Advanced High-Strength Steels

Science, Technology, and Applications

Mahmoud Y. Demeri

ASM International®
Materials Park, Ohio 44073-0002
asminternational.org

First printing, August 2013
Digital printing, December 2014

Prepared under the direction of the ASM International Technical Book Committee (2012–2013), Bradley J. Diak, Chair.

ASM International staff who worked on this project include Scott Henry, Senior Manager, Content Development and Publishing; Karen Marken, Senior Managing Editor; Sue Sellers, Product and Project Administrator; Madrid Tramble, Manager of Production; and Diane Whitelaw, Production Coordinator.

Library of Congress Control Number: 2013931685
ISBN-13: 978-1-62708-005-7
ISBN-10: 1-62708-005-8
SAN: 204-7586

ASM International®
Materials Park, OH 44073-0002
asminternational.org

Printed in the United States of America

Contents

Preface

Steel is used in many industries because of its favorable attributes and low cost relative to other materials. It is a globally dominant material and is a key driver of the world's economy. The auto industry is one of the biggest markets for steel products. In an era of weight reduction, steel was not perceived as a lightweight or a high-tech material because of its high density. This perception has changed with the introduction of new grades of high-specific-strength steels. The steel industry continues to develop new steel grades and innovative design concepts to meet challenges related to vehicle performance, affordability, safety, fuel efficiency, and the environment.

One key driver for steel development is the continuing increase in Corporate Average Fuel Economy (CAFE) standards from the current 27.5 mpg to 34.1 mpg by 2016 and to 54.5 mpg by 2025. A number of advanced high-strength steels (AHSS) that are highly formable and possess an excellent combination of strength, ductility, and strain hardening have been produced or are under development. Advanced high-strength steels are not lighter than traditional steels, but their high strength permits the use of thinner gages to reduce vehicle weight. The new steels extend the tensile strength limit of 800 MPa (116 ksi) provided by traditional high-strength steels (HSS) to over 1600 MPa (232 ksi). The superior properties of AHSS result from microstructural design that utilizes complex deformation and/or phase transformation processes.

The auto industry has adopted lightweighting as a greenhouse gas reduction strategy and automakers are now employing more AHSS in their vehicles to maintain affordability, reduce weight, lower emissions, and improve crash safety. Currently, AHSS account for 80 kg (175 lb) per vehicle and this is projected to double by 2020. Advanced high-strength steel panels and structures are becoming thinner, lighter, tougher, and stronger than those made from mild steel and traditional HSS. Studies have indicated that the economic and performance targets of automakers cannot be met without the significant use of steel. Advanced high-strength steels are currently the fastest growing class of materials offering solutions for current and future vehicles. It is believed that steel will continue to be the dominant material for the automotive market.

This book is a comprehensive guide to the grades, types, microstructures, thermal processing, deformation mechanisms, properties, performance, and applications of AHSS. The first objective of the book is to understand the drivers and solutions for building lighter, safer, efficient, and affordable cars. The second objective is to develop a rationale for using AHSS to accomplish weight reduction, fuel economy, crash safety, cost savings, and decrease in greenhouse gases. The third objective is to examine the complex interrelationship between composition, processing, microstructure, and mechanical properties. The fourth objective is to review the design guidelines, manufacturing hurdles, advanced forming, sustainability, and evolving grades of AHSS. The book is an essential reference/text for those who teach, study, develop, produce, use, design, or specify AHSS. A compelling and unique feature of the book is the extensive use of charts, figures, tables, and micrographs to provide a good perspective of the structure-property relationships and performance attributes of AHSS.

Acknowledgments

A number of figures and charts from sources whose permissions are acknowledged and appreciated are used at no cost. Special thanks are extended to WorldAutoSteel and Steel Market Development Institute for allowing the use of information and charts from their websites and photo galleries in the preparation of this book. Thanks are also due to William D. Callister, Jr. and David G. Rethwisch, authors of the 8th edition of the textbook *Materials Science and Engineering: An Introduction,* published in 2010 by Wiley & Sons. Their photos and figures are an essential part of Chapter 2, "Steel Fundamentals." I would also like to acknowledge the efforts of the following ASM International staff in the preparation of this book: Karen Marken, Senior Managing Editor; Sue Sellers, Product and Project Administrator; and Diane Whitelaw, Production Coordinator.

CHAPTER 1

Introduction

THE GLOBAL AUTOMOTIVE INDUSTRY is driven by consumer preferences, government regulations, and competitive pressures. Environmental, governmental, and customer demands to reduce fuel consumption, improve driver safety, ensure product reliability, and increase affordability have prompted the auto industry and material suppliers to develop a wide range of solutions to meet these requirements. The solutions include: optimization of product design, incorporation of lightweight materials, utilization of downgaging, and application of innovative manufacturing processes. All these solutions are interconnected and depend on the properties and attributes of the lightweight material.

New high-strength steel grades with superior attributes have been developed to compete with other lightweight materials on the basis of cost, performance, and manufacturability. At the core of this development is the advanced high-strength steel (AHSS) family, in which microstructures are manipulated to produce impressive mechanical properties such as very high tensile strength and remarkable ductility. Advanced high-strength steels are not intrinsically lighter than other steels, but they are strong enough that thinner gages can be used to reduce vehicle weight. Steel remains the dominant engineering alloy for building cars and structures because of its affordability, performance, manufacturability, recyclability, and wide range of applications. The typical 2010 light vehicle uses approximately 1080 kg (2390 lb) of steel. The versatility of steel results from its vast combinations of constituents, phases, microstructures, and thermal histories. These permutations impart to the steel properties that are desired for many applications. Also, the presence of the steel infrastructure and the knowledge acquired to manufacture steel products make it the material of choice for tomorrow's transportation products. Many auto manufacturers have aligned themselves with a lightweight strategy that is based on increasing AHSS content in their vehicles. Using AHSS presents manufacturing challenges in springback, die wear, press tonnage, and welding that must be resolved before any large-scale adoption of these materials is realized.

1.1 Drivers and Solutions

To ensure long-term success, automakers' business models, strategies, and activities have always been based on three major drivers:

- Consumer demands for lower cost, high quality, better performance, good reliability, higher safety, advanced features, and improved fuel economy
- Government regulations on Corporate Average Fuel Economy (CAFE) standards, crash safety, and gas emissions
- Competitive pressures on cost, quality, performance, and manufacturability

The automotive industry has identified four solutions to address these industry drivers and meet their business goals:

- Optimize product design
- Reduce vehicle weight
- Use low-cost materials
- Employ innovative manufacturing processes

All four solutions are based on the selection of lightweight materials that meet performance and cost requirements. The automotive industry, material producers, and part suppliers have been working for years to develop and deploy ferrous and nonferrous lightweight materials. It has been demonstrated that AHSS are the most affordable and best performing materials for lightweight applications. They can be used to reduce structural component weight by using thinner sections while maintaining the same performance characteristics. This "downgaging" leads to lighter vehicles and reduced costs.

One of the major environmental and societal challenges for the 21st century is the global increase in urban pollution—which results from the growing demand for petroleum consumption—and its adverse effect on the global climate. The estimated number of automobiles in the world will exceed 1.12 billion by 2015. According to the International Organization of Motor Vehicle Manufacturers (OICA), the world auto production for 2011 reached over 80 million vehicles. Over the next five years, the projected number of global vehicles produced annually will exceed 100 million. In 2011, the share for North America, which includes the United States, Canada, and Mexico, amounted to approximately 13 million vehicles. Of that annual production, the U.S. share is approximately 8.6 million units. Figure 1.1 shows historical and projected North American light vehicle production for the period 2006 to 2016. The forecast is for the production level to reach nearly 16 million vehicles by 2016.

With the number of vehicles produced globally on the rise, the demand for petroleum all over the world will continue to increase. According to the Transportation Energy Data Book, the world consumes 85.26 million petroleum barrels per day (M pbpd), of which the United States consumes 22.5%, or 19.15 M pbpd (Ref 1.2). Figure 1.2 shows a chart of the U.S. historical and projected petroleum production and consumption for all sectors of the economy for 1973 to 2035. The chart clearly shows that the transportation sector overwhelms other industry sectors in the consump-

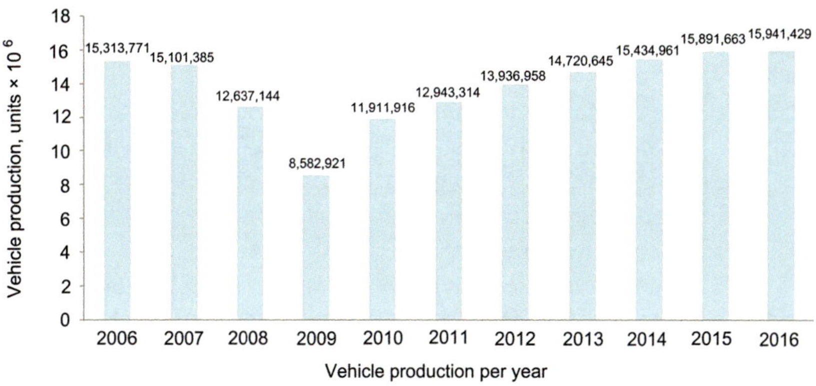

Fig. 1.1 Historical and projected North American light vehicle production from 2006–2016. Data is from 2011; data forecast from 2012 is 750,000 units higher per year than shown. Source: Global Insight, Ref 1.1

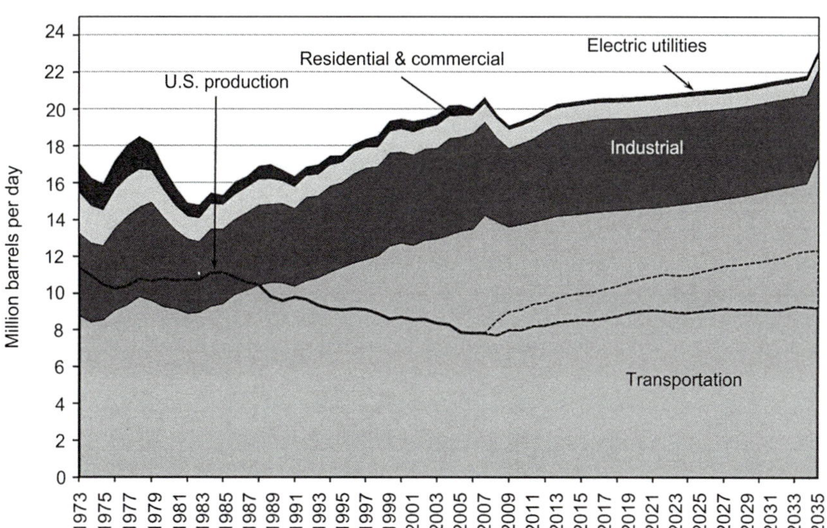

Fig. 1.2 United States petroleum production and consumption for all sectors of economy from 1973–2035. Source: Ref 1.2

tion of petroleum at all times. The U.S. production of petroleum is superimposed on the chart, which shows that the gap between petroleum production and consumption is increasing, and by 2035 the gap is expected to be approximately 10 M pbpd.

In the United States, there are over 235 million light vehicles, of which 135 million are cars and 100 million are light trucks. In addition, approximately 11 million heavy trucks were registered in the United States in 2009. The United States consumes 19.15 M pbpd, of which the transportation industry consumes 69.7%, or approximately13.4 M pbpd. Cars and light trucks account for 64%, or 8.6 M pbpd of U.S. transportation petroleum use (Ref 1.2). Figure 1.3 shows a chart of historical and projected U.S. petroleum production and consumption for the transportation sector for 1970 to 2035. The chart clearly shows that the automotive industry, which produces cars, light trucks, and heavy trucks, consumes the highest percentage of petroleum in relation to other industries such as air, rail, and marine. The chart also shows that by the year 2035, transportation petroleum consumption is expected to grow to more than 16 M pbpd. The gap between U.S. petroleum production and consumption for the transportation sector is also shown on the chart.

Figure 1.4 shows a more detailed plot of transportation sector fuel consumption per day for 1995 to 2035. It is clear that light and heavy duty vehicles will continue to dominate fuel consumption at a rate of 84% through 2035.

Increasing consumption of petroleum results in increasing emissions of greenhouse gases and adversely contributes to global climate change. Figure 1.5 shows the carbon dioxide (CO_2) emissions by sectors. Electricity

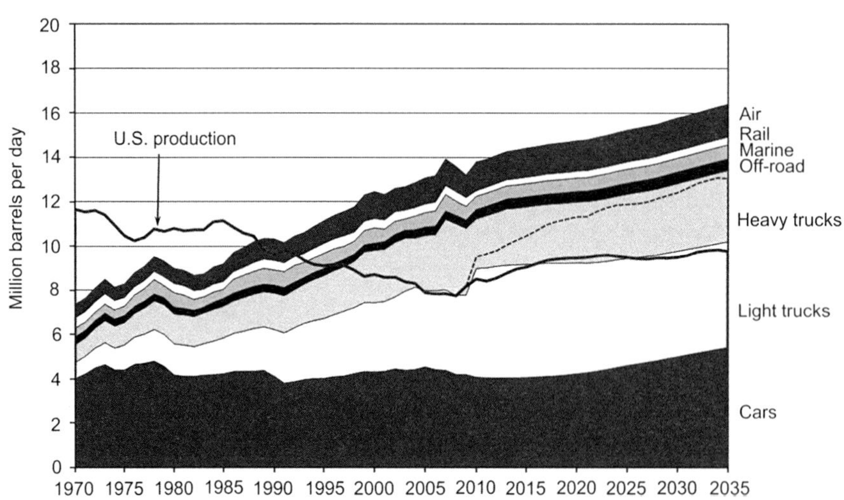

Fig. 1.3 United States petroleum production and consumption for the transportation sector from 1970–2035. Source: Ref 1.2

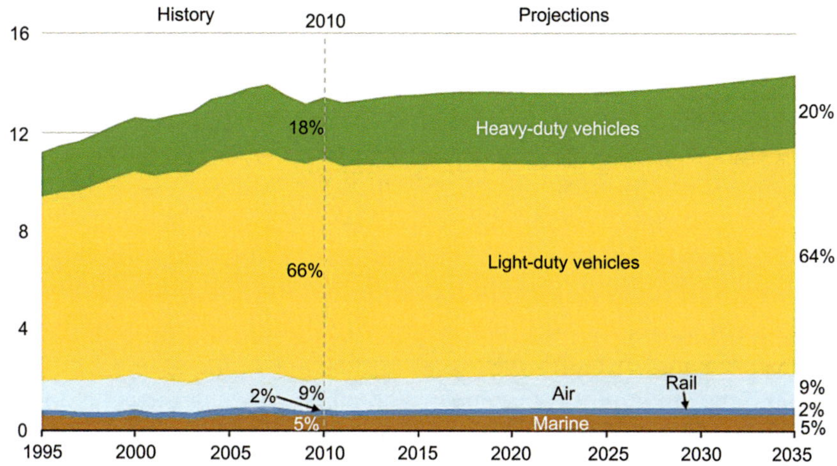

Fig. 1.4 United States petroleum consumption for the transportation sector from 1995–2035, million barrels per day equivalent. Source: Ref 1.3

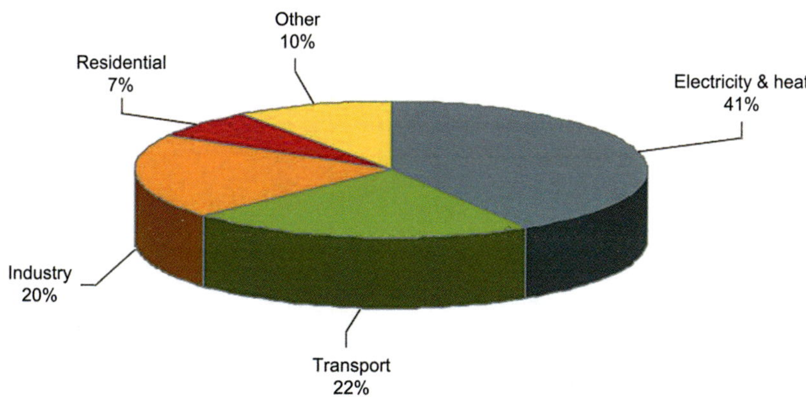

Fig. 1.5 Emissions of CO_2 by sectors. Source: Ref 1.4

and heat generation produces the highest percentage of CO_2 emissions, followed by the transportation sector.

Vehicles produce greenhouse gases (CO_2, CH_4, NO_x, and hydrofluorocarbons) that affect the environment and cause environmental impacts over the whole life cycle of the vehicle. Assessing greenhouse gas emissions of a vehicle requires understanding its life cycle stages, which includes all emissions from any process to produce, use, and retire the vehicle. Carbon dioxide accounts for the majority of greenhouse gases. In 2009, the transportation sector was responsible for 1757 million metric tons, which is approximately one-third of the total CO_2 emissions for that year. Most of the U.S. transportation sector CO_2 emissions come from petroleum fuels (98%). Table 1.1 lists the amount of CO_2 released into the atmosphere from a gallon of fuel.

Most of the emissions result from vehicle use (85%), while material production and vehicle manufacturing accounts for the rest (15%). The carbon footprint measures the impact of a vehicle on climate change in tons of CO_2 emitted annually. Table 1.2 lists the average annual carbon footprint for cars and light trucks between 1975 and 2010. The carbon footprint dropped 51.4% for cars and 42.2% for light trucks. This is a significant drop in carbon footprint for vehicles and it means that cars are becoming more fuel efficient.

Improvements in fuel economy and reduction of emissions and their effect on the environment became a national priority and the main objective of research and development in industrial, academic, and national research centers. The National Highway Traffic Safety Administration (NHTSA) and The Environmental Protection Agency (EPA) issued a joint rulemaking to establish a national program to regulate fuel economy and greenhouse gas emissions for model year 2012 to 2016 vehicles. Table 1.3 lists the average projected emissions compliance level for cars and light trucks for model years 2012 to 2016.

The fuel economy standards for model year 2012 to 2016 cars and light trucks are listed in Table 1.4. Also listed is the required fuel economy for the fleet average.

Table 1.1 Carbon dioxide emissions from a gallon of fuel

	CO_2 per gallon		
Fuel	g	kg	lb
Gasoline	8,788	8.8	19.4
Diesel	10,084	10.1	22.2

Source: Ref 1.2

Table 1.2 Average annual carbon footprint for light vehicles for 1975 and 2010

	CO_2, short tons		Change, %
Vehicles	1975	2010	1975–2010
Cars	11.8	5.7	–51.4
Light trucks	13.6	7.9	–42.2

Source: Ref 1.2

Table 1.3 Projected emissions compliance levels for 2012 to 2016 under the footprint-based carbon dioxide standards

	Average projected emissions compliance levels, grams/mile		
Year	Cars	Light trucks	Combined cars and light trucks
2012	263	346	295
2013	256	337	286
2014	247	326	276
2015	236	312	263
2016	225	298	250

Source: Ref 1.2

In August 2012, NHTSA and EPA released another proposed rulemaking to set stringent fuel and emissions requirements for model years 2017 to 2025. The proposed Energy Bill requires the U.S. auto industry to raise its CAFE standards to a fleetwide average of 34.1 mpg in 2016 to 54.5 mpg in 2025, up from 27.6 mpg in 2011. Proponents of the rules estimate that consumption will be reduced by 3.1 M pbpd. The EPA and NHTSA believe that the benefit of these rules to society will greatly offset the additional cost to industry and consumers.

Fuel consumption is measured in gallons per mile (gpm) and vehicle mileage is measured in miles per gallon (mpg). Both measures vary with car weight as shown in Fig. 1.6. Fuel consumption (gpm) varies linearly, while mileage (mpg) varies inversely with car weight. Simply put, reducing vehicle weight reduces its fuel consumption (gpm) and increases its mileage (mpg).

Automakers have been aware of the drawbacks of heavy cars and have been working diligently to find ways to reduce their weight. Figure 1.7 captures Henry Ford's observation about the heavy weight of cars.

Table 1.4 Fuel economy standards for 2012 to 2016

	Average required fuel economy, mpg		
Year	Cars	Light trucks	Combined cars and light trucks
2012	33.3	25.4	29.7
2013	34.2	26.0	30.5
2014	34.9	26.6	31.3
2015	36.2	27.5	32.6
2016	37.8	28.8	34.1

Source: Ref 1.2

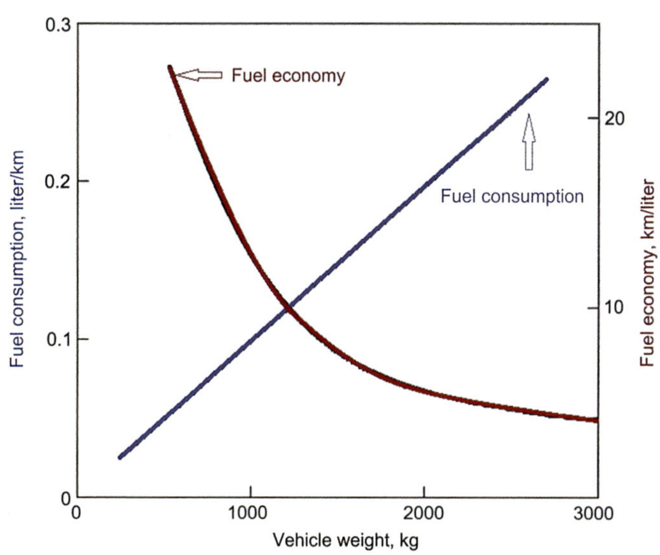

Fig. 1.6 Effect of car weight on fuel consumption and fuel economy. Source: Adapted from Ref 1.5

To achieve the fuel efficiency requirements, automakers are developing new strategies and advanced technologies to improve engines, drivetrains, transmissions, aerodynamics, tire rolling resistance, and vehicle weight. As Fig.1.8 shows, vehicle weight reduction is the most effective means for

Fig. 1.7 Henry Ford's observation regarding vehicle weight

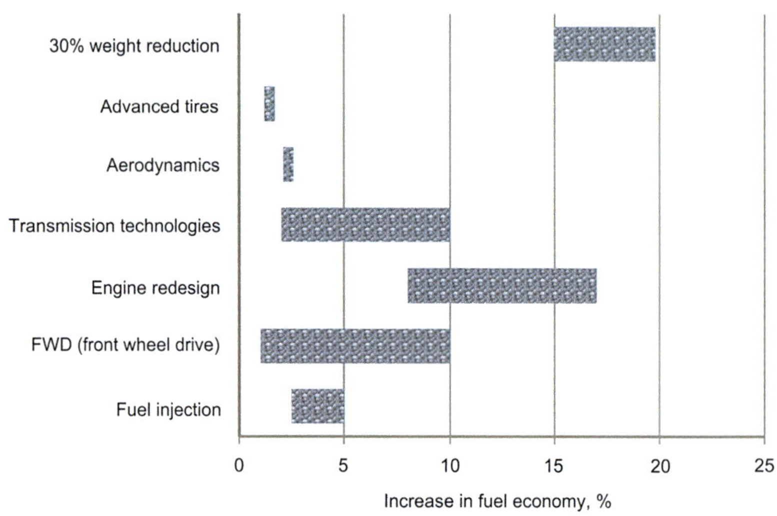

Fig. 1.8 Vehicle fuel economy improvement potential for various technologies. Source: Ref 1.6

improving fuel economy and reducing energy consumption. Reducing vehicle weight lowers the inertial forces that the engine has to overcome to decelerate and stop the vehicle. It also reduces the power required to move and accelerate the vehicle.

It has been suggested by many in the automotive industry that a 30% vehicle weight reduction will yield an acceptable target for improvement in fuel economy. The relationship between vehicle weight reduction and fuel economy improvement is complex and depends on many factors such as size, type, powertrain, speed, and driving cycles of the vehicle. As a rule of thumb, for every 10% of weight reduced from the average new car or light truck, fuel consumption is reduced by 6 to 8%.

The three strategies that can be used to reduce vehicle weight are:

1. Vehicle downsizing
2. Vehicle design changes
3. Lightweight/strong material substitution

Downsizing a vehicle to reduce its weight proves to be difficult because consumers prefer the comfort and functionality of larger vehicles. Vehicle design changes can produce nominal reduction in vehicle weight. Lightweighting through strong material substitution appears to be the only viable route to significant vehicle weight reduction. This can be achieved by replacing heavy steel components in body structures, closure panels, chassis, wheels, bumpers, and suspension parts with lightweight materials made from ferrous and nonferrous alloys, polymers, and composites. The overall weight of a car is distributed among its body (40%), chassis (25%), power train (15%), and equipment (20%). The body and chassis are the two major contributors to the weight of a car and therefore are the focus for lightweight design.

Lightweight materials include high-strength steels (HSS), aluminum alloys, magnesium alloys, titanium alloys, and various composite materials. By using lightweight materials, manufacturers can reduce the weight of a vehicle without sacrificing safety, durability, and comfort. The more weight that can be eliminated from a vehicle, the more fuel efficiency is achieved.

Ferrous alloys include all grades of steels and cast irons; nonferrous light alloys include aluminum and magnesium alloys. Nonmetallic lightweight materials are made of polymers and fiber reinforced polymer composites. The nonferrous and nonmetallic materials referred to have higher strength-to-weight ratios than HSS and could potentially be used for weight saving in automotive components. However, they are expensive, incompatible with existing manufacturing processes, and have higher production and manufacturing costs. These hurdles prevent such materials from being commonly used, especially in low-end but high-volume production cars.

Table 1.5 lists the weight savings and material and manufacturing relative cost per part resulting from replacing steel with different lightweight materials. The table shows clearly that conventional HSS maintain their cost advantage over other lightweight materials. This is because HSS and the first generation of AHSS are low-alloy steels; hence, expensive alloy cost is kept to a minimum.

While composites remain expensive due to high material cost and long production cycle times, HSS and aluminum are likely to remain popular substitutes for mild steel in occupant vehicles. Although aluminum use in vehicles has been rising modestly, the cost differential with steel is still significant and is responsible for its limited application.

Designers must weigh structural performance and life-cycle cost against material cost, which can vary widely and is only one factor in the overall cost of making a product. The total cost of a final product includes material, design, fabrication, and assembly costs. Product cost is usually a tradeoff between the various cost contributors. For instance, a more expensive material that meets performance requirements but is more durable and requires less processing may generate lower overall cost than a less expensive material. In most cases, using a different material leads to different performance and a different manufacturing process.

Figure 1.9 shows a plot of the additional manufacturing cost and the realized mass reduction ranges for various lightweight materials. While HSS show the least cost penalty of all other materials, their maximum weight reduction is limited to 20%. However, with increase in steel strength, as in AHSS, the weight reduction range can be significantly increased.

Cost-benefit analysis demonstrates that steel parts are stronger and cheaper than equivalent dimension parts made from other lightweight materials. When addressing cost, materials selection becomes the crucial factor, and competitive materials such as aluminum, magnesium, and fiber composites will be at a cost disadvantage compared to steel.

Because cost is a major driver for automakers and reducing it is the number one priority for success of any business, a more elaborate look at cost is justified. The Ultra-Light Steel Auto Body (ULSAB), Ultra-Light

Table 1.5 Weight savings and costs for lightweight automotive materials

Lightweight material	Material replaced	Mass reduction, %	Relative cost (per part), material and manufacturing
High-strength steel	Mild steel	10–25	1
Aluminum	Steel, cast iron	40–60	1.3–2
Magnesium	Steel or cast iron	60–75	1.5–2.5
Magnesium	Aluminum	25–35	1–1.5
Glass fiber reinforced polymer composites	Steel	25–35	1–1.5
Carbon fiber reinforced polymer composites	Steel	50–60	2–10+
Aluminum-matrix composites	Steel or cast iron	50–65	1.5–3+
Titanium	Alloy steel	40–55	1.5–10+
Stainless steel	Carbon steel	20–45	1.2–1.7

Source: Ref 1.8

Steel Auto Closure (ULSAC), and Ultra-Light Steel Auto Body—Advanced Vehicle Concepts (ULSAB-AVC) programs used cost models, developed by the Massachusetts Institute of Technology (MIT), to conduct economic analysis to assess the costs of body and door structures, as well as to evaluate the costs of all aspects of manufacturing of an entire vehicle (chassis, body, powertrain, and electric). Table 1.6 summarizes the costs associated with using HSS, tailor welding blanks, and part consolidation in the auto body structure. The results show that the body mass and cost can be reduced with respect to the reference vehicle.

The most important lightweight material competing with steel is based on aluminum alloys. Many studies have demonstrated the cost advantage of steel over aluminum alloys, and the results are (Ref 1.11):

- Raw material cost for aluminum is 3 times that of steel.
- Conversion cost for aluminum is 2 times that of steel (driven by the energy required to extract aluminum from raw material).

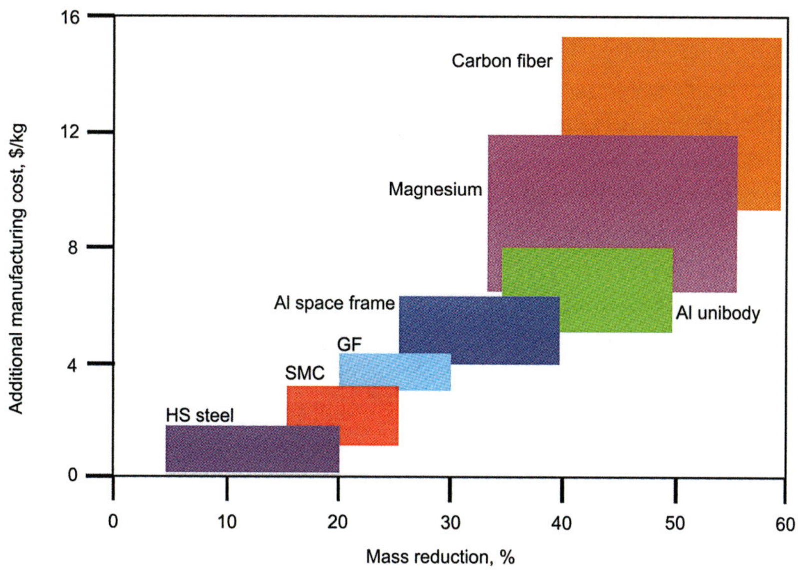

Fig. 1.9 Cost penalties for reducing mass of various lightweight materials. HS, high-strength; SMC, sheet molding compound; GF, glass fiber. Source: Ref 1.9

Table 1.6 Cost summary of Ultra-Light Steel Automotive Body (ULSAB) structure

	Reference vehicle	ULSAB	ULSAB-AVC(a)
Steel cost(b)	369	416	468
Forming cost(b)	282	250	213
Assembly cost(b)	328	281	291
Total body cost(b)	**979**	**947**	**972**
Tooling investment(b)	68.0 million	51.2 million	40.3 million
Part count	135	96	81
Body mass, kg (lb)	270 (595)	203 (448)	218 (481)

(a) Ultra-Light Steel Automotive Body—Advanced Vehicle Concept, designed to meet more stringent safety criteria than the listed alternatives. (b) U.S. dollars. Source: Ref 1.10

- Assembly cost of aluminum is 20 to 30 times that of steel.
- Body structure of aluminum costs 60 to 80% more than steel.

All indications show that steel has a big advantage in cost and performance compared to other lightweight materials.

In an era of weight reduction, traditional steels have been perceived negatively because of their high density. With the introduction of AHSS, the specific strength of those steels became comparable with those of aluminum and magnesium alloys, as shown in Fig. 1.10. Only carbon fiber reinforced polymers have better specific strength values, but their high cost and complicated manufacturing preclude them from being considered for high-volume vehicle production.

Although the densities of aluminum and magnesium alloys are lower than those of steel (2.7, 1.8 and 0.8 Mg/m^3 respectively), their yield strength and ductility combinations are much lower than those of other grades of steel. Figure 1.11 shows a comparison between the strength and ductility of aluminum alloys, magnesium alloys, and different grades of steels.

The figure shows the advantages that steel has over aluminum and magnesium alloys in two key properties: strength and ductility. These properties are extremely important for the manufacturing and performance of automotive structures and components. Also, it is important to realize that because aluminum and magnesium alloys suffer from low ductility, they require special forming processes such as warm forming and unconventional die designs.

1.2 Importance of Steel

Steel is one of the most common materials in the world. It has fueled the industrial revolution and is central to the transport system, infrastruc-

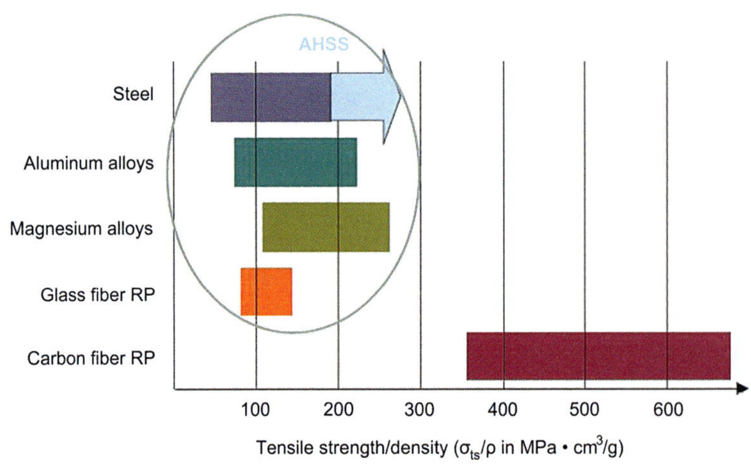

Fig. 1.10 Specific strength of various lightweight materials. RP, reinforced polymer. Source: Ref 1.12

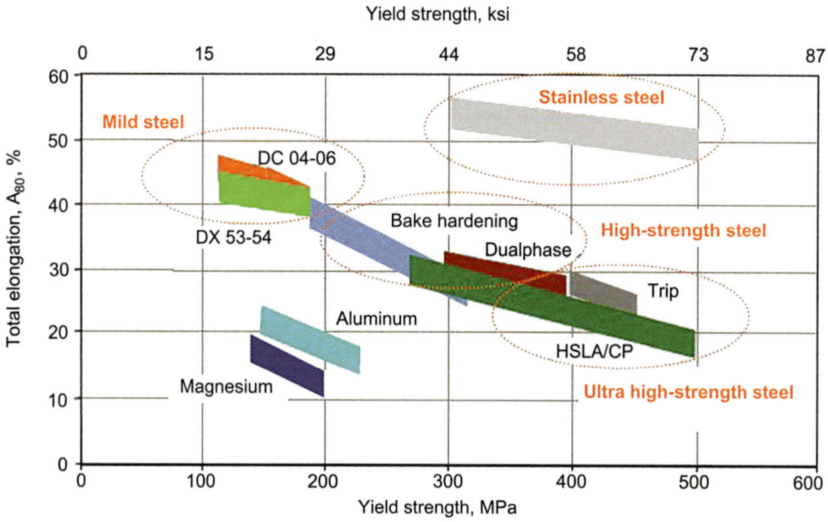

Fig. 1.11 Yield strength and ductility for various metal alloys. HSLA/CP, high-strength, low-alloy/[insert definition of CP, complex phase; TRIP, transformation-induced plasticity steels. Source: Ref 1.13

ture, housing, and manufacturing. Steel producers reported that 1518 million tons of steel were produced in 2011. Nearly everything that is man-made is either made of steel or manufactured by equipment made of steel. Steel has been used in vehicles since the inception of the automotive industry. It has been commonly used because of its strength, formability, and low cost relative to other metals.

Steel has evolved over the years from mild steel in early 1900s to high-strength low-alloy (HSLA) steels in the late 1970s and the first generation of AHSS in the 1990s. There are approximately 30 steel grades that are used in automobiles. In 2010, steel made up approximately 60% of the light vehicle total weight. This amounts to approximately 1084 kg (2390 lb) of steel in a typical car.

Recent approaches to improving the properties of steel have been based on developing steel microstructures of higher strength in order to reduce component section size and weight. The last two decades have seen dramatic changes in the compositions, microstructure, strength, ductility, and processing approaches to steel manufacture. Figure 1.12 shows HSS development over time. The strengthening mechanisms and microstructural features for the different steel grades are also listed.

Figure 1.13 shows a spectrum of the steel categories that were used throughout the automotive industry history. The chart reflects the traditional inverse relationship between strength and ductility. At the low end of the strength spectrum are interstitial-free (IF) and mild (MILD) steels. These steels are very soft and formable and they are mostly used in outer auto body applications. In the middle are the conventional grades of HSS that include interstitial-free, high-strength (IF-HS), isotropic (IS), bake-

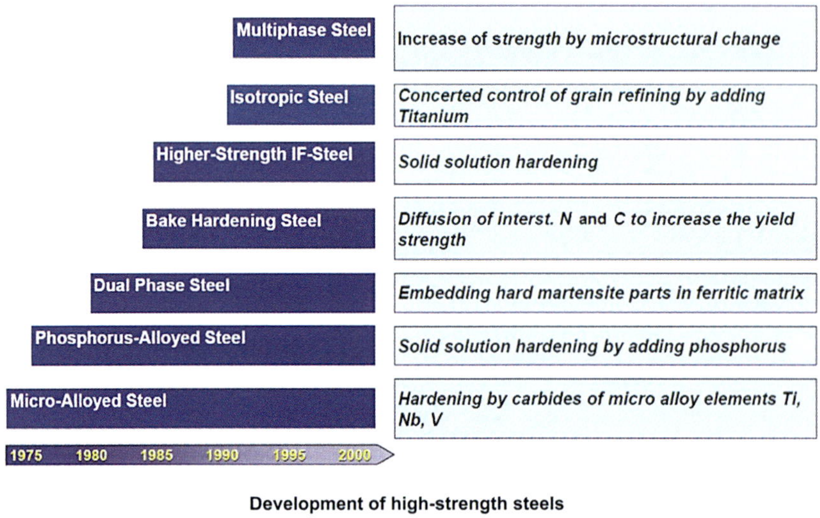

Fig. 1.12 Development of high-strength steels over time. Source: Thyssen, Ref 1.14

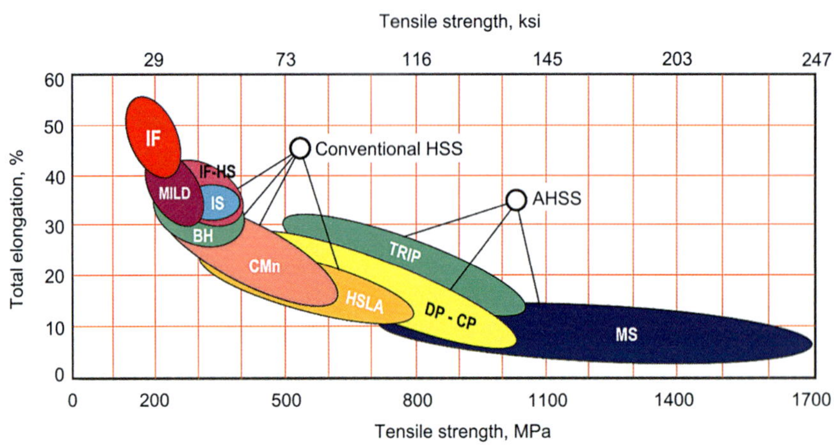

Fig. 1.13 Location of conventional high-speed strength (HSS) and first-generation advanced high-strength steel (AHSS) in the strength-elongation space. IF, interstitial-free; IF-HS, interstitial-free, high-strength; IS, isotropic steel; BH, bake-hardenable; CMn, carbon manganese; HSLA, high-strength, low-alloy; TRIP, transformation-induced plasticity steels; DP-CP, dual-phase, complex-phase; MS, martensitic. Source: Ref 1.15

hardenable (BH), and high-strength, low-alloy (HSLA) steels. These steels have higher strength compared to traditional steels but their ductility is much lower. At the high end are the AHSS that include dual-phase (DP), complex-phase (CP), transformation-induced plasticity (TRIP), and martensitic (MS) steels. These steels have very high strength and, except for the MS grade, have good formability and are being used in many automotive structural applications.

Figure 1.13 also shows that steel grades overlap in some ranges, giving the designer possible options for steel grade selection for specific applications. For example, if part specification requires a material with a minimum of 1000 MPa (145 ksi), the designer can choose among four grades of steel that meet the specific requirement.

References to HSS usually include steels with tensile strength >300 MPa (>44 ksi). Such steels include IF, BH, HSLA, and AHSS (TRIP, DP, CP, and MS). Figure 1.14 shows different grades and types of HSS superimposed on the strength-elongation chart.

With the increasing demand for more fuel efficient and safer cars, steel has continued to evolve to meet the demands and to compete with other lightweight materials, such as aluminum, magnesium, titanium, and fiber reinforced polymers. The major difference between conventional HSS and AHSS is their microstructure. Conventional HSS are single-phase ferritic steels while AHSS are multiphase steels that contain ferrite, martensite, bainite, and/or retained austenite. The first generation of AHSS derive their strength, to a large extent, from manipulation of the microstructure and not from alloying. Therefore, they can be produced for approximately the same price as traditional HSS.

The major attribute of AHSS is their extremely high strength. This means that thinner sheets of AHSS are used to replace current thick steel components. In addition, advances in optimization methods and manufacturing processes have increased the mass savings achievable in AHSS. These improvements in technology have enabled AHSS to be at the same level with aluminum concerning lightweighting applications.

In section 1.1, "Drivers and Solutions," the cost advantage of steel over other lightweight materials was established. Table 1.7 gives approximate comparative cost data for some types of steel that can be used to advantage by the automotive designer. Austenitic stainless steels have a high

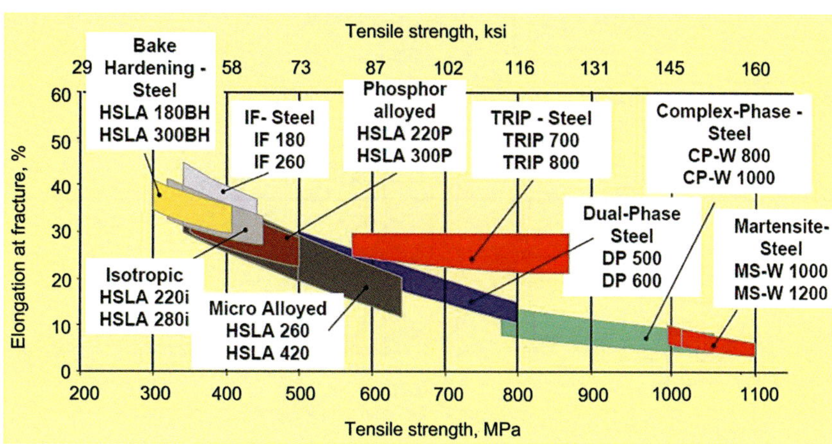

Fig. 1.14 Grades and types of HSS. HSLA, high-strength, low-alloy; IF, interstitial-free; TRIP, transformation-induced plasticity. Source: Ref 1.14

relative cost because they contain a high percentage of expensive alloying elements. Conventional HSS and first-generation AHSS grades have slightly higher relative cost compared to cold rolled carbon steel because they are considered to be low-alloy steels and hence keep expensive alloy cost to a minimum.

The importance of steel in industry is revealed by the amount of steel products shipped to manufacturers. The total North American steel shipments for 2010 were over 113 million tons, of which approximately 15 million tons of flat rolled steel went into the production of approximately 12 million new light vehicles. Figure 1.15 shows that in 2010, HSS and AHSS accounted for approximately 40% of the types of flat rolled steel delivered to vehicle manufacturers. Flat rolled AHSS applications include body structures, closures, suspensions, subframes, crossmembers, cradles, bumpers, and wheels.

Figure 1.16 shows the 2010 flat rolled steel content in North American light vehicles. It can be seen that 58% of the steel is used in the body and closure subsystems. Specifically, 42% of the steel is used in body-in-white

Table 1.7 Approximate relative costs of various sheet steels

Steel grade	Approximate relative cost
Hot rolled carbon	0.80
Cold rolled carbon	1.00
Bake-hardenable	1.10
Hot dipped galvanized	1.12
Aluminized	1.21
Electrogalvanized	1.35
High-strength, low-alloy	1.15
Dual-phase	1.40
Martensitic	1.50
Austenitic stainless type 304	5.70

Source: Ref 1.16

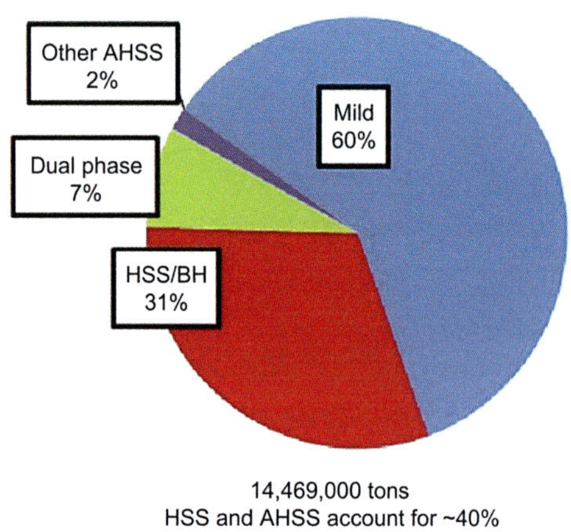

Fig. 1.15 Percentage of different types of flat rolled steel delivered in 2010. AHSS, advanced high-strength steel; HSS, high-strength steel; BH, bake-hardenable. Source: Ref 1.1

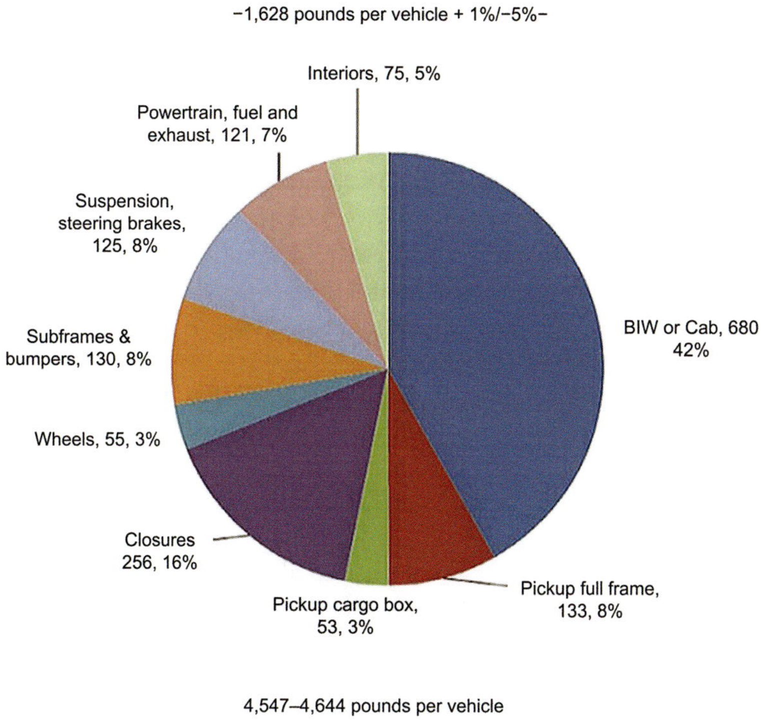

Fig. 1.16 Distribution of flat rolled steel content in the 2010 North American light vehicles. BIW, body-in-white. Source: Ref 1.1

(BIW), followed by 16% for closures applications. This means that the greatest potential for weight savings is in the body structure and closure panels. The rest of the steel is used in chassis subsystems, wheels, bumpers, exhaust, and other applications.

1.3 Advanced Steels

Many years and vast amounts of resources were invested by the steel and automotive industries in research and development to bring about advanced steels that are stronger, safer, lighter, greener, fuel efficient, and cost competitive. These new generations of steels possess extremely high strength while maintaining the high formability required for manufacturing. New technologies such as continuous casting and thermomechanical processing have resulted in the development of several new grades of steels. Advanced high-strength steels (AHSS) are multiphase steels consisting of hard islands of martensite, bainite, and/or retained austenite dispersed in a ductile ferrite matrix, in quantities and combinations sufficient to produce a variety of desired mechanical properties. The AHSS family includes dual-phase (DP), complex-phase (CP), transformation-induced plasticity (TRIP), twinning-induced plasticity (TWIP), and martensitic

(MS) steels. They show superior strength compared to the high-strength, low-alloy steel (HSLA) grades with the same ductility. The TWIP grades, in particular, exhibit excellent formability that is comparable to that of mild steel but at an incredibly higher tensile strength. The strengthening mechanisms involved in AHSS include solid-solution strengthening, precipitation hardening, grain refinement strengthening, and phase transformation from soft ferrite to hard martensite. Advanced high-strength steels are the fastest growing materials in today's automotive industry and are designed to leverage steel as the optimal automotive material. These advanced steels are best suited for the manufacture of crash-resistant auto body structures, where strength and energy absorption are required.

Advanced high-strength steels are classified into two categories based on the levels of their tensile strength and percent elongation. A chart giving the location of the two groups of AHSS in the strength-elongation space is shown in Fig. 1.17. The first group is known as the first generation of AHSS, and it includes DP, CP, TRIP, and MS steels. The second group is known as the second generation of AHSS, and it includes twinning-induced plasticity (TWIP), lightweight induced-plasticity (L-IP), and austenitic stainless steel (AUST SS).

The first generation of AHSS is ferrite-based while the second generation has an austenite-based microstructure. The austenite matrix provides better formability but the high austenite stabilizer content of up to 20 wt% manganese and nickel increase the cost of the second-generation AHSS.

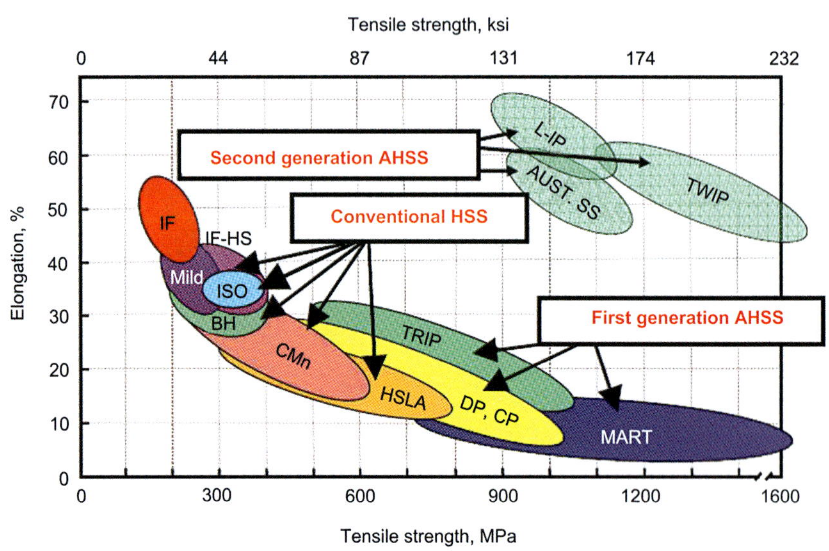

Fig. 1.17 Location of the first and second generation of AHSS. IF, interstitial-free; IF-HS, interstitial-free, high-strength; ISO, isotropic; BH, bake-hardenable; CMn, carbon manganese; HSLA, high-strength, low-alloy; TRIP, transformation-induced plasticity steels; DP-CP, dual-phase, complex-phase; MART, martensitic; L-IP, lightweight induced-plasticity; AUST. SS, austenitic stainless steel; TWIP, twinning-induced plasticity. Source: Ref 1.17

In the quest for safer and lighter cars, automotive engineers are specifying increased amounts of AHSS for auto body structures to achieve greater strength without a corresponding increase in weight. Advanced high-strength steels have been incorporated into nearly every new car, and many of today's vehicles contain as much as 40% of those materials. They are predicted to replace approximately 60% of the currently used conventional steels, including many of the HSLA steel grades. The currently used AHSS have reduced vehicle body structure mass by at least 25%. The third-generation AHSS, under development, are expected to achieve more than 35% in structural mass reduction. This will significantly improve fuel efficiency and reduce greenhouse gas emissions.

Figure 1.18 shows historical and projected increase in use of AHSS per vehicle from 1975 to 2015. Also included in the figure are data for the total amount of HSS content, which includes AHSS, HSS, and BH steel grades.

The best prediction for the use of AHSS comes from the steel industry's FutureSteelVehicle (FSV) program, which developed a steel body structure 35% lighter than a benchmarked vehicle. Approximately 97% of the steels used are HSS and AHSS, and approximately 50% of those steels have strengths in the giga pascal range. Using this amount of HSS is expected to reduce the total lifetime emissions by approximately 70%. The program claims that weight reduction is achieved at no cost penalty (Ref 1.19).

Advanced high-strength steels are multiphase steels with complex microstructures that contain phases such as ferrite, martensite, bainite, austenite, and/or retained austenite. Understanding the structure-property re-

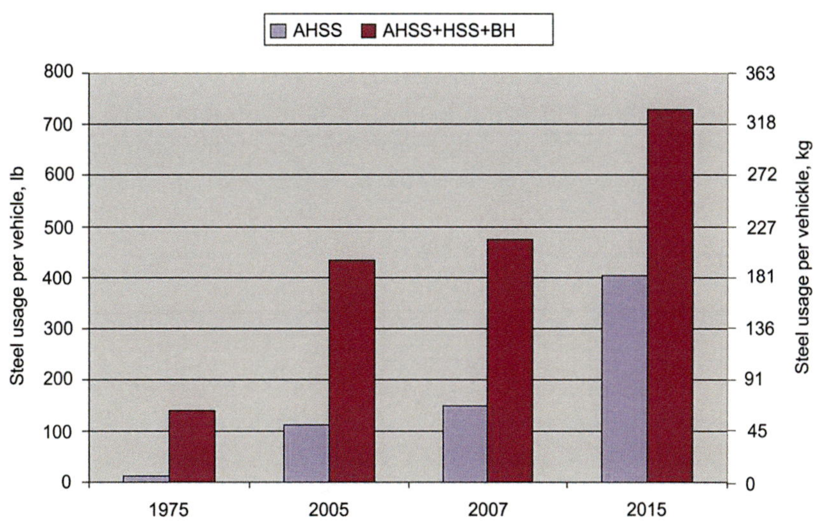

Fig. 1.18 Historical and projected content of AHSS, HSS, and bake-hardenable (BH) steels in North American vehicles from 1975–2015. Source: Ref 1.18

lationship in AHSS requires a basic knowledge about steel compositions, phases, microstructures, strengthening mechanisms, and thermal processing. This basic knowledge is provided in Chapter 2, "Steel Fundamentals," in this book

1.4 Significance of High Strength

The biggest advantage of AHSS over other lightweight materials is their wide range of strengths, extending from approximately 500 to 1700 MPa (70 to 245 ksi). In most cases, high strength allows thick-gage components to be replaced by thin-gage material without any loss of performance. The significance of strength manifests itself in weight reduction, increased fuel efficiency, enhanced crashworthiness, lower material cost, and reduced environmental impact.

Weight Reduction and Fuel Efficiency. It has been demonstrated, by the early ULSAB program, that vehicle weight reduction is highly dependent on the strength of the new steels replacing the traditional ones. Table 1.8 shows the dependence of weight reduction on the yield strength of AHSS. The potential for weight reduction increases with increase in the yield strength of the material.

Vehicle weight reduction is a key factor for fuel efficiency and is accomplished by downgaging components made from AHSS. Reducing vehicle weight lowers the inertial forces that the engine has to overcome and reduces the power needed to move the vehicle. It is estimated that a 10% reduction in vehicle weight results in 6 to 8% improvement in fuel economy.

Crash Performance and Energy Absorption. A large part of the energy produced in a collision is absorbed by the steel front end structure as it deforms under impact. The structure is designed to collapse like an accordion to absorb the impacting force. The ability of a material to absorb an impact depends on its toughness, which is the product of strength and ductility. High strength levels of 1800 MPa (260 ksi) are indicated as a future requirement for anti-intrusion applications. Also, a high strain-hardening rate means that the material becomes stronger on impact and will absorb more of the impact energy.

Material and Cost Savings. When AHSS are used, the mass reduction due to downgaging results in less steel used per car. This reduces the amount of steel that needs to be produced, hence decreasing the demand on natural resources and the energy needed to convert iron ore into steel.

Table 1.8 Dependence of weight reduction on yield strength of AHSS

Yield strength, MPa (ksi)	Weight reduction, %
400 (58)	15–25
700 (102)	35–45
900 (131)	45–50
1100 (160)	50–55

Source: Ref 1.20

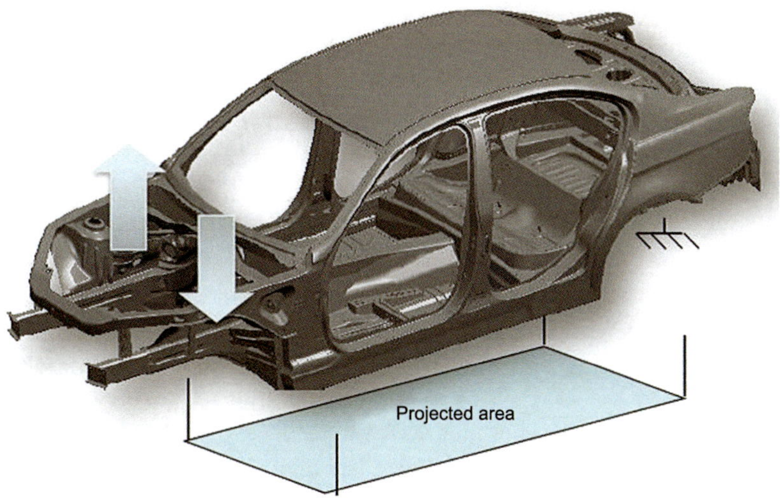

Fig. 1.19 Determination of mass efficiency coefficient. Source: Ref 1.21

It also reduces emissions from steelmakers' smoke stacks and vehicle tail pipes. Using less steel per car reduces material and fuel cost and benefits the environment. Simply put, if steel strength is increased from 500 to 1000 MPa (70 to 145 ksi), steel consumption would be reduced by about half.

Mass Efficiency. Europeans use a mass efficiency measure to rate the performance of vehicle lightweight body-in-white designs. The basis of the measure is illustrated in Fig. 1.19.

The efficiency measure is called the lightweight design coefficient (LDC) and it is defined as the body-in-white mass divided by the product of the projected area and the torsional stiffness of the vehicle:

$$\text{LDC} = \text{BIW mass (kg)} / [\text{Area (m}^2\text{)}.\text{Torsional stiffness (N-m/deg)}]$$

Lightweight design coefficient values reported by carmakers for their BIW designs ranged from 2.3 to 5.5. Smaller LDC numbers indicate more efficient BIW design.

REFERENCES

1.1 R. Schultz, Ducker Worldwide, personal communication, Nov 2012

1.2 S. Davis, S. Diegel, and R. Boundy, *Transportation Energy Data Book: Edition 30,* Oak Ridge National Laboratory, June 2011

1.3 H. Gruenspecht, "Annual Energy Outlook 2012—Early Release Reference Case," AEO2012 Early Release Rollout Presentation, Jan 23, 2012 (Washington, DC) Johns Hopkins University

1.4 E. Basson, "The Steel Industry in a Sustainable Society," SBB Green Steel Strategies Conference, April 19, 2012 (Berlin), Platts

1.5 M. Ashby, H. Shercliff, and D. Cebon, *Materials: Engineering, Science, Processing and Design*, 2nd ed., Elsevier, 2010

1.6 M.F. Ashby and D.R.H. Jones, *Engineering Materials 1,* 4th ed., Elsevier, 2012

1.7 F. Stodolsky, A. Vyas, and R. Cuenca, Lightweight Materials in the Light-Duty Occupant Vehicle Market: Their Market Penetration Potential and Impacts, *Proc. The 2nd World Car Conference,* University of California, Riverside, March 1995

1.8 W.F. Powers, *Adv. Mater. Process.,* May 2000, p 38–41

1.9 C. Fine and R. Roth, "Lightweight Materials for Transport: Developing a Vehicle Technology Roadmap for the Use of Lightweight Materials," presented at Innovation Roundtable II: The Future of Manufacturing Innovation—Advanced Technologies, March 29, 2010, Massachusetts Institute of Technology

1.10 F. Field, R. Kirchain, and R. Roth, Process Cost Modeling: Strategic Engineering and Economic Evaluation of Materials Technologies, *JOM,* Oct 2007

1.11 World Steel Association, Sustainability Indicator, www.worldsteel.org, 2012

1.12 G. Coates, "Importance of Materials and Manufacturing Emissions for Future Vehicle Considerations," The Phoenix Group Presentation at Great Designs in Steel Seminar, May 16, 2012 (Livonia, MI), AUTOSTEEL

1.13 M. Mehrkens, "Advanced High Strength Steel Technology in the Porsche Cayenne," Porsche Engineering presentation at Great Designs in Steel Seminar, Feb 18, 2004 (Livonia, MI), AUTOSTEEL

1.14 K. Siegert and M. Vulcan, "Tool and Die Design for Deep Drawing AHSS," IFU presentation at Great Designs in Steel Seminar, March 9, 2005 (Livonia, MI), AUTOSTEEL

1.15 "Advanced High Strength Steels Applications Guidelines," Version 4.1, WorldAutoSteel, www.worldautosteel.org, June 2009

1.16 "Automotive Steel Design Manual," Revision 6.1, American Iron and Steel Institute/ASP, Aug 2002, www.autosteel.org

1.17 R.H. Wagoner, Advanced High Strength Steel Workshop, Oct 22–23, 2006 (Arlington, VA), sponsored by the National Science Foundation, Department of Energy, and the Auto/Steel Partnership

1.18 B. Yan, "Progress and Challenges in Forming AHSS," Workshop on Forming of AHSS, Ohio State University, April 29, 2009

1.19 WorldAutoSteel, "FutureSteelVehicle (FSV)," Final Engineering Report, Steel Market Development Institute, May 2011, www.worldautosteel.org

1.20 Uddeholm Tooling and SSAB Swedish Steel, "Tooling Solutions for Advanced High Strength Steels," presented at Uddeholm Swedish Rally, 2005

1.21 W.J. Parsons, "Lightweighting the 2013 Cadillac ATS Body Structure," General Motors presentation at Great Designs in Steel Seminar, May 16, 2012 (Livonia, MI), AUTOSTEEL

CHAPTER 2

Steel Fundamentals

STEELS are among the most widely used metals and have a wide variety of applications. Traditionally, steels are used to build cars, appliances, machines, and structures. More advanced specialty steels are used for corrosion resistance (stainless steel), high-temperature applications (super alloys), and lightweighting (high-strength steel, HSS). New grades of advanced high-strength steels (AHSS) have been developed or are still under development for use in applications where strength, safety, recyclability, and cost are essential. Steel has the widest range of compositions, microstructures, and properties of most metal alloys and, therefore, is suited to a wide range of applications.

2.1 Compositions and Metallurgical Phases

Steel is a versatile metal alloy that is basically iron and carbon with varying small amounts of other alloying elements such as manganese, chromium, nickel, and molybdenum. Some of the elements are residual while others are added in order to impart useful properties such as strength, ductility, hardness, toughness, wear resistance, machinability, and weldability. In carbon steel, up to 2% C interacts with iron to produce two principal phases: a single-phase solid solution of body-centered cubic (bcc) iron that is soft and ductile, called *ferrite* (α), and another phase composed of a chemical compound that is hard and strong, called *cementite* or iron carbide (Fe_3C). Cementite forms in the presence of high carbon concentrations. A two-phase mixture called *pearlite* forms at lower carbon content and is made up of alternating layers of soft ferrite and hard cementite. *Martensite* is a very hard phase that results from the very rapid quenching of *austenite* (γ), which is the high-temperature phase of the Fe-C solid solution. *Bainite* is a strong phase that has the same composition as that of pearlite but forms when austenite is cooled more quickly than necessary to form pearlite, but not quickly enough to form martensite.

The basic guide to understanding the behavior of carbon steel is the Fe-C phase diagram. This diagram is fundamental to understanding the relationship between heat treatment, microstructure, and mechanical properties of iron-carbon alloys. Figure 2.1 shows part of the phase diagram for compositions of carbon steels with <2% C and temperatures below 1000 °C (1830 °F). This range defines the wrought steels that are significant for most engineering applications. The diagram shows how temperature and carbon content combine to give a variety of metallurgical structures. Carbon atoms dissolve in iron as interstitial impurities to form the solid solution known as ferrite. When the solubility limit of carbon in iron is exceeded, the chemical compound cementite, or iron carbide (Fe_3C), forms. Of special interest in the Fe-C phase diagram is the eutectoid composition that occurs at 0.76% C. At this composition, the high-temperature austenite will undergo the eutectoid reaction (austenite → ferrite + cementite) at 720 °C (1330 °F). The eutectoid steel is composed of 100% pearlite whose microstructure consists of alternate layers of ferrite and cementite.

At room temperature, the stable phases of steel are ferrite and cementite. Ferrite is a common constituent in steels and has a less densely packed bcc structure. Cementite is a very hard and brittle chemical compound that enhances the strength of steels. All steel alloys, up to 2%, must cool through the γ phase, austenite. Austenite is a high-temperature phase and has a close-packed face-centered cubic (fcc) structure. Heating carbon steel into the austenite region is called austenitization. Figure 2.1 shows that the austenitization temperature decreases with carbon content up to the eutectic composition of 0.76% C, then increases for higher carbon content up to 2% C. Austenite is nonmagnetic and can be retained at room temperature by the addition of appropriate alloying elements such as manganese and nickel.

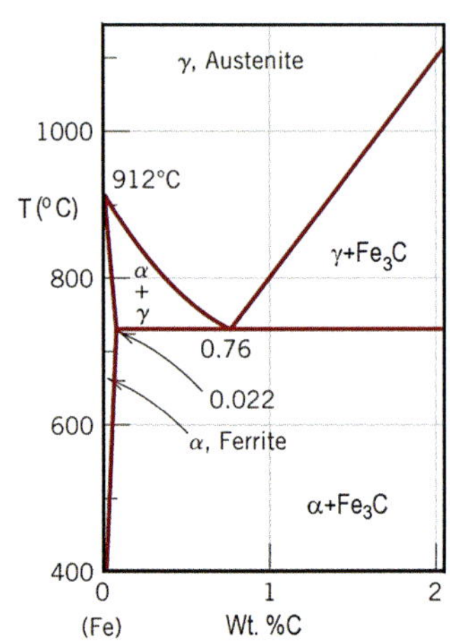

Fig. 2.1 Portion of the Fe-Fe_3C phase diagram for carbon steel

Ferrite is iron containing a very small amount of carbon. The maximum solubility of carbon in iron is 0.022% at 725 °C. The iron-carbon phase diagram describes what happens when a certain grade of carbon steel is slowly heated or cooled. For instance, eutectoid steel transforms completely to austenite on heating above the eutectoid temperature of 725 °C and transforms completely to pearlite on cooling below this temperature. The carbon content provides many different steel compositions and microstructures and, therefore, is the major factor in determining the properties that can be developed in steel. For example, strength increases with carbon content up to the eutectoid composition but then starts to drop as a grain-boundary network of brittle cementite is formed.

Most of the carbon in steel is present in the microstructure as iron carbide. Hence, the wide range of properties in steel that has not been heat treated is determined primarily by the relative amounts of iron and iron carbide in the microstructure. Increasing the fraction of iron carbide in steel, while holding microstructural features constant, will produce harder and stronger steel. Although carbon is the element that transforms iron into steel, other elements are also added to produce a variety of desirable properties. The effect of the alloying elements on the Fe-C phase diagram is to shift its phase boundaries and the shapes of the phase areas. Figure 2.2 shows the dependence of the eutectoid temperature on alloy concentration for several alloying elements in steel. The figure shows that the eutectoid reaction, which occurs at 725 °C for unalloyed steel, is greatly influenced by the type and concentration of the alloying elements.

2.2 Microstructure and Heat Treatment

The three major variables in steel processing are alloying, work hardening, and thermal processing. Alloying refers to the ability of iron to form

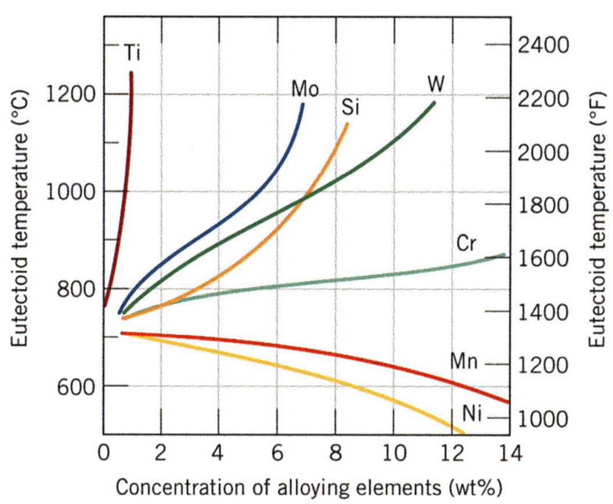

Fig. 2.2 Effect of alloy concentration on the eutectoid temperature in steel. Source: Ref 2.1

solid solutions and compounds with other metals. Work hardening refers to the tendency of steel to become stronger when subjected to plastic deformation. Thermal processing refers to the ability of steel to change its mechanical properties by heat treating while in the solid state. Although chemistry and plastic deformation are important to steel manufacture, heating and cooling are essential for phase transformation and the development of microstructures. Steel properties and performance depend on its microstructure, which is a function of processing.

Heat treatment involves moving atoms around within the crystalline structure of metal by means of solid-state diffusion. The mechanism of diffusion relies on the presence of vacancies in the crystal lattice. The basis of heat treatment relies on the fact that the solubility of carbon in austenite (fcc) is much higher than in ferrite (bcc) because the size of interstices that can accommodate carbon atoms are larger in the fcc structure. At room temperature, all carbon steels have two phases (ferrite and iron carbide). Austenitizing steel produces a single phase (α phase) and this is used as the starting point to control microstructure by heat treating procedures. For carbon steels, cooling transforms the single phase to two phases of widely different compositions. For alloy steels that have more sophisticated chemical compositions, more complex phases are produced.

Iron carbide particles embedded in the ferrite grains form a microstructure that can be observed under a microscope. Figure 2.3 shows the microstructure of eutectoid steel consisting of alternating lamellae of two phases (α and Fe_3C).

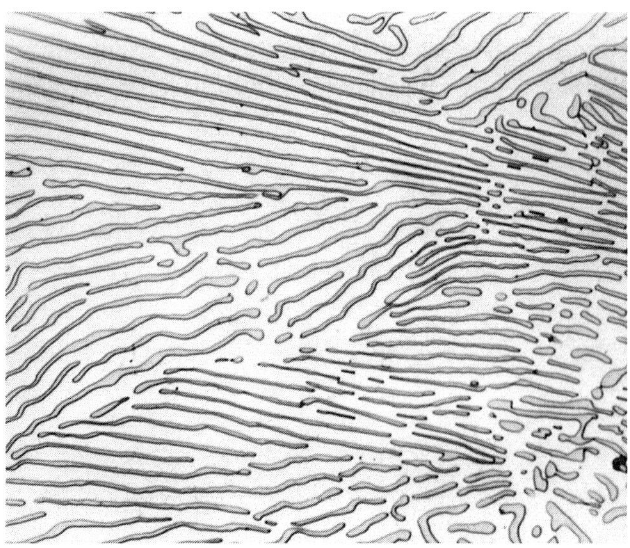

Fig. 2.3 Microstructure of pearlite in eutectoid steel consisting of alternating layers of α-ferrite (light phase) and iron carbide (dark layers). Source: Ref 2.1

Microstructural features include not only the relative amounts of the constituents but also their nature, size, shape, and distribution. For a given carbon content, these features are brought about by heat treatment and can result in significant changes in the properties of carbon steel. The role of heat treatment is to produce a final microstructure in steel that will result in desired properties for the intended application. Microstructure development can be achieved by heat treating cycles aimed at controlling the precipitation of iron carbide from a solid solution of austenite.

Austenite is the high-temperature phase of steel that decomposes on cooling to form the final microstructure of the steel. For a specific steel composition, this microstructure depends on the temperature at which the precipitation occurs. Under equilibrium conditions, austenite transforms to pearlite at a temperature just below the critical temperature of 725 °C (1340 °F).

If nonequilibrium conditions exist (rapid or slow cooling), the transformation microstructure varies markedly with the temperature of the transformation. The transformation of austenite is studied by observing the constant temperature transformation behavior at a series of temperatures below the critical temperature of 725 °C. The results of these observations are presented as a chart where the times at which transformation begins and at which it goes to completion, at a given temperature, are plotted. Such a diagram is referred to as a time-temperature-transformation (TTT) diagram and is shown in Fig. 2.4 for eutectoid steel.

Figure 2.4 also shows how the TTT diagram is generated from individual percent transformation versus logarithm of time plots obtained after the rapid cooling of a sample to a specified constant temperature. The shape of the curve results from a diffusion-limited transformation at low temperatures and an energy-limited transformation at high temperatures. At high temperature, diffusion is fast and coarse pearlite forms. At low temperatures, diffusion is slow and fine pearlite forms, but only near the "knee" of the TTT diagram. Below the "knee," transformation produces a feathery microstructure called bainite.

Figure 2.5 shows a superimposed isothermal heat treatment curve (*ABCD*) where austenite begins its transformation at $T > 725$ °C. It is then rapidly cooled to $T = 625$ °C (1160 °F) and then held at constant temperature of 625 °C until decomposition takes place. This isothermal treatment produces the fine pearlite structure shown in Fig. 2.5. The temperature around 550 °C (1020 °F) gives the shortest time (incubation period) for starting the austenite-to-pearlite transformation and is referred to as the "knee" of the curve, *N*, as shown in Fig. 2.6. This location also defines the critical cooling rate of the steel.

A more complete isothermal transformation diagram for austenite-to-pearlite (*A-P*) and austenite-to-bainite (*A-B*) transformation regions is shown in Fig. 2.6. While pearlite forms above the "knee" of the curve, *N*, bainite, which consists of a very fine combination of ferrite and carbide,

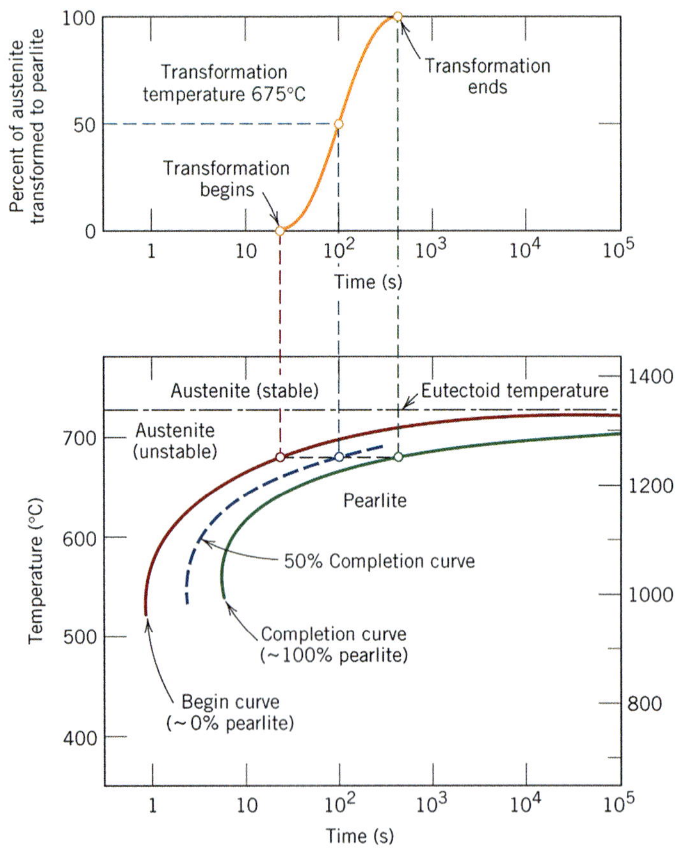

Fig. 2.4 Generation of time-temperature-transformation (TTT) diagram from percent of transformation vs. log time plots. Source: Ref 2.1

forms below it. Pearlite forms in the region from the eutectoid temperature 725 °C (1340 °F) down to approximately 550 °C (1020 °F). Below that temperature, pearlite does not form and, instead, bainite forms. In general, the transformed microstructure becomes more fine grained as the transformation temperature is decreased. Increasing the carbon and alloy content decreases the transformation rate and affects the shape of the TTT diagram by shifting it to the right. This means that the transformation at all temperature levels starts later and takes more time to completion.

Most heat treating processes for steel involve continuous cooling to room temperature rather than isothermal heat treatment. For continuous cooling, the time required to start the transformation is delayed. Therefore, the isothermal transformation curves will be slightly shifted to longer times and lower temperatures. Such curves are referred to as continuous cooling transformation (CCT) diagrams. Figure 2.7 shows the superposition of the CCT diagram on the isothermal cooling (TTT) diagram.

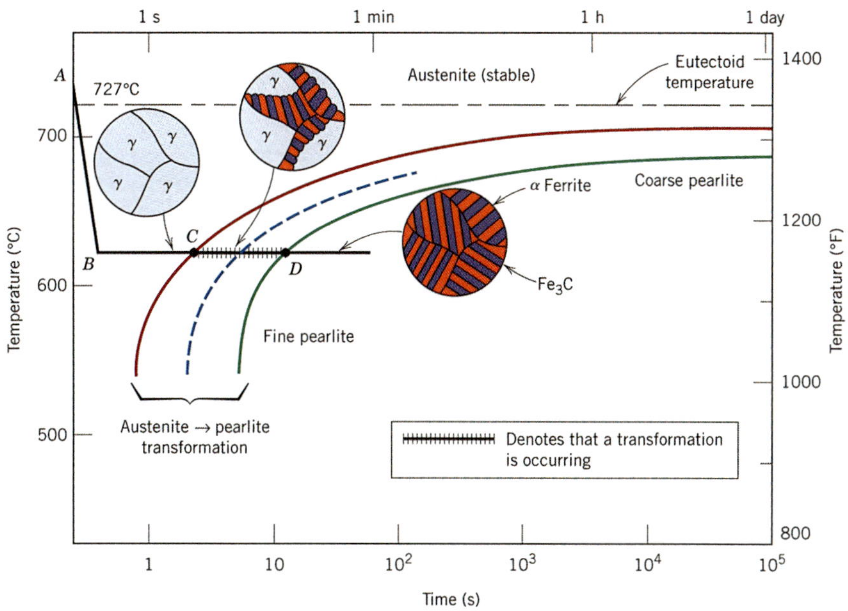

Fig. 2.5 Isothermal transformation of austenite to pearlite in eutectoid carbon steel. Source: Ref 2.1

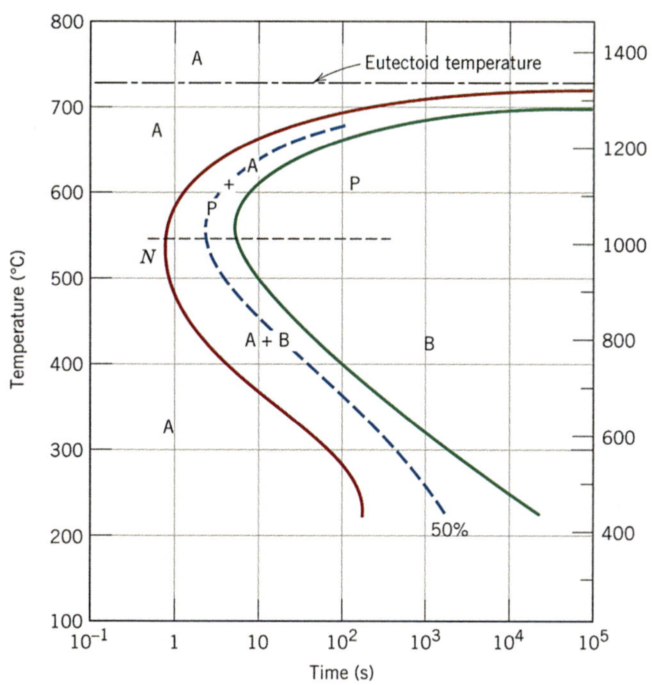

Fig. 2.6 Isothermal transformation of eutectoid steel from austenite to pearlite (*A-P*) and austenite to bainite (*A-B*). Source: Ref 2.1

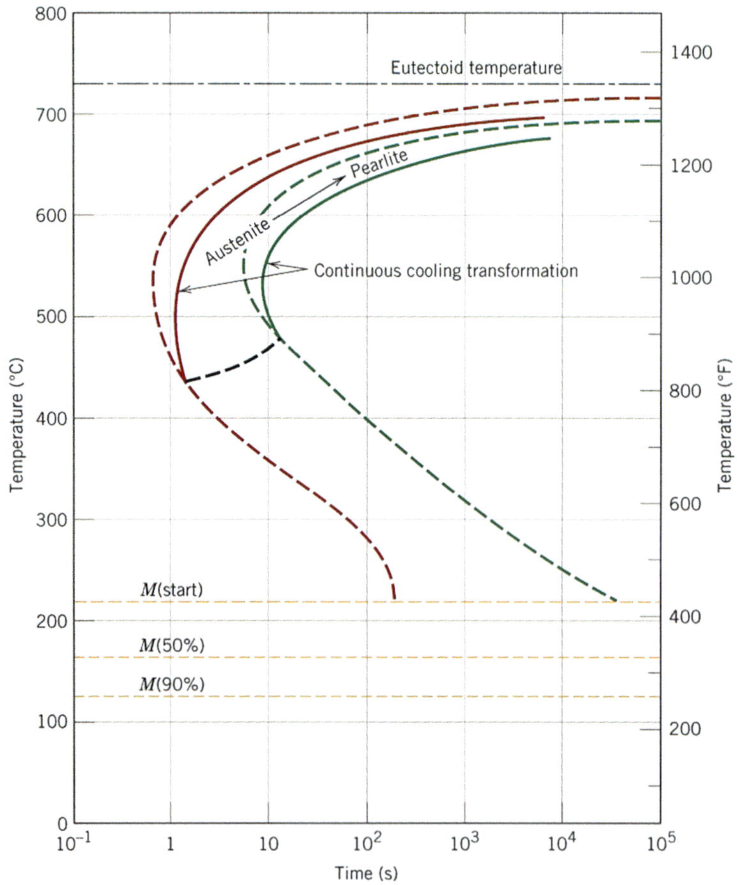

Fig. 2.7 Relationship between isothermal cooling diagram (TTT) and continuous cooling diagram (CCT) for eutectoid steel. Source: Ref 2.1

A schematic representation of the temperature histories for an isothermal cooling diagram (TTT) and a CCT diagram is shown in Fig. 2.8. The temperature histories for the TTT diagram in Fig. 2.8(a) show two quench and hold isothermal treatments, *A* and *B*, to room temperature, *RT*. At *A*, 50% of the transformation from austenite to bainite occurred, while at *B* the transformation is 100% complete. For the CCT diagram in Fig. 2.8(b), three continuous cooling treatments for three different cooling rates are shown. The middle cooling rate is known as the critical cooling rate (CCR) and is defined as the minimum rate that avoids the start of any diffusion-controlled transformation.

Figure 2.9 shows cooling curves for moderately fast (10 min) and slow (15 h) cooling times superimposed on a CCT diagram of eutectoid steel. The fast cooling rate produces a microstructure of fine pearlite and the slow cooling rate produces one with coarse pearlite. The explanation for this is based on the fact that low nucleation rates and high diffusion rates

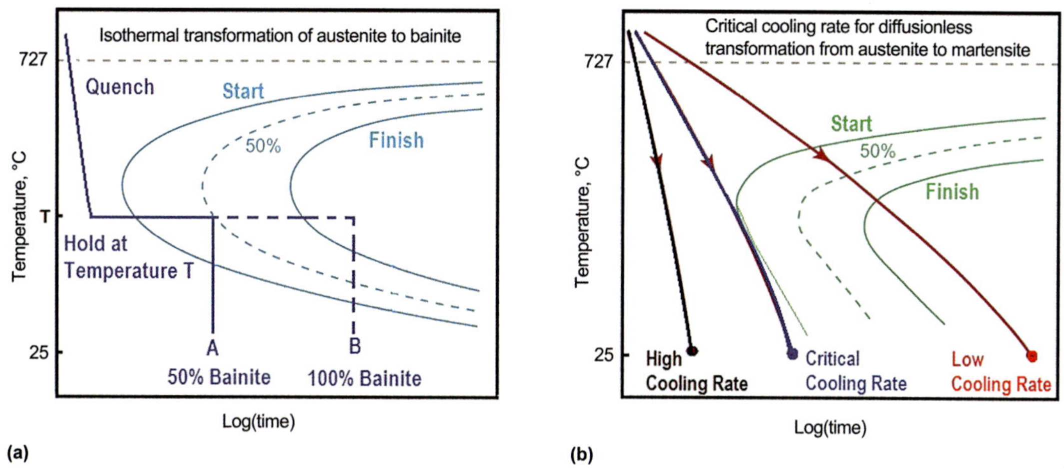

Fig. 2.8 Schematics showing the temperature histories of (a) an isothermal cooling diagram (TTT), and (b) a continuous transformation diagram (CCT). Source: Adapted from Ref 2.2

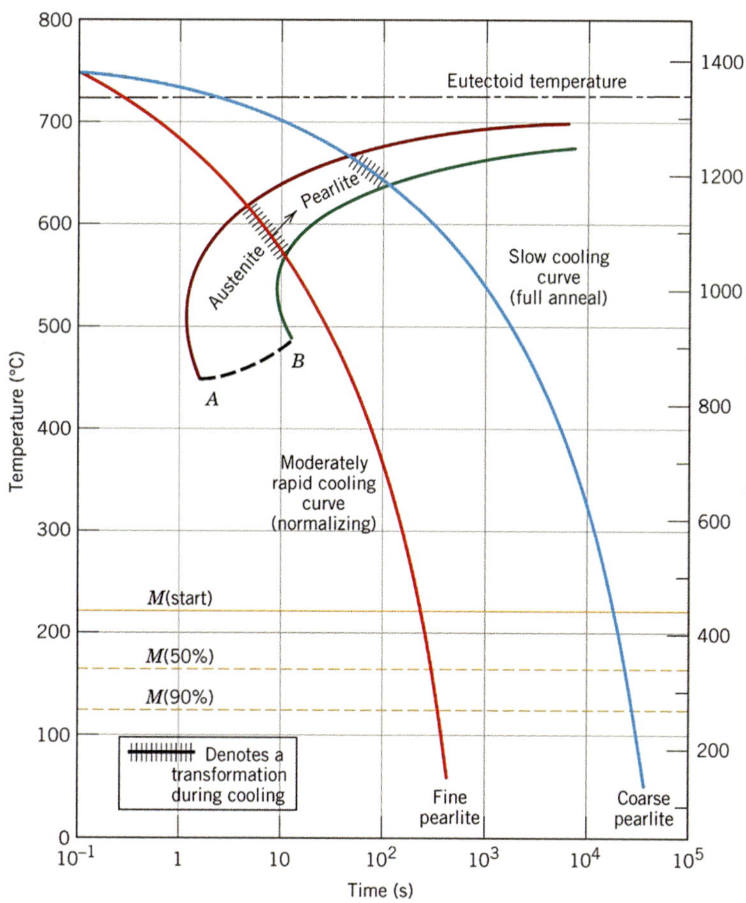

Fig. 2.9 Microstructures of eutectoid steel resulting from fast and slow continuous cooling rates. Source: Ref 2.1

lead to coarse structures while high nucleation rates and low diffusion rates lead to fine structures.

The microstructures of coarse and fine pearlite for eutectoid steel are shown in Fig. 2.10. Coarse pearlite is characterized by thick and widely spaced lamellae of ferrite and cementite while the lamellae of fine pearlite are thin and closely packed. Fine pearlite is harder and less ductile than coarse pearlite.

In many situations, the objective of heat treatment is to cool the sample rapidly enough so that transformation does not occur at all. Close inspection of the CCT diagram in Fig. 2.7 indicates that no information is available below approximately 250 °C (480 °F). Three horizontal lines are added below that temperature to show the transformation of quenched austenite to martensite. Martensite is the microstructure obtained by the rapid quenching of austenite to bypass the "knee" of the CCT diagram. Martensite is important because it is the hardest structure obtained from the transformation of austenite. Figure 2.11 shows a micrograph of martensitic microstructure. As formed, martensite is of little use and must be softened for metalworking. Softening is accomplished by reheating or tempering martensite at temperatures below the eutectoid temperature of 725 °C (1340 °F). Properties of tempered martensite depend on the tempering temperature and the hold time at that temperature.

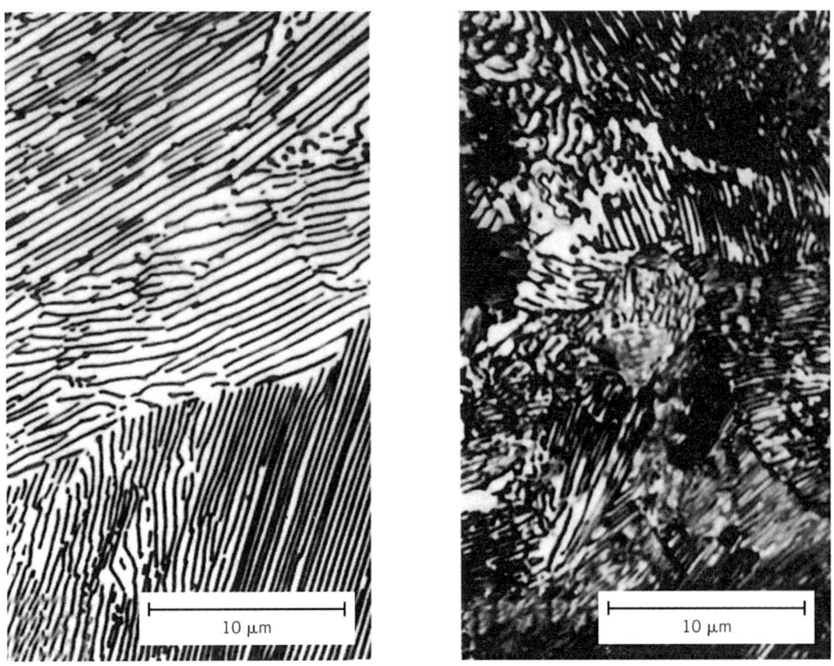

(a) (b)

Fig. 2.10 Micrographs of (a) coarse pearlite and (b) fine pearlite of eutectoid steel. Source: Ref 2.1

Fig. 2.11 Martensite microstructure (needle-shaped grains) and retained austenite (white matrix). Source: Ref 2.1

Martensitic transformation involves the reorientation of carbon and iron atoms from the fcc solid solution of austenite to a distorted body-centered tetragonal (bct) solid solution of martensite. The distortion is caused by trapped atoms of carbon and the distorted lattice is resistant to dislocation glide and hence increases its strength and hardness. Transformation to martensite occurs at temperatures below 250 °C, and the percentage of transformation depends only on the temperature to which it is cooled. Transformation to martensite occurs by a diffusionless process, and it differs from transformation to pearlite or bainite in that it is not time dependent, but occurs instantly during cooling. Figure 2.12 shows the CCR for eutectoid steel superimposed on a CCT diagram. It also demonstrates the dependence of the final microstructure on the cooling rate. Cooling rates greater than the critical cooling rate will produce martensite only. Lower cooling rates than the critical will produce both martensite and pearlite, and for the lowest cooling rates only pearlite forms.

The transformation from austenite to martensite produces a brittle microstructure with low fracture toughness and diminished ability for deformation processing of the material. Ductility is regained by a reheating process called tempering. On tempering, the excess carbon in the deformed bct lattice combines with iron atoms to form Fe_3C precipitates.

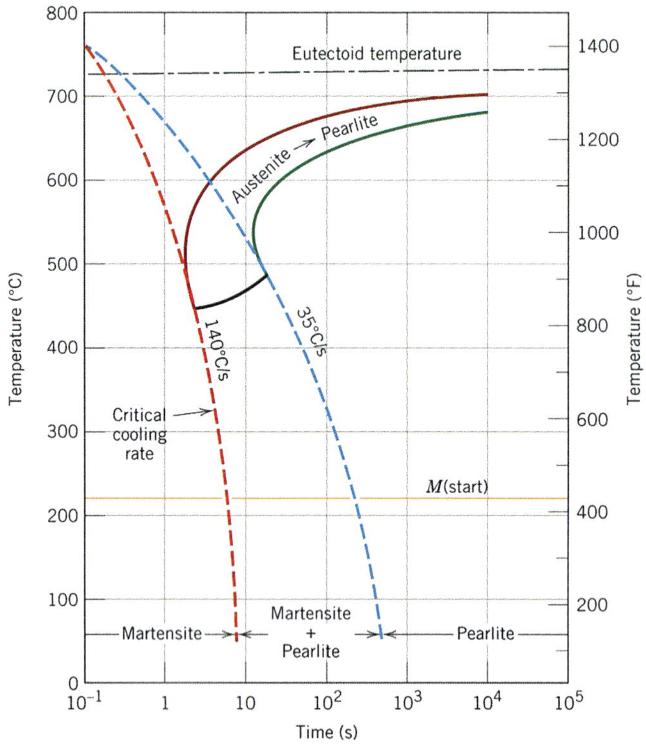

Fig. 2.12 Effect of cooling rate on development of final microstructure. Source: Ref 2.1

The strengthening mechanism in tempered martensite shifts from resistance to dislocation glide in a distorted lattice to resistance to dislocation glide by precipitates. The resulting yield stress evolves with tempering time because the process of nucleation and growth of the Fe_3C precipitates is time dependent.

Figure 2.13 shows (a) quenching of austenite to martensite on a CCT diagram and (b) evolution of yield stress during tempering of quenched carbon steel. Tempering reduces the yield strength to levels between quenched martensite (2500 MPa, or 363 ksi) and normalized ferrite-pearlite microstructure (350 MPa, or 51 ksi). Manipulating properties of steel by using quench and temper treatment is a widespread technique for controlling the toughness of steel.

A block diagram that summarizes the effect of cooling rates on the microstructure of steel is shown in Fig. 2.14.

A summary of the microstructures and mechanical attributes of carbon steel is listed in Table 2.1.

A major concern in heat treatment is that a low enough cooling rate cannot be practically achieved to avoid the "knee" of the CCT diagram; hence, 100% martensite cannot be produced. This problem can be solved

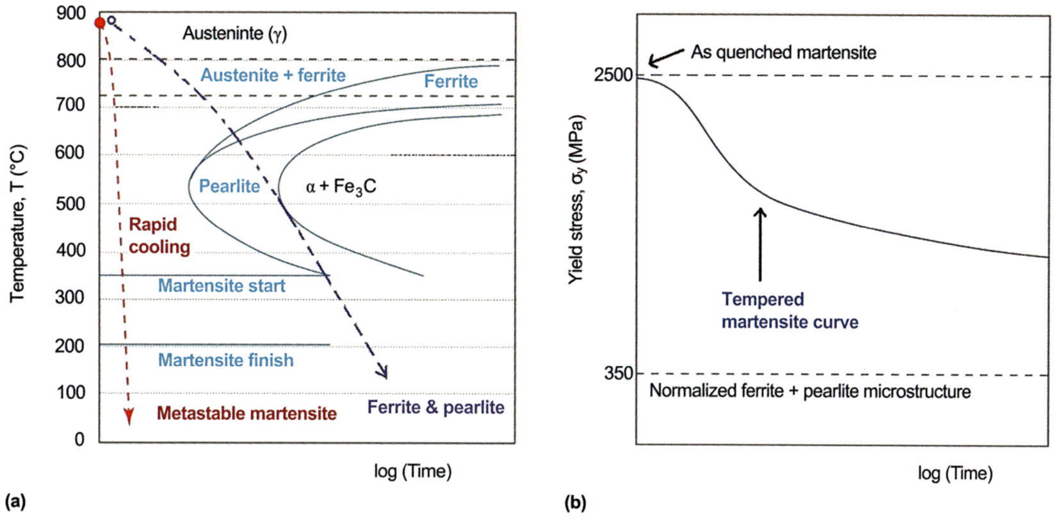

Fig. 2.13 (a) Martensite formation from austenite. (b) Evolution of yield stress during tempering of martensite. Source: Adapted from Ref 2.2

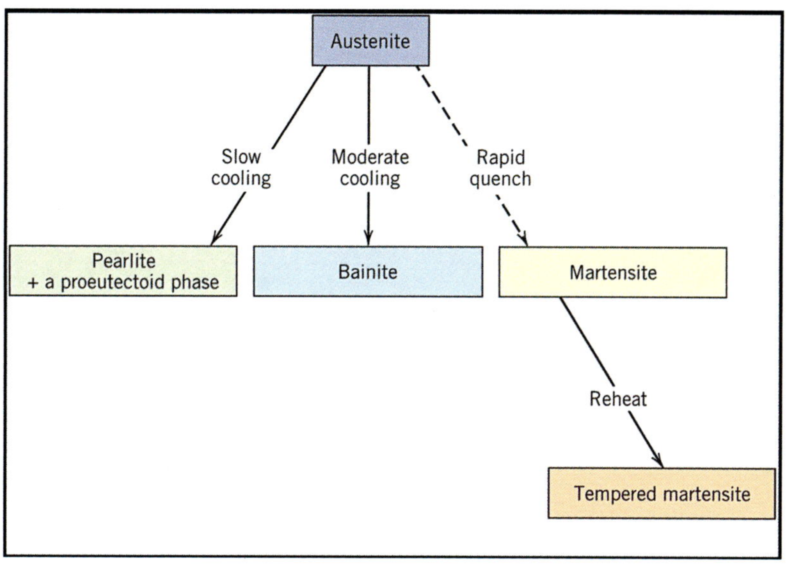

Fig. 2.14 Block diagram of the effect of cooling rate on development of final microstructure. Source: Ref 2.1

Table 2.1 Microstructure and attributes of steel

Microconstituent	Phases present	Mechanical attributes
Pearlite	α ferrite + Fe_3C	Soft and ductile
Bainite	α ferrite + Fe_3C	Stronger and less ductile than pearlite
Martensite	Single-phase bct structure	Stronger and less ductile than bainite
Tempered martensite	α ferrite + fine Fe_3C	Strong but more ductile than martensite

bct, body-centered tetragonal. Source: Compiled from Ref 2.1

by shifting the CCT diagram to the right to achieve longer times for its "knee." The shift can be accomplished by increasing the carbon content and adding small amounts of alloying elements. Figure 2.15 shows a shifted CCT diagram for high-strength, low-alloy (HSLA) steel AISI/SAE 4340 obtained by increasing carbon content to 0.40% and adding alloying elements nickel, chromium, and molybdenum. Alloying elements affect diffusion and limit austenite transformation at high temperature to the extent that a double "knee" is formed. Figure 2.15 shows that the "knee" of the CCT diagram shifted from a low cooling time of approximately 1 second for low carbon steel (Fig. 2.9) to a high cooling time of approximately 10 seconds for HSLA steel (Fig. 2.15).

The properties of steel can be controlled by heat treating procedures through the control of microstructure. Heat treating involves heating the steel to a temperature where austenite is formed and then cooling it at an appropriate rate to transform it to the desired microstructure. Heat treat-

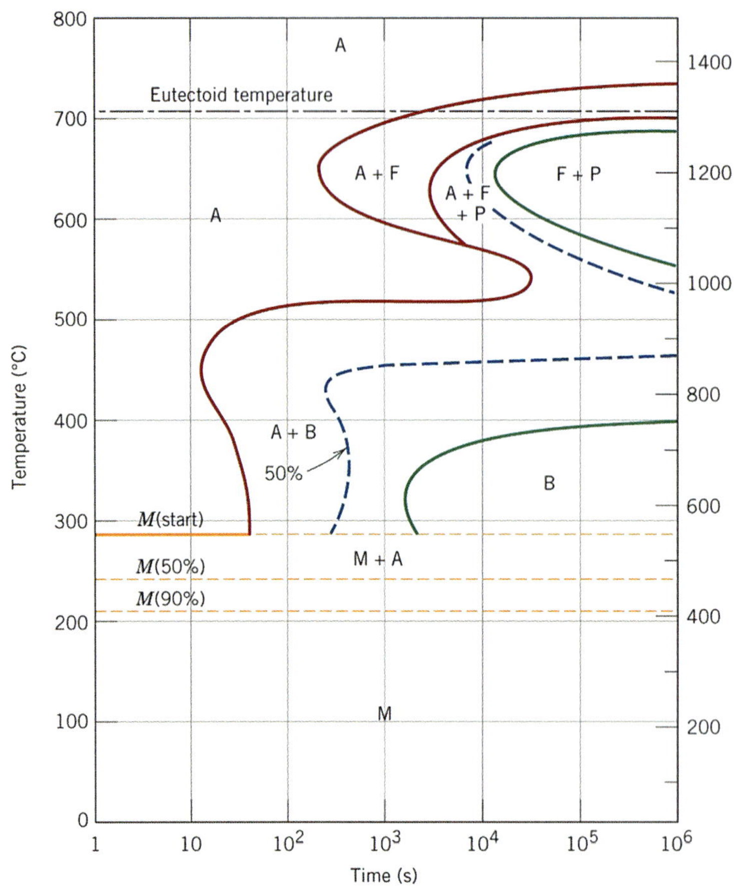

Fig. 2.15 Continuous cooling transformation (CCT) diagram for high-strength, low-alloy steel AISI/SAE 4340. Source: Ref 2.1

ment usually includes reheating and cooling to relieve residual stresses and to modify the microstructure to control the strength and ductility of the steel.

Conventional heat treating procedures include annealing, normalizing, quenching and tempering, stress relieving, martempering, and austempering. Figure 2.16 shows time-temperature diagrams of the major heat treating procedures and products produced by decomposition of austenite. The heat treating diagrams are superimposed on a typical isothermal transformation diagram to illustrate graphically the heating and cooling operations involved in each heat treating process. Each heat treating procedure produces a microstructure that determines the final mechanical properties of the steel. The first step in the heat treating procedures that are based on the decomposition of austenite requires that the steel be heated to a temperature within the austenitizing range of 790 to 915 °C (1455 to 1680 °F). Following is a summary of the four major transformation cycles:

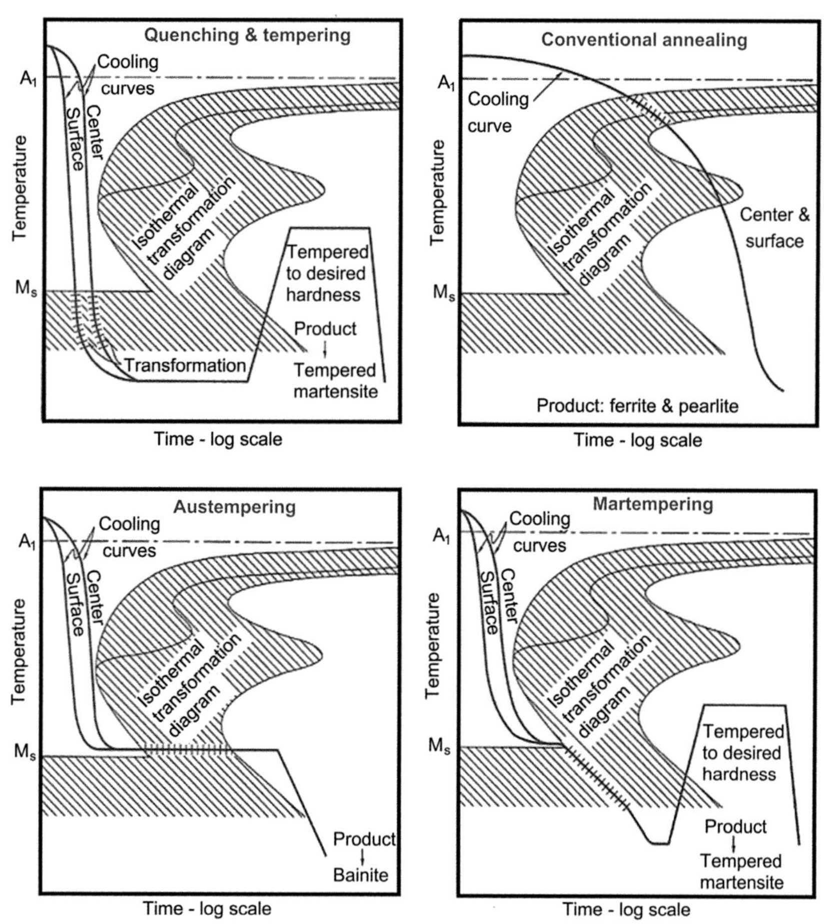

Fig. 2.16 Major heat treating procedures superimposed on TTT diagram. Source: Ref 2.3

1. *Conventional Quenching and Tempering:* Austenitizing steel disperses the carbon into solution. Quenching to room temperature traps the carbon in a supersaturated solution. The phase transformation of iron from fcc to bcc results in a metastable phase called martensite. The extra carbon in the bcc lattice causes it to distort into the bct lattice of martensite. Martensite has very high yield strength but very low ductility. It is then tempered to desired hardness by heating at temperatures in the range of 210 to 650 °C (410 to 1200 °F) to transform martensite to ferrite and cementite. The microstructure of tempered martensite is different from that of annealed ferrite and cementite.
2. *Conventional Annealing:* Austenitized steel is cooled slowly to form ferrite and pearlite.
3. *Austempering:* Austenitized steel is quenched in a bath maintained at a constant temperature range of 260 to 400 °C (500 to 750 °F). The steel is allowed to transform isothermally to bainite in this bath, then cooled to room temperature.
4. *Martempering:* Austenitized steel is quenched below the "knee" of the TTT diagram. It is then slow cooled to form martensite and tempered to desired hardness by transforming martensite to ferrite and ce-mentite.

2.3 Structure-Property Relationships

The mechanical properties of metals that are of most interest to engineers are strength, ductility, and toughness. Strength determines the stress that can be applied to a structure without resulting in plastic deformation. For most applications the yield strength is important to prevent the onset of plastic flow. For other applications the tensile strength defines the ability for a structure to carry or resist mechanical forces or stresses. Ductility measures the ability of a metal to be formed and is defined as the amount of plastic flow that can occur beyond the yield and before fracture. It is measured in terms of percent elongation or reduction in area. Toughness is related to the resistance of the metal to fracture under conditions of stress concentration resulting from cracks, notches, or defects. It is measured in terms of the energy required to fracture a standard test specimen.

Strength, ductility, and toughness are related because all three are governed by the same deformation mechanism of dislocation glide. In general, as strength increases, ductility and toughness decrease. These properties are microstructure-sensitive because they are influenced by the amount and distribution of constituents and the presence of imperfections such as solutes, inclusions, precipitates, dislocations, twins, grain boundaries, and cracks in the metal.

Mechanical and performance properties of steels are determined by three main factors:

- Composition of constituents
- Microstructures resulting from heat treatment

- Loading mode (tension, compression, and shear)

Engineering components can have complex shapes and can be loaded by a variety of forces. The stresses in small elements within the components can be a combination of three basic modes of loading: tension, compression, and shear. Loading effects, other than tension, are not considered in this chapter. Accordingly, the structure-property relationship in steel is based on the effects of composition and microstructure. Variations in mechanical properties due to steel composition are shown in Fig. 2.17. Increasing the carbon content increases the yield and tensile strengths as well as the hardness of plain carbon steel. Ductility, as measured by elongation or reduction in area, and toughness decrease with increase in carbon content of plain carbon steels.

The microstructure of steel has a significant effect on its strength. For steel with a certain composition the microstructure can be altered by varying the heat treating process. Table 2.2 shows how different yield strengths can be achieved by obtaining different microstructures for a specific steel composition of 0.15 wt% C and 1.0 wt% Mn.

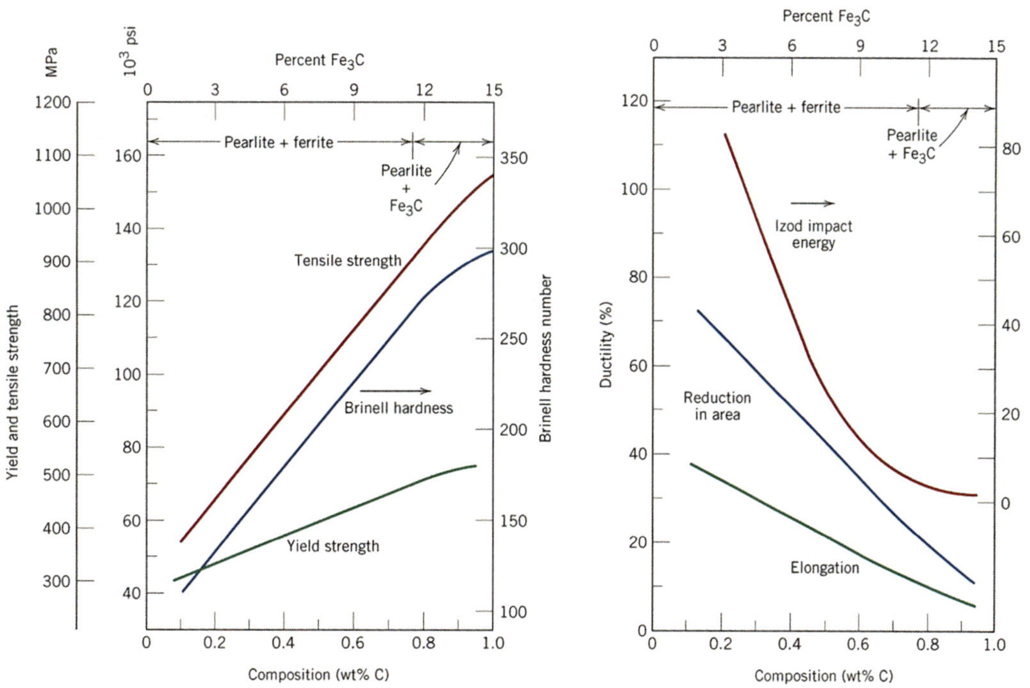

Fig. 2.17 Effect of carbon content on the mechanical properties of plain carbon steels. Source: Ref 2.1

Table 2.2 Effect of microstructure on yield strength of steel

Heat treating process	Microstructure	Yield strength, MPa (ksi)
Water quench	Martensite	450 (65)
Oil quench	Bainite	300 (44)
Air cool	Ferrite + pearlite	150 (22)

Figure 2.18 shows the effect of microstructure on the hardness and ductility of different plain carbon steel compositions. Alloys containing pearlitic microstructures have greater strength and hardness than those containing ductile spheroidite.

Figure 2.19 shows the effect of carbon content on the tensile strength and hardness of carbon steels. High-carbon steels can also be heat treated to form martensite, making the steel harder and stronger.

Different thermal processing schemes produce different room-temperature microstructures. Slow cooling leads to soft ferrite microstructure while quenching produces hard martensite. Table 2.3 shows the microstructural effects on mechanical properties of various grades of steel.

The effect of quenching and tempering on fracture toughness of plain carbon steels of varying carbon content is shown in Fig. 2.20. Fracture toughness decreases with increase in the yield strength of steel. Quenched steel (martensite) has the highest yield strength and the lowest fracture toughness compared to the other tempered and normalized steels.

Increasing carbon content increases strength and reduces fracture toughness of steel. Quenching the 0.4% C steel produces martensite, which has high strength but extremely low fracture toughness, below the threshold of 15 MPa (2.2 ksi) for structural applications. Tempering the 0.4% C steel reduces its yield strength but increases its fracture toughness.

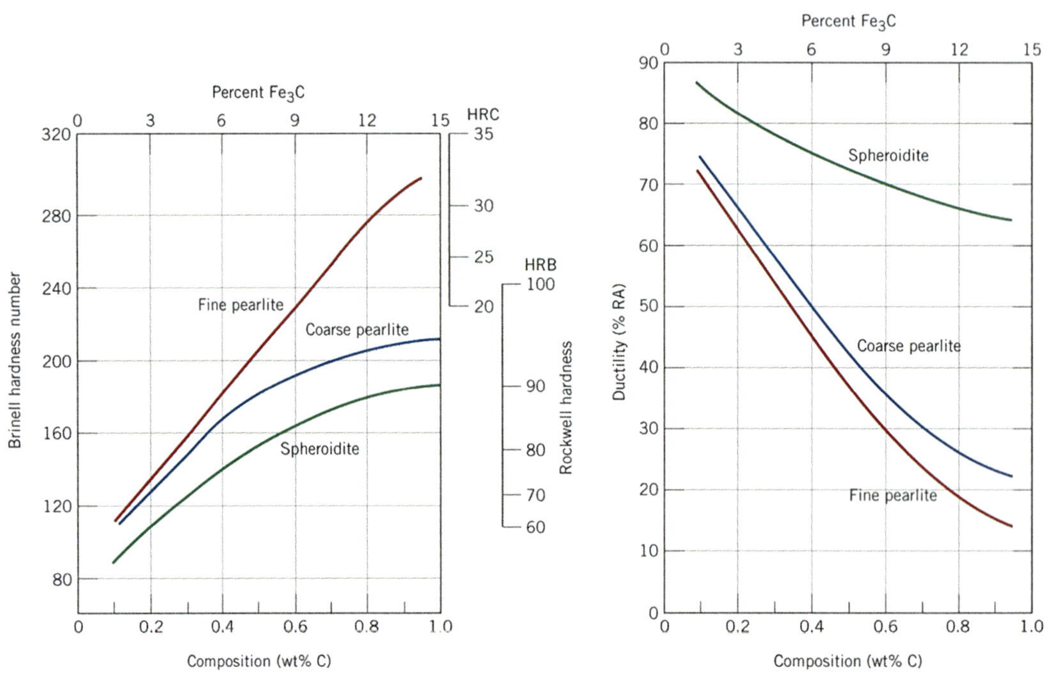

Fig. 2.18 Effect of microstructure on mechanical properties of different plain carbon steel compositions. Source: Ref 2.1

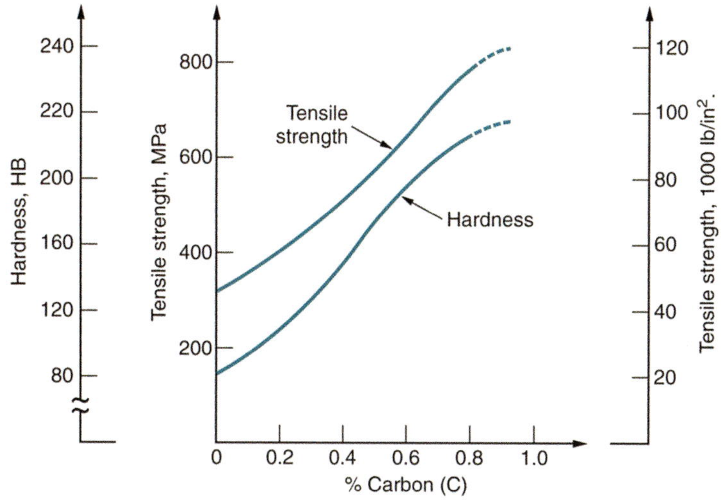

Fig. 2.19 Effect of carbon content on the tensile strength and hardness of carbon steels. Source: Ref 2.1

Table 2.3 Microstructural effects on mechanical properties of various grades of steel

Steel grade	Microstructure	Tensile strength range, MPa (ksi)	Ductility range, %
Interstitial-free (IF)	Ferrite	150–250 (22–36)	40–55
Mild steel	Ferrite + pearlite	180–300 (26–44)	30–45
High-strength, low-alloy (HSLA)	Ferrite + pearlite + alloy precipitates	300–800 (44–116)	10–25
Dual-phase (DP)	Ferrite + martensite	450–1000 (65–145)	6–30
Complex-phase (CP)	Fine ferrite + higher fraction of martensite and pearlite	450–1200 (65–174)	6–30
Transformation-induced plasticity (TRIP)	Soft ferrite + hard bainite + retained austenite	500–1200 (73–174)	10–33
Martensitic (MS)	100% martensite	720–1650 (104–239)	3–12
Twinning-induced plasticity (TWIP)	Retained austenite	1100–1650 (160–239)	43–63
Austenitic stainless steel (AUST SS)	Retained austenite	900–1150 (131–167)	45–63

The higher the tempering temperature, the higher is the fracture toughness and the lower is the yield strength.

Figure 2.21 compares the microstructures of various phases of steel grades. Ferrite and martensite have single-phase microstructures. Dual-phase (DP) steel has a two-phase ferrite and martensite, while transformation-induced plasticity (TRIP) steel has a multiphase ferrite, bainite/martensite, and retained austenite microstructure.

2.4 Categories of Steels

Within a steel category, a distinction should be made as to whether or not a significant amount of alloying elements, other than carbon, is used. A composition of 5% alloying elements is an arbitrary boundary between low-alloy and high-alloy steels. Various elements are added to steels to

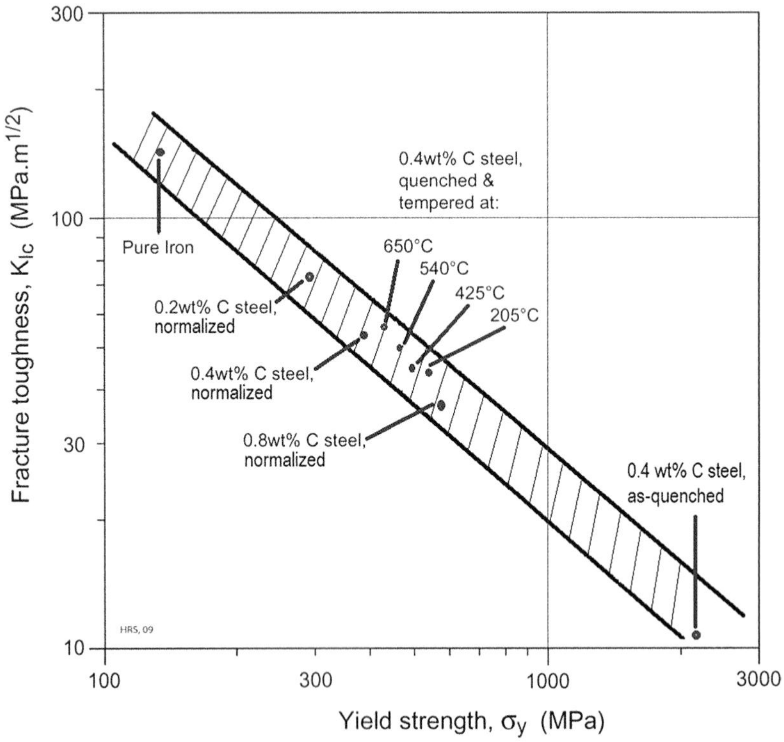

Fig. 2.20 Effect of quenching and tempering on the fracture toughness of plain carbon steels. Source: Adapted from Ref 2.2

produce beneficial properties in strength, hardness, toughness, wear resistance, formability, weldability, and machinability. Alloy additions increase material costs and are justified only by essential improvements in properties for certain applications.

A number of arbitrary definitions of the classes of steels have been developed. The American Iron and Steel Institute (AISI) and the International Ultra-Light Steel Auto Body (ULSAB) Consortium define the strength levels of mild steel, HSS, and AHSS (Table 2.4).

It is important to realize that some HSS may be designated by minimum tensile strength and not by minimum yield strength. To provide a consistent nomenclature, the ULSAB Consortium adopted a standard practice that defines both yield strength (YS) and ultimate tensile strength (UTS).

Original equipment manufacturers (OEMs) use a different categorization scheme for flat rolled steels that is based on the tensile strength of the steel (Table 2.5).

The AISI/SAE and UNS designation systems and composition ranges for plain carbon and low alloy steels are listed in Table 2.6.

Following is a brief description of relevant steel grades (Ref 2.5).

Carbon steels are the most widely used metals and have a wide range of applications. They are characterized by good strength, ductility, weld-

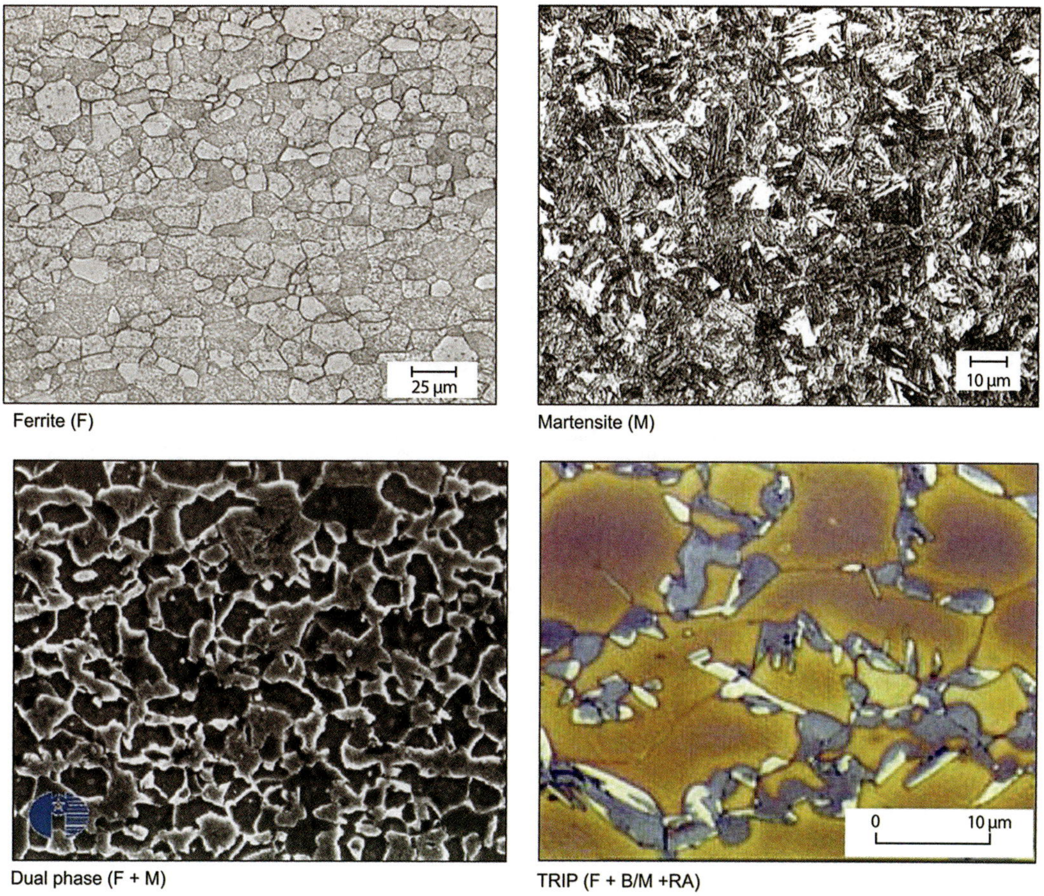

Fig. 2.21 Microstructures of various phases of steel. Source: Ref 2.4

Table 2.4 AISI and ULSAB definitions of mild, high-strength, and advanced high-strength steels

Steel category	Yield strength, MPa (ksi)
Mild steel	140– <210 (20– <30)
High-strength steel (HSS)	210–550 (30–80)
Advanced high-strength steel (AHSS)	>550 (>80)

AISI, American Iron and Steel Institute. ULSAB, International Ultra-Light Steel Auto Body Consortium. Source: Ref 2.5

Table 2.5 Segmentation of flat rolled steel grades by original equipment manufacturers

Steel category	Tensile strength, MPa (ksi)
Mild steel	<270 (39)
Conventional high-strength steel (HSS)	350–590 (51–86)
Advanced high-strength steel (AHSS)	>700 (102)

Table 2.6 Designation systems and composition ranges for plain carbon and low-alloy steels

		Composition ranges (wt% of alloying elements in addition to C)(b)			
AISI/SAE designation(a)	UNS designation	Ni	Cr	Mo	Other
10xx, plain carbon	G10xx0				
11xx, free machining	G11xx0				0.08–0.33 S
12xx, free machining	G12xx0				0.10–0.35 S, 0.04–0.12 P
13xx	G13xx0				1.60–1.90 Mn
40xx	G40xx0			0.20–0.30	
41xx	G41xx0		0.80–1.10	0.15–0.25	
43xx	G43xx0	1.65–2.00	0.40–0.90	0.20–0.30	
46xx	G46xx0	0.70–2.00		0.15–0.30	
48xx	G48xx0	3.25–3.75		0.20–0.30	
51xx	G51xx0		0.70–1.10		
61xx	G61xx0		0.50–1.10		0.10–0.15 V
86xx	G86xx0	0.40–0.70	0.40–0.60	0.15–0.25	
92xx	G92xx0				1.80–2.20 Si

(a) The carbon concentration, in weight percent ×100, is inserted in the place of "xx" for each specified steel. (b) Except for 13xx alloys, Mn concentration is <1.00 wt%. Except for 12xx alloys, P concentration is <0.35 wt%. Except for 11xx and 12xx alloys, S concentration is <0.04 wt%. Except for 92xx alloys, Si concentration varies between 0.15–0.35 wt%. Source: Ref 2.1

ability, and low cost. Carbon steels are classified according to their carbon content. Low-carbon steel has <0.3% C, medium-carbon steel has 0.30 to 0.6% C, and high-carbon steel has >0.6% C. The higher the carbon content of the steel, the higher are its strength and hardness and the lower its ductility, toughness, and weldability.

Alloy steels are basically carbon steels with certain chemical elements added to improve the properties of the metal for specific applications. Alloying elements include manganese, silicon, copper, sulfur, phosphorus, nickel, selenium, molybdenum, vanadium, and chromium in amounts totaling less than ~5% by weight. Their mechanical properties, such as strength, hardness, creep resistance, and fatigue resistance are superior to plain carbon steels. One of the reasons for adding alloying elements is to enhance the metal hardenability. Heat treatments, such as quenching and tempering, normalizing, annealing, and stress relieving, are often required to achieve these improved properties. Alloy steels are mainly used in the construction and transportation industries.

Bake-hardenable (BH) steels have a good combination of strength and formability. This combination makes them ideal for dent-resistant auto applications such as hoods, doors, and fenders. The increase in yield strength in this class of steel results from work hardening and strain aging during paint-baking after the stamping process. Bake-hardening produces an increase in yield strength ranging from 28 to 56 MPa (4 to 8 ksi) after a low-temperature heat treatment cycle similar to a paint-bake schedule (175 °C, or 350 °F, for 20 to 30 min). The bake-hardening effect is possible only if the steel has undergone low plastic deformation of <5% and has been exposed to a baking temperature of at least 150 °C (300 °F). These steels are appropriate for some low-deformation body panels.

High-strength, low-alloy (HSLA) steels have low carbon content (<0.3% C) and approximately 1% or less of alloying elements such as manganese, phosphorus, chromium, nickel, silicon, or molybdenum. Their microstructure consists of a fine-grain ferrite phase and a hard martensite and austenite phase. The high strength of the HSLA steels results from carefully controlled microalloying and thermal processing. These steels are used mainly in the automotive industry where strength-to-weight ratio applications are important.

High-strength, low-alloy steels have small amounts of alloying elements such as phosphorus, manganese, or silicon added to low-carbon (0.02 to 0.13% C) steels to attain high strength levels. In these steels, higher strength is achieved by rapid cooling to produce a very fine ferrite grain size, by solid-solution strengthening with phosphorus, nitrogen, silicon, and manganese and by formation of carbides or carbonitrides with vanadium, nickel, and titanium. These steels have better mechanical properties, corrosion resistance, and weldability than mild steels. High-strength, low-alloy steels are produced as hot rolled products and cold rolled sheet. These steels can achieve yield strengths up to 485 MPa (70 ksi) without heat treatment. This class of steel has low formability and is used for shallow drawn parts.

Stainless steels are iron-base alloys containing a minimum of 12 to 30% Cr and >50% Fe. Carbon is usually present in amounts ranging from <0.03% to >1.0% in certain martensitic grades. Other alloying elements such as nickel, molybdenum, copper, titanium, manganese, cobalt, aluminum, nitrogen, and sulfur are added to enhance specific characteristics. Stainless steels are characterized by their corrosion resistance, high strength, and ductility. Chromium forms a thin impermeable oxide film that protects the surface from corrosion. Nickel is another alloying ingredient in certain stainless steels that increases corrosion protection. The amount of nickel determines whether the stainless steel is heat treated or remains austenitic. Carbon is used to strengthen and harden stainless steels, but high carbon content reduces corrosion protection because chromium carbide forms and this reduces the availability of the corrosion-resisting free chromium. Stainless steels types and compositions are classified according to the predominant phase present at ambient temperature. They are commonly divided into five groups:

Martensitic stainless (MS) steels are alloys of iron, chromium, and carbon. They contain as much as 18% Cr but no nickel. Carbon content may exceed 1%. They are magnetic and are heat treatable. Heat treatment produces martensite that is hard and strong. They are mainly used where strength, hardness, and wear resistance are required.

Ferritic stainless steels are iron alloys containing approximately 15 to 20% Cr, low carbon, and no nickel. They have a bcc crystal structure and are not heat treatable. They are used for decorative trim, sinks, and automotive exhaust systems.

Austenitic stainless steels (AUST SS) are iron alloys containing between 16 to 26% Cr and approximately 8% Ni. They are nonmagnetic and have a fcc crystal structure. Austenite is retained because of the large amounts of austenitizing elements used, such as nickel, manganese, and nitrogen.

They cannot be hardened by heat treatment and are hardened only by cold working. The austenite phase is tough and ductile. Because of their high chromium and nickel content, they are the most corrosion resistant of the stainless group. Also, they have excellent mechanical properties. Typical applications include flatware and sinks, pressure and cryogenic vessels, and food and chemical processing equipment.

Duplex (ferritic-austenitic) stainless steels are alloys of iron containing roughly equal amounts of bcc ferrite and fcc austenite. The percentage of each phase depends on the composition and heat treatment. The primary alloying elements are chromium and nickel. Duplex stainless steels have greater tensile and yield strengths, but lower toughness than austenitic stainless steels. They are more resistant to stress-corrosion cracking than austenitic alloys, yet tougher than ferritic alloys.

Precipitation-hardening stainless steels contain chromium and nickel and may be either austenitic or martensitic in the annealed condition. They develop high strength during heat treatment by utilizing precipitation hardening rather than phase transformation. Originally developed for aerospace applications, they are used in many applications.

Table 2.7 compares the major relative properties of four grades of stainless steel. Austenitic stainless steels have the best formability, corrosion resistance, and weldability compared to the other grades.

Stainless steels are significantly more expensive than plain carbon or low-alloy steels because they contain large amounts of expensive alloying metals. They have been used for many industrial, architectural, chemical, and consumer applications. Austenitic stainless steels have excellent formability with good strength and this makes them potential material for automotive application. Austenitic stainless steels are covered in more detail in Chapter 10 in this book, "Austenitic Stainless Steels (AUST SS)."

Advanced high-strength steels are multiphase steels that contain a ferrite matrix and islands of martensite, bainite, austenite, and/or retained austenite in quantities and combinations sufficient to produce desired mechanical properties. They are produced by nontraditional thermal cycles and exhibit a superior combination of high strength with good formability and high strain-hardening capacity. Advanced high-strength steels fall in

Table 2.7 Comparison of relative properties of various stainless steel grades

	Stainless steel grade			
Property	Ferritic 51400	Martensitic 51400	Austenitic 30300	Precipitation hardening
Heat treatable	No	Yes	No	Yes
Magnetic	Yes	Yes	No	Yes
Microstructure	Ferrite	Martensite	Austenite	Martensite
Formability	B	D	A	C
Strength	D	A	C	B
Corrosion resistance	C	D	A	B
Oxidation resistance	A	D	B	D
Weldability	B	D	A	D

A = highest, D = lowest. Source: Ref 2.6

two categories: first generation and second generation. Members of the first generation include dual-phase (DP), complex-phase (CP), transformation-induced plasticity (TRIP), and martensitic (MS) grades. Members of the second generation include twinning-induced plasticity (TWIP), light steel with induced plasticity (L-IP), and austenitic stainless steel (AUST SS). Following is a brief review of the AHSS steel grades.

Dual-Phase (DP) Grades. The duplex microstructure of DP steels is comprised of a soft ferrite matrix and between 20 and 70% volume fraction of martensite. The volume fraction of martensite determines the strength level of this steel. Special heat treating practices that involve quenching and tempering are used to generate the martensite phase. The dual-phase structure is produced by quenching low-carbon steels from the $\alpha + \gamma$ phase region to form a microstructure of martensite islands in a ferrite matrix. This microstructure gives DP steels their high strain-hardening capability and better formability compared to HSLA grades. Dual-phase steels can also exhibit a bake-hardening effect, which is the increase in yield strength resulting from prestraining and elevated-temperature aging. The extent of the bake-hardening effect in AHSS depends on the specific chemistry and thermal histories of the steels. Dual-phase steels are designed to provide ultimate tensile strengths of 600 to 1000 MPa (87 to 145 ksi).

Complex-Phase (CP) Grades. Complex-phase steels consist of a very fine microstructure of ferrite and a higher volume fraction of hard phases that are further strengthened by fine precipitates. They use many of the same alloy elements found in DP and TRIP steels, but additionally have small quantities of niobium, titanium, and/or vanadium to form fine strengthening precipitates. Complex-phase steels provide ultimate tensile strengths of 800 MPa (115 ksi) and greater. They are characterized by high deformability and high energy absorption, which make them ideal candidates for automotive crash applications, such as bumper and B-pillar reinforcements.

Transformation-Induced Plasticity (TRIP) Grades. Transformation-induced plasticity steels consist of two phases: a ferrite/bainite matrix and a 5 to 20% volume fraction of metastable retained austenite, which progressively transforms to martensite during plastic deformation. The transformation provides significant dispersion hardening and resistance to localized deformation. This combination of phases gives TRIP steels the high formability of austenite during the initial stages of the stamping process followed by the high strength of martensite at the end of the forming process. Transformation-induced plasticity steels are characterized by high tensile strengths and high strain-hardening rates. The excellent formability of TRIP steels combined with their high strength make them attractive candidate materials for lightweight applications. Because the martensite formation occurs at extremely fast rates, the TRIP phenomenon occurs even at high deformation rates, as encountered during vehicle crash conditions. Therefore, TRIP steels are also being considered for applications requiring maximization of crash energy absorption.

Martensitic (MS) Grades. Martensitic steels contain a high fraction of martensite. The microstructure is low-carbon martensite, with the carbon content determining the strength level of the material. The austenite that exists during hot rolling or annealing of the material is transformed to martensite during quenching and/or post-forming heat treatment. Martensitic steels are often subjected to post-quench tempering to improve ductility and can provide good formability even at extremely high ultimate tensile strengths of up to 1500 MPa (215 ksi). Carbon is added to martensitic steels to increase hardenability and also to strengthen the martensite. Elements such as manganese, silicon, chromium, molybdenum, boron, vanadium, and nickel are also used in various combinations to increase hardenability.

Twinning-Induced Plasticity (TWIP) Grades. Twinning-induced plasticity steels are iron-manganese alloys that contain 17 to 20% Mn. They possess a fully austenitic structure at room temperature, mainly due to the high content of manganese. Austenite is retained because of the large amount of the austenitizing element manganese. Twinning-induced plasticity steels derive their exceptional mechanical properties, such as strength, ductility, and work hardening, from a strengthening mechanism known as twinning, in addition to the classical deformation mechanism of dislocation glide. The mechanical twins appear during straining and they act as planar obstacles to dislocation glide. The steels are fully austenitic, with no phase transformation. The formation of mechanical twins during deformation generates high strain-hardening rates, delaying localized deformation and thus retaining high ductility. Unlike TRIP steels that have high strength and moderate ductility, TWIP steels have both high strength and high ductility. Increase in strength enables weight reduction, whereas increase in ductility allows complex parts to be formed. Twinning-induced plasticity steels are, therefore, particularly attractive for many automotive applications.

Austenitic stainless steels (AUST SS) contain between 16 to 26% Cr and 8% Ni. They have the fcc crystal structure of austenite that is retained at room temperature because of the large amounts of austenitizing elements, such as nickel, manganese, and nitrogen, used in making the steel alloy. Austenitic stainless steels cannot be hardened by heat treatment and are hardened only by cold working. Because of their high chromium and nickel content, they are the most corrosion resistant of the stainless group. Also, they have excellent mechanical properties, such as very high tensile strength and excellent ductility. The higher strength is caused by deformation-induced transformation from austenite to martensite. Austenitic stainless steels show superior crash performance due to the energy absorption capability that results from their high toughness.

Figure 2.22 shows the relationship between tensile strength and elongation percent for the steel categories and groups listed in Table 2.4. Mild steel, conventional HSS, and first-generation AHSS obey the inverse relationship between strength and ductility. However, second-generation AHSS do not conform to the traditional inverse relationship because of the unique characteristic of having elevated ranges of both strength and ductility.

Table 2.8 summarizes the steel categories, gives their tensile strength ranges, and lists the steel groups for each category. The table shows the extremely high strength and excellent ductility of the second generation of AHSS.

The steels currently used in the automotive industry are those of the first three categories in Table 2.8. The last category is still under commercial development and some of those steels have been used in automotive applications on a limited basis. Figure 2.23 shows a ductility-strength-

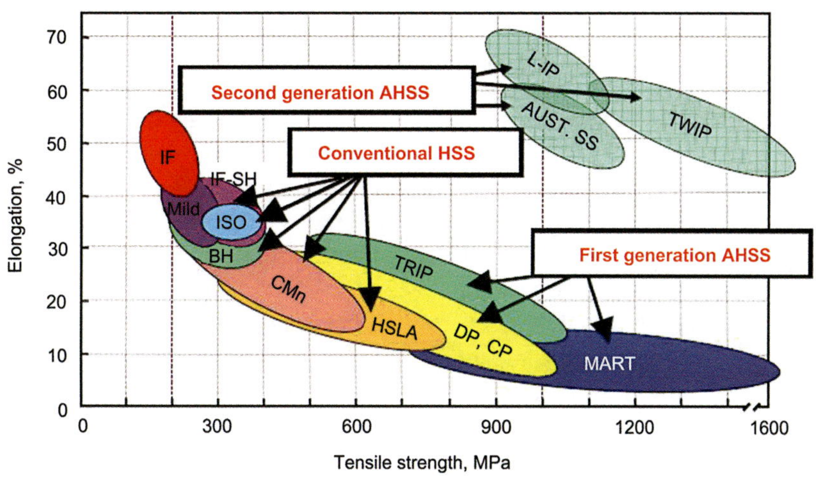

Fig. 2.22 Strength-elongation relationships for various categories of steel. Source: Ref 2.7

Table 2.8 Steel categories with tensile strength ranges and steel groups in each category

Category	Tensile strength range, MPa (ksi)	Ductility, %	Grade
Mild steels	<300 (<44)	30–45	Low-carbon, interstitial-free (IF)
Conventional high-strength steels	300–800 (44–116)	10–45	Interstitial-free, high-strength (IF-HS), isotropic (ISO), bake-hardenable (BH), carbon-manganese (CMn), high-strength, low-alloy (HSLA)
Advanced high-strength steels (1st generation)	500–1600 (73–232)	5–30	Dual-phase (DP), Transformation-induced plasticity (TRIP), complex-phase (CP), martensitic (MS)
Advanced high-strength steels (2nd generation)	900–1600 (131–232)	45–70	Twinning-induced plasticity (TWIP), austenitic stainless steel (AUST SS), light steel with induced plasticity (L-IP)

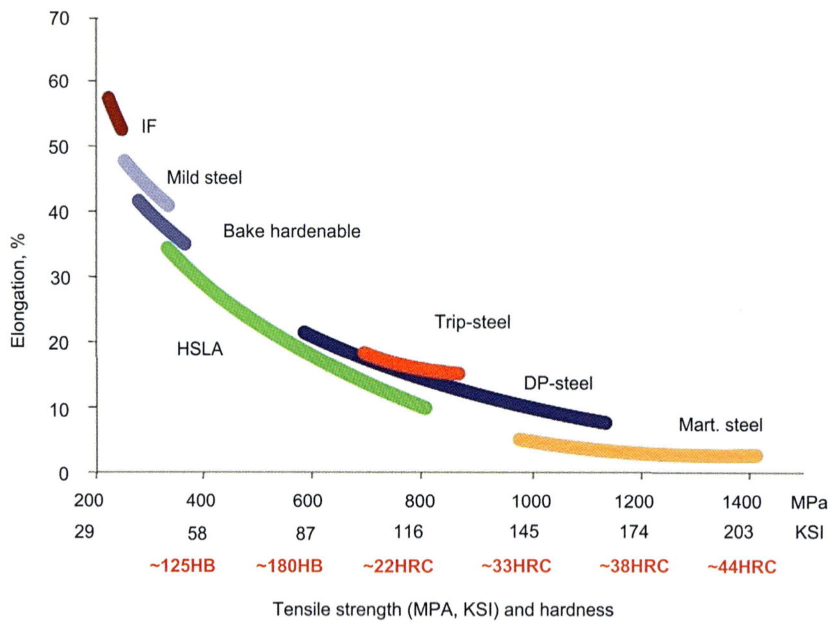

Fig. 2.23 Elongation-strength-hardness chart for currently used steel grades. Source: Adapted from Ref 2.8

hardness chart of the currently used steel grades in the automotive industry. Ductility decreases with increase in strength and hardness. Note the change in the hardness scale on the chart.

Advanced high-strength steels are increasing their share of incorporation into new vehicles at the expense of traditional mild and HSLA steels. These steels and their mechanical properties are discussed in more detail in the following chapters.

2.5 Deformation Mechanisms

Metals are made of crystals that have imperfections such as vacancies, interstitials, dislocations, grain boundaries, and twin boundaries. Properties of metals and alloys depend to a large extent on the presence of such defects.

Deformation by Dislocation Glide. Dislocations are simply extra half planes in metal crystals. They are basic to understanding the mechanical behavior of metals and it is assumed that the concept of dislocations is clear to the reader. Special x-ray techniques and transmission electron microscopy (TEM) revealed their existence and theories about them have been experimentally verified. Now the presence of dislocations in metals is universally accepted and recognized. Dislocations are responsible for plastic deformation in metals by their glide inside the metal crystal and they do so at a stress level that is less than the ideal strength of the metal, estimated to be equal to $E/15$,where E is the elastic modulus of metal. Figure 2.24 shows how a dislocation moves across the slip plane producing a unit slip step of shear strain, γ.

A dislocation moves across the metal crystal, one plane at a time, by re-shuffling the bonds between the surrounding atoms. This movement across the plane eventually causes the top half of the crystal to move with respect to the bottom half. Because only a small fraction of the bonds are broken at any given time, a much smaller force is required to cause slip in metal crystals.

Atoms around a dislocation are displaced from their lattice positions and this leads to lattice distortion and associated high-strain-energy field. To lower this energy, a dislocation tends to get shorter by dissociating into smaller segments.

The driving forces for plastic deformation in metals and alloys are of two types:

- *Mechanical:* A shear stress exerts a high enough force on a dislocation to move it through the metal crystal or a shear stress is used to produce a twin.
- *Thermodynamic:* The formation of mechanical twins and phase transformation of austenite to martensite is determined by the chemical free energy difference between parent and product phases, which depend on both composition and temperature.

Stacking fault (SF) results from the dissociation of a dislocation into two segments, called partials. The SF is created between the two partials, and this structure has a higher Gibbs free energy than the equilibrium fcc structure because it is not thermodynamically stable under the imposed conditions. This process is driven by a tendency for the dislocation to reduce its strain energy. The stacking fault energy (SFE) is related to the separation distance between the two segments; the larger the separation, the lower the SFE. The formation of mechanical twins and ε-martensite transformation originate from stacking faults that rely on the ease of dis-

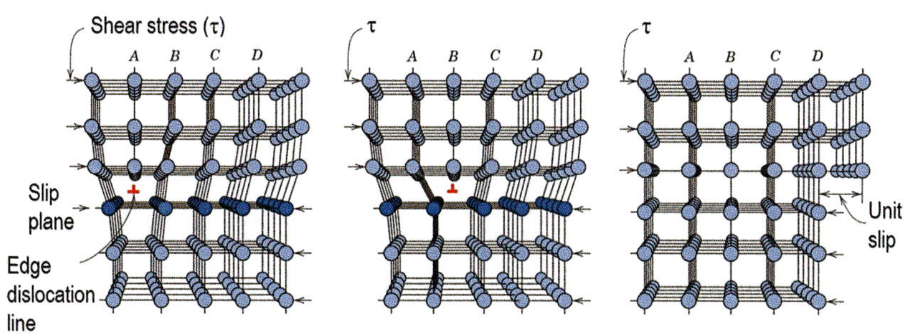

Fig. 2.24 Dislocation line sweeps through the slip plane to produce a shear strain, γ. Source: Ref 2.1

sociation of dislocations. Transmission electron microscopy revealed that there are two possible processes that are responsible for the formation of ε-martensite: regular and irregular overlapping of stacking faults.

Temperature and composition were found to affect the SFE. The effect of alloying elements is generally to decrease SFE. Many researchers concluded that the operating deformation mechanism, whether it is mechanical twinning or martensitic transformation, depends on composition and temperature because of their effect on the SFE values. Early studies on austenite-to-martensite phase transformation showed that SFE diminishes with decreasing temperature.

The effect of SFE on deformation mechanisms of Fe-22Mn-0.6C steel was studied by changing the deformation temperature to vary the SFE values of the alloy. At high temperatures of 400 °C (750 °F), dislocation glide was the only mechanism for plastic deformation. At room temperature, mechanical twinning and martensitic transformation occurred. This verified the dependence of deformation mechanisms on the SFE (Ref 2.9). Such results led to the conclusion that the deformation mechanisms in high-manganese steels are highly dependent on SFE. When SFE is high, dislocation glide is the only deformation mechanism operating in metals. For intermediate SFE, mechanical twinning (TWIP effect) becomes active, and for low SFE, transformation-induced plasticity (TRIP effect) dominates.

Deformation by Twinning. Twins are formed when a shear force, τ, produces atomic displacement in a crystal where atoms on one side of the twin plane form a mirror image of atoms on the other side. Figure 2.25 shows (a) initial position of atoms in a crystal lattice and (b) formation of a deformation twin.

Deformation or mechanical twinning is the second most important mechanism of plastic deformation in metals. It is not as common as dislocation glide. The stress necessary to form twins is generally greater than

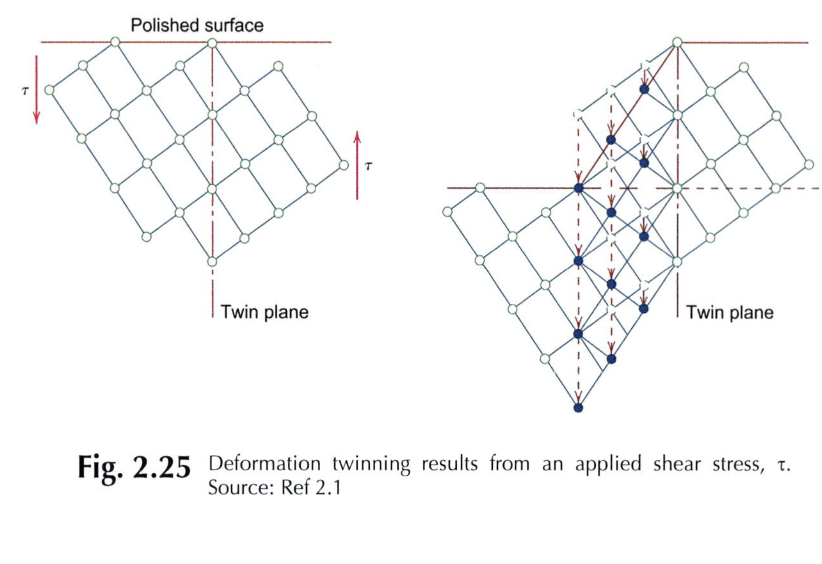

Fig. 2.25 Deformation twinning results from an applied shear stress, τ. Source: Ref 2.1

that for dislocation glide. Deformation twinning occurs when the applied stress is high due to work hardening or low temperature. This deformation mechanism is important in hexagonal close-packed (hcp) and bcc, but not in fcc metals unless they deform at very low temperatures and at high rates of loading.

Dislocation glide and twinning are competing plastic deformation mechanisms. Twinning becomes active when dislocation glide is restricted by the availability of few operable slip systems. Twinning is generally favored at high strain rates or low temperatures. Unlike the thermally activated dislocation glide, twinning is not a thermally activated deformation mechanism.

2.6 Strengthening Mechanisms

The favorable combination of strength and ductility in steels comes not from their perfect crystal structures, but rather from imperfections that they contain. Such imperfections are caused by the presence of impurities, precipitates, grain boundaries, and dislocations. Plastic deformation occurs when dislocations move within and exit the lattice. The presence of impurities, precipitates, and grain boundaries create strains and distortions that tend to impede the movement of dislocations and causes strength to increase and ductility to decrease. Accordingly, strengthening mechanisms are attributed to the reduction of the dislocation mean free path.

The strength of a metal determines the magnitude of force it can withstand before it fails. For metals, the design criterion is based on the yield strength, which comes from two different sources: intrinsic and extrinsic. Intrinsic factors are related to the resistance of the crystal structure to plastic shear and this depends on the strength of the metallic bond in the material. Extrinsic factors involve the introduction of obstacles to slip. Deformation in metals including steels results from dislocation motion on specific slip systems in crystalline materials. Strength in metals including steels results from different mechanisms acting to limit dislocation motion. The harder it is for dislocations to move inside crystal lattices the stronger the metal gets. Dislocation motion is impeded by obstacles in the form of solutes, grain boundaries, dislocation interactions, grain boundaries, and, in the case of steels, changes in crystalline structures, known as phase transformations. Twin boundaries that separate two crystalline regions that are, structurally, mirror images of each other are also considered planar defects and can produce additional obstacles to dislocation motion. Each type of obstacle exerts a pinning force on the dislocation line (a resisting force per unit length of dislocation). Strengthening mechanisms add up and the total strength of a metal is directly proportional to the sum of the contributing strength of the individual obstacles. This means that to manipulate the strength of steels, one or more of the following strengthening mechanisms must be employed.

Solid-Solution Strengthening (Alloying). Solute atoms in solid solutions may be interstitial or substitutional (Fig. 2.26). Interstitial atoms are relatively small and fit in the open spaces of the host metal, while substitutional atoms are large and occupy the host lattice sites. Solute atoms are deliberately added to form barriers to dislocation motion. Alloying elements are generally bigger than those of the host metal and this causes distortion in the lattice and roughens the slip plane on which the dislocation moves. Impurities tend to concentrate at dislocations; hence, they reduce their mobility and increase the strength of the alloy steel. Solid-solution strengthening increases with increase in solute concentrations.

Mechanical Working (Strain Hardening). Although the presence of dislocations lowers the shear stress required to cause slip, the dislocations can become entangled and interfere with each other. Plastic deformation generates more dislocations and as the number of dislocations increase, their interactions increase and the stress needed to move them also increases. This leads to an overall increase in the strength of the metal. Work hardening is used extensively to strengthen metals in metalworking processes. Examples of work hardening include strengthening wire by reducing its cross section by drawing, cold forging the head of a nail, and cold drawing of sheet metal by flat rolling.

Dispersion and Precipitation Hardening. An effective way to impede dislocation motion is to disperse and precipitate small solid and hard particles in their path. In dispersion, small solid particles of high melting points are dispersed into a liquid metal (Fe-SiC metal-matrix composite). In precipitation, solutes are dissolved in a metal while both are molten, then they precipitate as small particles when cooled. The extent of hardening depends on factors such as particle size and shape, volume fraction, chemical and mechanical properties, and coherency with the matrix. These factors determine the extent of particle-dislocation interaction. Precipitation hardening provides a greater hardening effect than the other strengthening mechanisms.

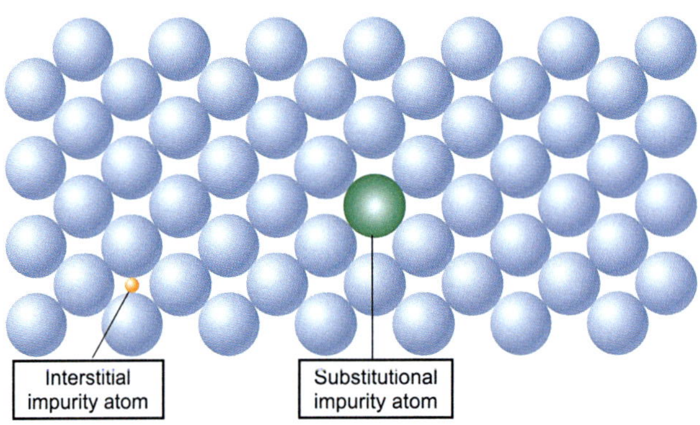

Fig. 2.26 Interstitial and substitutional atoms in a metal crystal lattice. Source: Ref 2.1

Grain-Boundary/Grain-Refinement Hardening. Grain boundaries are surfaces that separate individual grains. The crystallographic orientation of grains changes abruptly from one grain to the next across the grain boundaries. Grain boundaries affect the strength and ductility of steels because they interfere with the movement of dislocations. A dislocation in one grain cannot slide into the next grain because slip planes in the two grains do not line up (Fig. 2.27). Barrier strength increases with increasing the angle of misorientation between the adjacent grains.

Grain size has an important effect on the mechanical properties of steels. The smaller the grains, the larger the number of grain boundaries and, hence, the more barriers to dislocation slip. Large grains are, therefore, generally associated with lower strength because they reduce the number of grain boundaries.

Grain sizes are determined by the rate of solidification from the liquid phase and also by plastic deformation followed by appropriate heat treatment. In general, the enhancement in strength due to grain refinement is usually accompanied by a deterioration of ductility. However, it was shown in many studies, that the uniform and total elongations are only slightly affected by a decreasing ferrite grain size in some AHSS.

Quench Hardening (Phase Transformation). Quench hardening involves heat treating and deformation-induced austenite-to-martensite transformation. Austenitizing steel changes its bcc structure into fcc structure, which holds more carbon atoms than the bcc structure. Rapid cooling prevents equilibrium and transforms austenite to a nonequilibrium phase called martensite. The extra carbons are trapped in the bcc structure and cause it to transform into the bct structure of martensite. The extreme hardness and brittleness of martensite results from the lattice strain created by the trapped carbon atoms, thus providing a barrier to dislocation

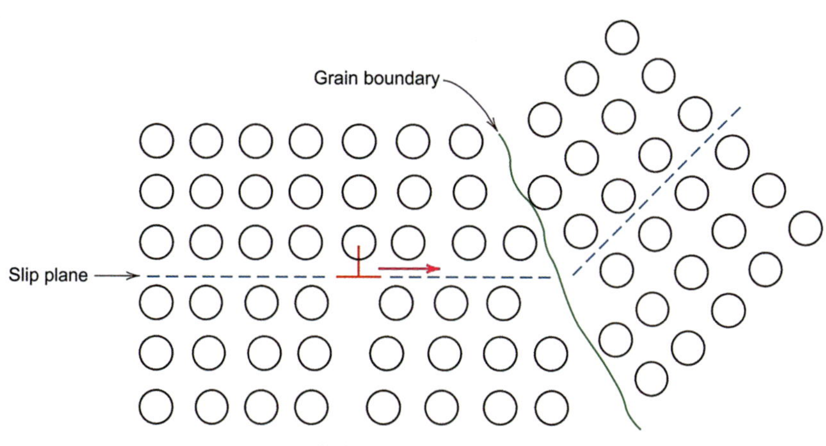

Fig. 2.27 Grain boundaries act as barriers to slip by dislocations. Source: Ref 2.1

slip. Martensite composition is the same as the austenite from which it was quenched. Although dislocation glide is the primary mechanism for plastic deformation, phase transformations can produce additional obstacles to dislocation glide and, therefore, increase the strength of the metal.

Twin Boundary Hardening. Twin boundaries are planar defects that separate two crystalline regions that are, structurally, mirror images of each other. Figure 2.28 shows a schematic diagram of a twin plane and twin boundary. Twinning occurs on certain crystallographic systems that depend on the crystal structure of the metal. Crystal twins are produced during crystal growth from solid, liquid, or vapor, while mechanical twins are produced in bcc metals during plastic deformation, and annealing twins are produced in fcc metals after heat treatment following plastic deformation. Twins produce additional obstacles to dislocation glide and thus increase the strength of the metal.

These strengthening methods are extremely important in designing alloys to meet the properties desired for the intended application. The conventional mechanisms to increase the strength of mild steels and HSLA steels are grain size, solid solution, and precipitation strengthening and these are usually accompanied by a drastic drop in ductility. Advanced high-strength steels, such as DP, CP, TRIP, and TWIP steels, have additional mechanisms that rely on transformation and twinning to increase their strength by providing additional obstacles to dislocation glide.

2.7 Deformation Processing and Annealing

Continuous casting or hot rolling of carbon steels leads to an initial solid grain size as shown in Fig. 2.29 for a typical metal (brass alloy). Cold rolling causes the metal grains to elongate and work hardening increases dislocation density within those grains, causing strengthening and loss of ductility of the metal. Further deformation processing of the metal

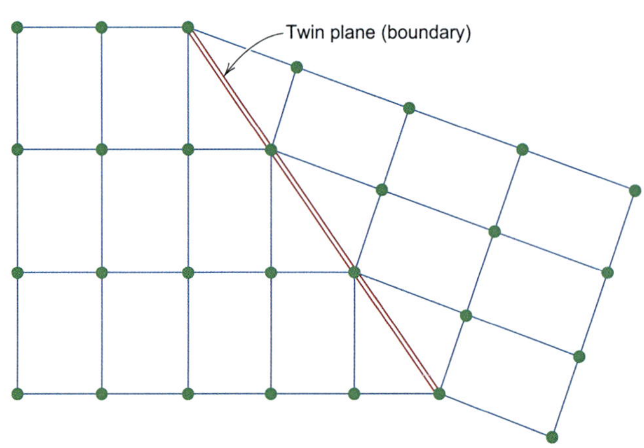

Fig. 2.28 Schematic diagram of a twin plane. Source: Ref 2.1

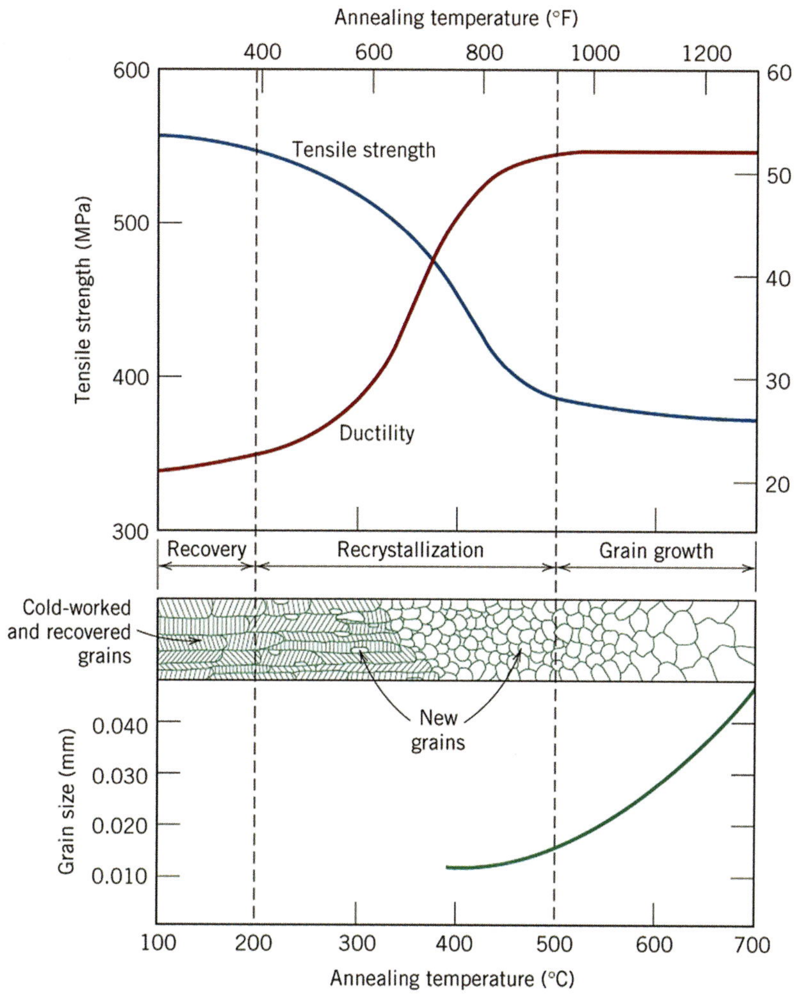

Fig. 2.29 Evolution of grain structure in cold rolling and annealing. Source: Ref 2.1

requires softening it to lower its strength and to increase its ductility. This is done by a heat treating process called annealing that involves two mechanisms: recovery and recrystallization. The mechanisms are solid-state changes that occur during hot forming or applied after cold rolling. Figure 2.29 shows the evolution of grain structure in cold rolling and annealing. Recovery forms subgrains within the grains, and recrystallization eliminates most of the dislocations by reforming the grain boundaries and creating new equiaxed grains.

REFERENCES

2.1 W.D. Callister, Jr. and D.G. Rethwisch, *Materials Science and Engineering: An Introduction,* 8th ed., Wiley, 2010

2.2 M. Ashby, H. Shercliff, and D. Cebon, *Materials: Engineering, Science, Processing and Design,* 2nd ed., Elsevier, 2010
2.3 D. Collins and T. Vasilos, *Industrial Materials: Metals and Alloys,* Vol 1, Prentice Hall, 1995
2.4 S. Bhat, Advances in High Strength Steels for Automotive Applications, Arcelor Mittal Presentation, Great Designs in Steel Seminar, 2008 (Livonia, MI), AUTOSTEEL
2.5 M.Y. Demeri, Forming of Advanced High Strength Steels, *Metal Working: Sheet Forming,* Vol 14B, *ASM Handbook,* ASM International, 2006, p 530–538
2.6 "Automotive Steel Design Manual," Revision 6.1, American Iron and Steel Institute/ASP, Aug 2002, www.autosteel.org
2.7 R.H. Wagoner, Advanced High Strength Steel Workshop, Oct 22–23, 2006 (Arlington, VA), sponsored by the National Science Foundation, Department of Energy, and the Auto/Steel Partnership
2.8 E. Severson, "Better Tool Steels for Better Dies: Balanced Properties for Optimum Performance," Bohler-Uddeholm United States presentation at Great Design in Steels Seminar, March 8, 2006 (Livonia, MI), AUTOSTEEL
2.9 S. Allain, J.-P. Chateau, O. Bouaziz, S. Migot, and N. Guelton, Correlations between the Calculated Stacking Fault Energy and the Plasticity Mechanisms in Fe-Mn-C Alloys, *Mat. Sci. Eng. A,* Vol 387–389, 2004, p 158–162

CHAPTER 3

Advanced High-Strength Steels

ADVANCED HIGH-STRENGTH STEELS (AHSS) are primarily steels with tensile strengths ≥500 MPa (73 ksi) and complex microstructures of different phases such as ferrite, martensite, bainite, and retained austenite. Some types of AHSS have a higher strain-hardening capacity resulting in a strength-ductility balance superior to conventional steels. Many AHSS were developed for weight reduction, where the higher strength allows downgaging while maintaining sufficient formability to make a successful stamping. Typical applications of AHSS include rocker panels, B-pillars, A-pillars, and roof rails. Most of these are energy-absorbing structural components used to increase the level of vehicle impact protection and reduce intrusion into the occupant compartment.

3.1 Nomenclature

Advanced high-strength steels are relatively new materials and a consistent nomenclature was not available until the Ultra-Light Steel Automotive Body—Advanced Vehicle Concept (ULSAB-AVC) Consortium adopted a new AHSS identification standard. The standard specifies both yield strength (YS) and ultimate tensile strength (UTS). Accordingly, AHSS are identified by the system:

XX aaa/bbb

where
XX = type of steel
aaa = minimum (YS) in MPa
bbb = minimum (UTS) in MPa

As an example, a classification of DP 500/800 refers to dual-phase steel with 500 MPa (73 ksi) minimum yield strength and 800 MPa (116 ksi) minimum ultimate tensile strength. The steel grade designator uses the classification shown in Table 3.1.

3.2 Generations

Advanced high-strength steels are relatively new to the materials world, but historically speaking, they are divided into three generations. The first generation includes steel grades that have been completely developed and fully used in many applications throughout the automotive industry. This group has four steel grades that include dual-phase (DP), complex-phase (CP), transformation-induced plasticity (TRIP), and martensitic (MS) steels. Table 3.2 lists the microstructure and tensile strength ranges for the four steel grades of the first-generation AHSS.

Figure 3.1 shows the location of the first generation of AHSS on a tensile strength versus elongation chart. Included in the chart, as a reference, are the mild steels and the conventional high-strength steels (HSS) groups. This chart gives an overview, showing where steel groups and their subgroups lie in the strength-ductility space. Some of the steel grades overlap and give the same strength-elongation combination. This provides the designer several options for the selection of a steel grade and type that is suitable for the manufacture of a specific component.

The second-generation AHSS includes the newer steel grades that have been developed and are being transitioned to application. This group has three steel grades that include twinning-induced plasticity (TWIP), lightweight steel with induced plasticity (L-IP), and austenitic stainless steel

Table 3.1 High-strength steel (HSS) and advanced high-strength steel (AHSS) grade classification

HSS grades	AHSS grades
BH, bake-hardenable	DP, dual-phase
HSLA, high-strength, low-alloy	CP, complex-phase
	MS, martensitic
	TRIP, transformation-induced plasticity
	TWIP, twinning-induced plasticity
	AUST SS, austenitic stainless steel

Source: Ref 3.1

Table 3.2 Microstructure and strength ranges for steel grades of the first generation of AHSS

First-generation AHSS	Microstructure	Strength range, MPa (ksi)
DP	Ferrite + martensite	400–1000 (58–145)
CP	(Ferrite + bainite) matrix + small amounts of pearlite, martensite and retained austenite	400–1000 (58–145)
TRIP	Ferrite + martensite/bainite + austenite	500–1000 (73–145)
MS	Martensite	700–1600 (102–232)

DP, dual-phase; CP, complex-phase; TRIP, transformation-induced plasticity; MS, martensitic

(AUST SS). Table 3.3 lists the microstructure and tensile strength ranges for the three steel grades of the second-generation AHSS.

Figure 3.2 shows the location of the second generation of AHSS on a tensile strength versus elongation chart. Included in the chart, for comparison, are the mild steels, the conventional HSS, and the first generation of AHSS groups. The chart clearly shows that the second-generation AHSS provide great potential for structural applications due to their excellent tensile strength-ductility property combination.

Although the second-generation AHSS offer outstanding combinations of strength and ductility, their widespread adoption is hampered because of the high cost of alloying elements, such as chromium, nickel, and manganese, used in making these alloys.

The third-generation AHSS is the newest group of advanced steels and is still under research and development. Figure 3.3 shows a chart of the strength-ductility space for all currently used steel groups. It also includes the potential location of the third-generation AHSS. This chart can be used as a lead-in to envision the kind of steel that can be developed by filling the empty regions in the strength-ductility space. A beneficial region for future development is circled and is referred to as future opportunities for

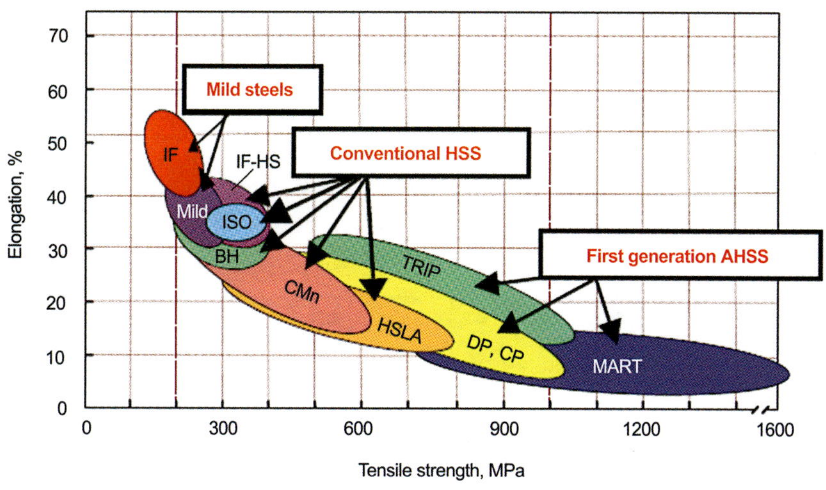

Fig. 3.1 Location of first-generation advanced high-strength steel (AHSS) in the strength-ductility space. Source: Ref 3.2

Table 3.3 Microstructure and strength ranges for steel grades of the second generation of AHSS

Second-generation AHSS	Microstructure	Strength range, MPa (ksi)
TWIP	Single-phase retained austenite	1100–1650 (160–239)
L-IP	Single-phase retained austenite	850–1150 (123–167)
AUST SS	Single-phase retained austenite	900–1150 (131–167)

TWIP, twinning-induced plasticity; L-IP, lightweight steel with induced plasticity; AUST SS, austenitic stainless steel

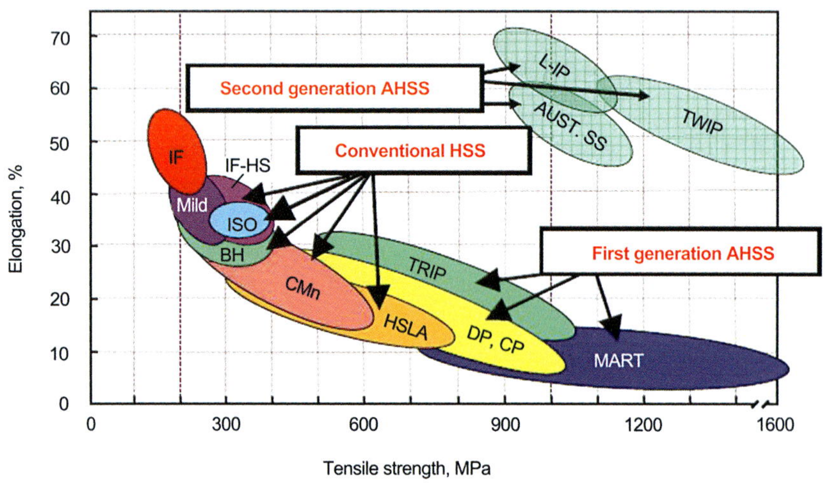

Fig. 3.2 Location of second-generation AHSS in the strength-ductility space. Source: Ref 3.2

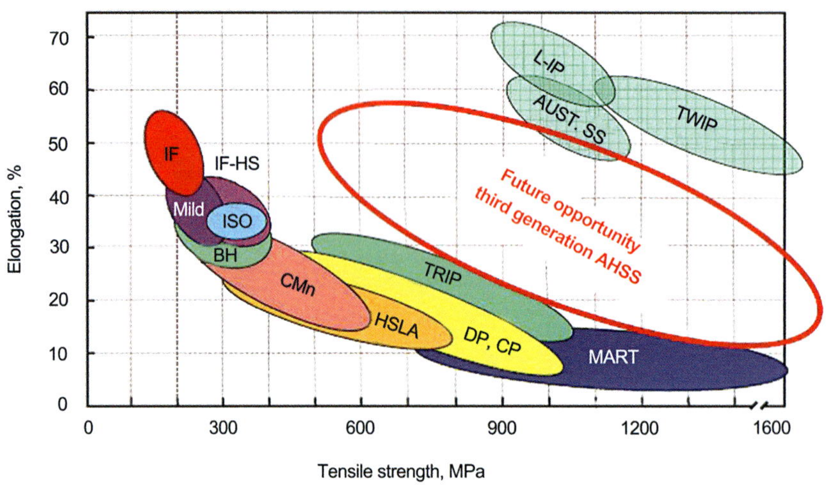

Fig. 3.3 Location of future third-generation AHSS in the strength-ductility space. Source: Ref 3.2

third-generation AHSS. The proposed development region involves material property combinations that bridge the gap between first and second generations of AHSS. These future steel grades will improve the ductility of the first generation and the affordability of the second-generation AHSS.

It is important to note that the second- and future third-generation AHSS groups increase both strength and ductility at the same time. This behavior is contrary to the well-established trade-off relationship between strength and ductility and is bound to change attitudes toward material development.

3.3 Thermomechanical Processing

Advanced high-strength steel grades are produced by controlling the cooling rate from the austenite phase or the ferrite-plus-austenite two-phase region to room temperature. In production, this heat treatment is done on the run-out table of the hot mill (for hot rolled products) or in the cooling section of the continuous annealing furnace (for continuously annealed or hot dip coated products).

Steels are produced based on annealing the material in the two-phase intercritical temperature region where the carbon content in austenite increases with lowering the intercritical temperature as shown in Fig. 3.4. Steel with carbon content (C_{st}) in the austenite region will have a higher carbon content in austenite if the temperature in the two-phase region is reduced from T to T1 (i.e., $C_{\gamma 2} > C_{\gamma 1}$). This means that the closer the annealing temperature (T_{an}) is to the Ac_1 line, the higher the carbon content is in austenite (C_γ) and the higher is its hardenability. Also, lowering the annealing temperature in the two-phase ($\alpha + \gamma$) region lowers the cooling rate required for martensite transformation while avoiding pearlite and/or bainite formation.

The high carbon content in austenite, after intercritical annealing, results in a significant shift of the pearlite forming region toward lower temperature and slower cooling rates. Intercritical annealing can be used to enrich austenite by carbon from ferrite and this stabilizes and retains austenite at room temperature. Carbon redistribution from ferrite to austenite occurs at slow cooling rates and under paraequilibrium conditions.

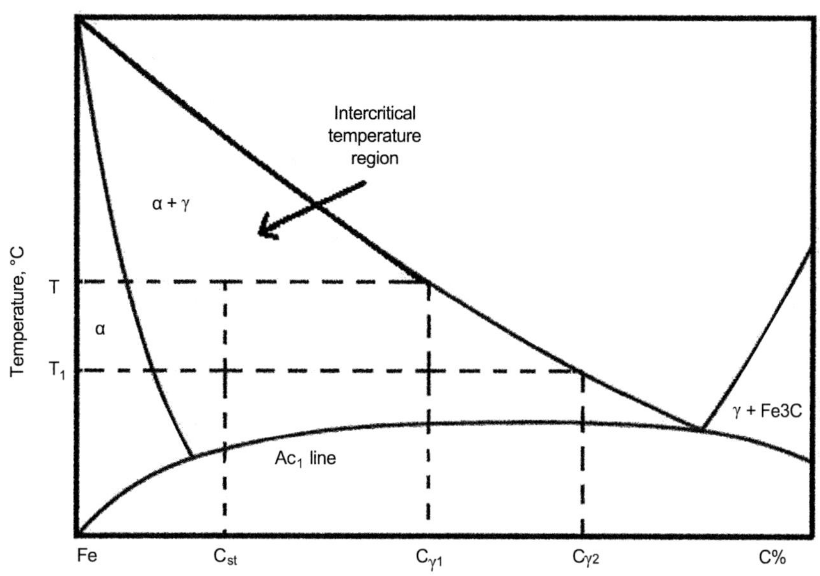

Fig. 3.4 Effect of the intercritical temperature on the carbon content of austenite. Source: Ref 3.2

The following is a brief description of the processing schemes for AHSS grades (Ref 3.1):

- *Martensitic* steels are produced from the austenite phase by rapid quenching to transform most of the austenite to martensite.
- *Dual-phase* steels are produced by controlled cooling from the austenite phase (in hot rolled products) or from the two-phase ferrite-plus-austenite region (in hot rolled products or continuously annealed and hot dip coated products) to transform some of the austenite to ferrite before rapid cooling to transform the remaining austenite to martensite.
- *TRIP* steels require the use of an isothermal hold at an intermediate temperature to produce some bainite. The presence of high silicon and carbon content in TRIP steels produces significant volume fractions of retained austenite.
- *Complex-phase* steels follow a cooling pattern similar to that of TRIP steels, however, the chemistry is adjusted to produce less retained austenite and to include fine precipitates to strengthen the martensite and bainite phases.

Advanced high-strength steels achieve their mechanical properties through engineering of the volume fractions of microconstituents of ferrite, bainite, austenite, and martensite. This is achieved by thermomechanical processing steps and/or by adding significant amounts of alloying elements such as manganese. Optimal alloys with high strength and high ductility are designed by controlling the following processes:

- *Intercritical annealing process:* Ferrite-pearlite mixture is transformed into a ferrite/austenite structure. Phase composition can be controlled by changing the temperature and holding time.
- *Bainite isothermal transformation treatment:* Carbon atoms enrich the newly generated austenite phase. The heat treatment parameters can be decided by the time-temperature-transformation (TTT) diagram.
- *Ultrafine-grained alloy:* The grain size is refined by the equal channel angular extrusion (ECAE) method.

3.4 Microstructure Development

Microstructure plays an important role in the properties and performance of steels. The development of microstructure in AHSS occurs by controlling the cooling rate from the austenite or austenite plus ferrite for hot rolled products or by regulating the cooling section of the continuous annealing furnace for continuously annealed or hot dipped coated products. The mechanical properties in steels are determined by their chemical composition and microstructure. Table 3.4 shows the mixture of microstructure and the resulting properties of the first generation of AHSS.

Table 3.4 Microstructure and resulting properties of the first generation of AHSS

Steel subgroup	Microstructure	Properties
Dual-phase (DP)	Ferrite matrix + second-phase martensite	High strength, good ductility, high bake hardenability
Complex-phase (CP)	Fine ferrite + higher fraction of martensite and pearlite	High strength, good ductility
Transformation-induced plasticity (TRIP)	Soft ferrite + hard martensite and/or bainite + >5% retained austenite	Highest combination of strength and ductility
Martensitic (MS)	100% martensite	High strength, low ductility, high bake hardenability

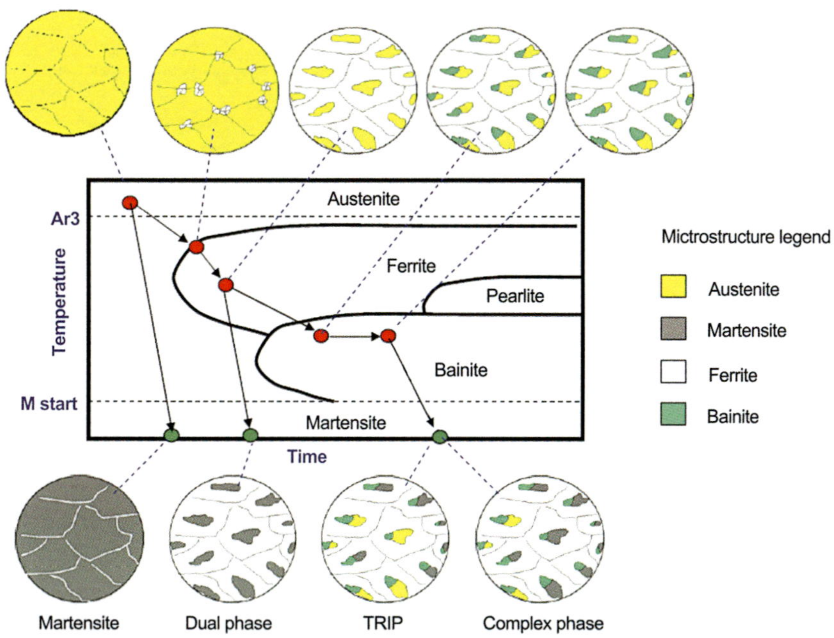

Fig. 3.5 Cooling routes and microstructure development in AHSS. Source: Adapted from Ref 3.3

Figure 3.5 shows the thermal processing routes for obtaining the final microstructures for four grades of AHSS: MS, DP, TRIP, and CP steels. The effect of the alloying elements in the austenite shifts the lobes of the continuous cooling transformation (CCT) diagram to help achieve the desired microstructure. Different cooling schemes produce different room-temperature microstructures. Slow cooling leads to an all-ferrite microstructure while quenching produces martensite. Slow cooling followed by quenching produces different phase mixes.

3.5 Property Trends

The combination of high strength and ductility of AHSS allows the use of thinner gage components for automotive and construction applications. Additionally, their high energy absorption capabilities improve the crash safety of automobiles. The following is a review of the mechanical property trends of AHSS.

Table 3.5 lists mechanical properties of typical HSS and AHSS. The table shows that except for TRIP- and TWIP-grade steels, all other categories of steel follow the traditional inverse relationship between strength and ductility. The main advantage of TRIP and TWIP steels is that they have much higher ductility than other steels with similar strength.

For a perspective on the relative magnitude of the two major attributes of steel, tensile strength and ductility, a composite plot that includes most of the steel grades used by the automotive industry is shown in Fig. 3.6. The high-manganese alloys show extraordinary ductility compared to other grades of steel.

Table 3.5 Typical values of mechanical properties for various grades of steels

Steel grade	Yield strength, MPa (ksi)	Ultimate tensile strength, MPa (ksi)	Total elongation, %	*n*-value, 5–15%	R-bar	*K*-value, MPa (ksi)
BH 210/340	210 (30)	340 (49)	34–39	0.18	1.8	582 (84)
BH 260/370	260 (38)	370 (54)	29–34	0.13	1.6	550 (80)
DP 280/600	280 (41)	600 (87)	30–34	0.21	1.0	1082 (157)
DP 300/500	300 (44)	500 (73)	30–34	0.16	1.0	762 (111)
HSLA 350/450	350 (51)	450 (65)	23–27	0.14	1.1	807 (117)
DP 350/600	350 (51)	600 (87)	24–30	0.14	1.0	976 (142)
DP 400/700	400 (58)	700 (102)	19–25	0.14	1.0	1028 (149)
TRIP 450/800	450 (65)	800 (116)	26–32	0.24	0.9	1690 (245)
DP 500/800	500 (73)	800 (116)	14–20	0.14	1.0	1303 (189)
CP 700/800	700 (102)	800 (116)	10–15	0.13	1.0	1380 (200)
DP 700/1000	700 (102)	1000 (145)	12–17	0.09	0.9	1521 (221)
TWIP 400/1000	400 (58)	1000 (145)	45	0.40(a)	...	...
MS 950/1200	950 (138)	1200 (174)	5–7	0.07	0.9	1678 (243)
MS 1250/1520	1250 (181)	1520 (220)	4–6	0.065	0.9	2021 (293)

(a) *n* increases to a value of 0.4 at 30% strain and remains constant after that. Source: Ref 3.1

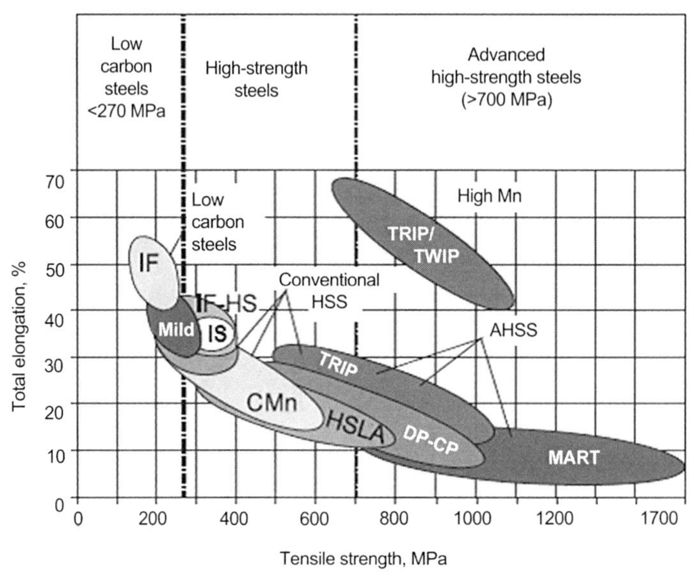

Fig. 3.6 Relationship between tensile strength and elongation for various grades of steels. Source: Ref 3.2

Figure 3.7 shows engineering stress-strain curves for high-strength, low-alloy (HSLA), DP, and TRIP steels of approximately similar yield strengths. The TRIP steel has a lower initial work-hardening rate than the DP steel, but continues to increase at higher strains. Also, the total elongation for TRIP steel is much higher than those of HSLA and DP steels.

A comparison between typical engineering stress-strain curves of various grades of cold rolled, hot dip galvanized HSS (HSLA340, DP600, and TRIP700) and stainless steel 301LN is shown in Fig. 3.8. The figure shows

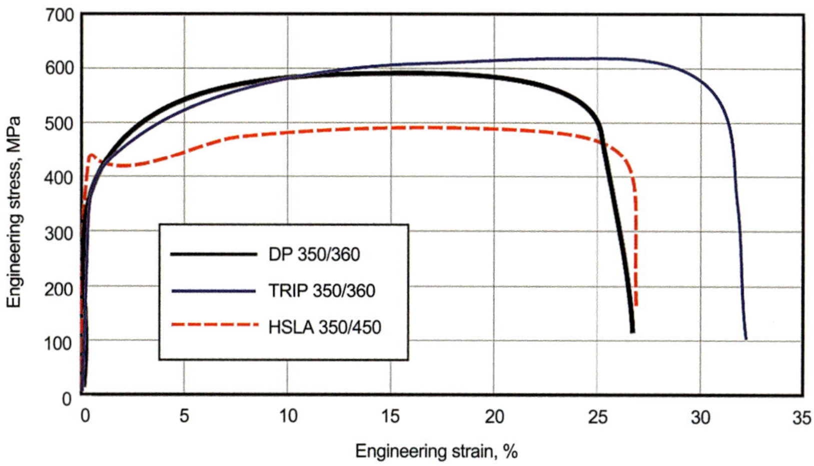

Fig. 3.7 Comparison of engineering stress-strain curves for three grades of steel. Source: Ref 3.4

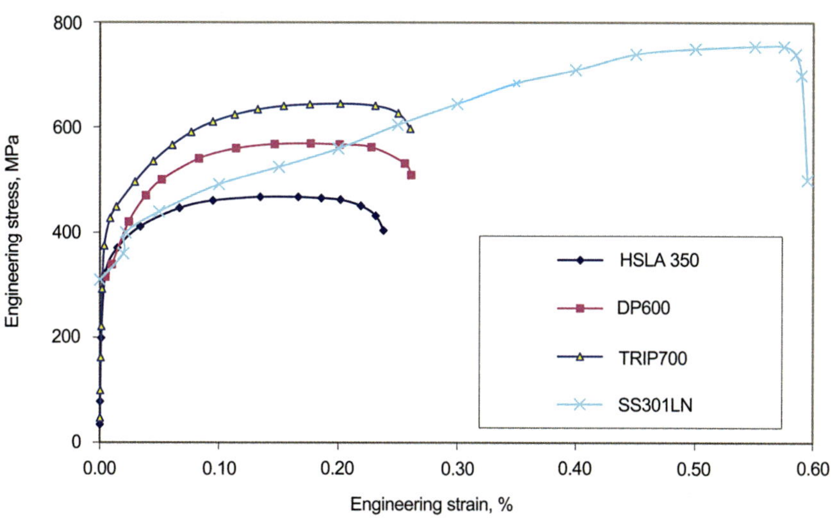

Fig. 3.8 Stress-strain curves of various grades of high-strength steel (HSS) and SS301LN. Source: Ref 3.5

that austenitic stainless steel 301LN exhibits very high tensile strength and excellent ductility compared to the other grades of steel. The material also shows superior crash performance due to the energy absorption capability that results from its high toughness. A comparison between typical strength levels of various grades of cold rolled, hot dip galvanized HSS (BH180, HSLA220, IF260, DP600, TRIP700, and CP1000) and stainless steel 301LN is shown in Fig. 3.9.

A comparison between the ductility of the same grades of HSS and stainless steel 301LN is shown in Fig. 3.10. The superior ductility of austenitic stainless steel 301LN in comparison to other steel grades is very clear.

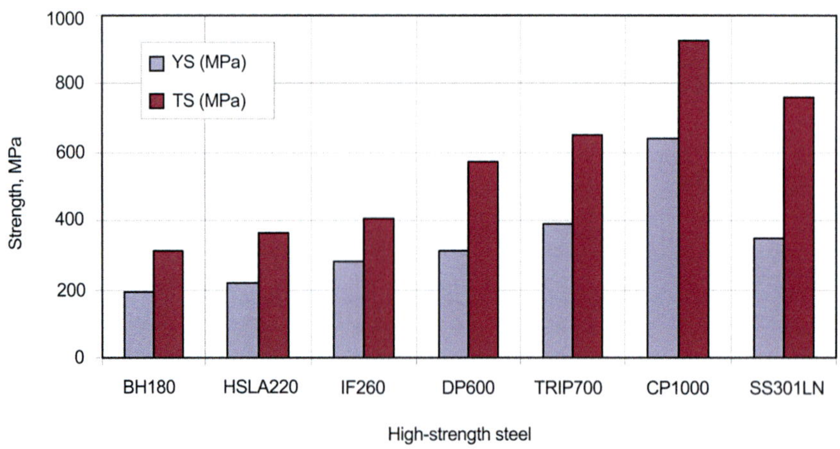

Fig. 3.9 Strength levels of various grades of HSS and SS301LN. Source: Ref 3.5

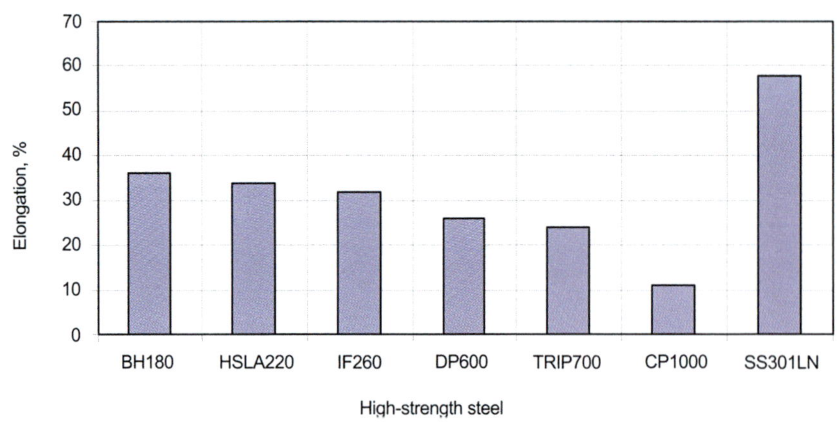

Fig. 3.10 Variation of percent elongation of various grades of HSS and SS301LN. Source: Ref 3.5

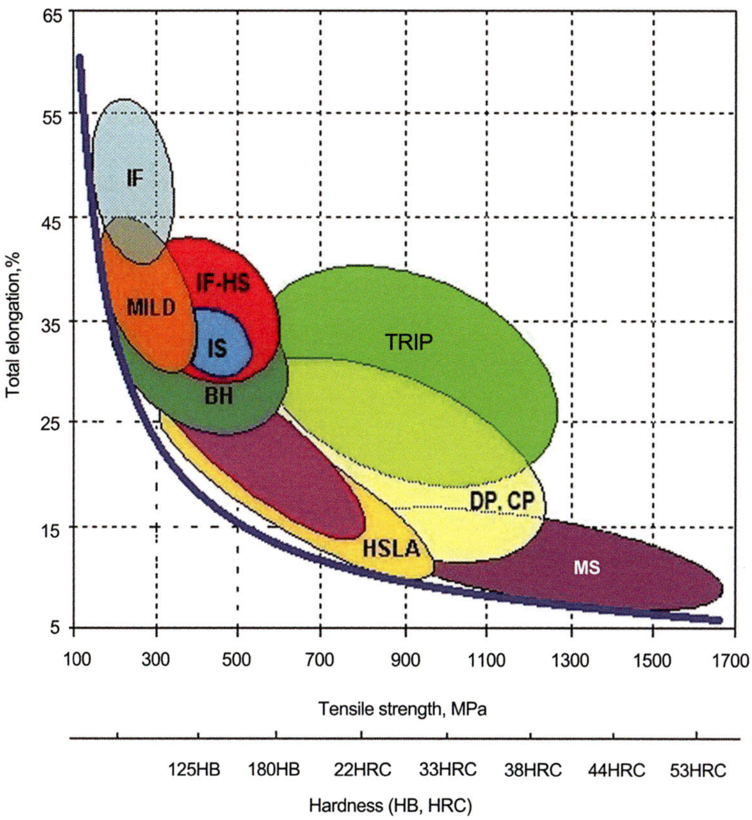

Fig. 3.11 Relationships between strength or hardness and formability for various steel grades.

An interesting plot of tensile strength and hardness versus formability for various steel grades is shown in Fig. 3.11. The traditional inverse relationship between strength or hardness and ductility is followed by the listed steel grades.

The formability of two grades of AHSS (DP600 and TRIP700), HSLA340, SS304, and mild steel, as determined by the forming limit diagram (FLD), is shown in Fig. 3.12. It can be seen that only stainless steel SS304 has, for the most part, a higher formability than mild steel. The formability of the other grades of HSS follow the general trend of decreasing with increase in the strength of the material.

REFERENCES

3.1 "ULSAB-AVC Consortium, Advanced Vehicle Concepts," Technical Transfer Dispatch #6, WorldAutoSteel, www.worldautosteel.org, May 2001

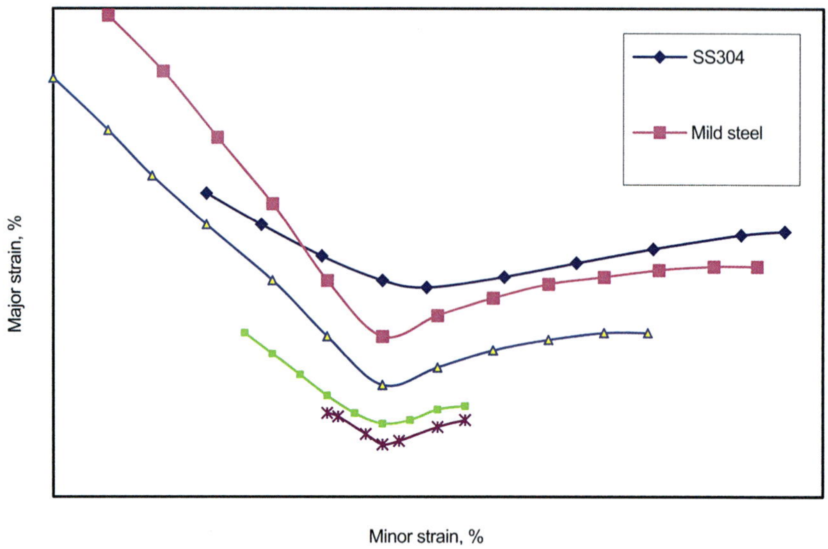

Fig. 3.12 Comparison of forming limit diagrams for various steel grades. Source: Ref 3.6

3.2 J. Zrnik et al., Recent progress in high strength low carbon steels, Comtes FHT, Ltd., Plzen, Czech Republic, ISSN 0543-5846, METABK 45 (4) 323–331 (2006)
3.3 S.P. Bhat, "Advances in High Strength Steels for Automotive Applications," ArcelorMittal presentation at Great Designs in Steel Seminar, April 9, 2008 (Livonia, MI), AUTOSTEEL
3.4 "Advanced High Strength Steels Applications Guidelines," Version 4.1, WorldAutoSteel, www.worldautosteel.org, June 2009
3.5 M.Y. Demeri, High Strength and Stainless Steels for Automotive Applications, *Innovations in Processing and Manufacturing of Sheet Materials,* M.Y. Demeri, Ed., Feb 11–15, 2001 (New Orleans, LA), TMS, Warrendale, PA, 2001, p 83–95

CHAPTER 4

Attributes of Advanced High-Strength Steels

STEELS ARE GENERALLY USED as structural members and, therefore, their mechanical properties such as stiffness, strength, ductility, and toughness are of most interest for engineering applications. Hardness is related to strength and is important for applications involving wear or abrasion resistance. The performance attributes of advanced high-strength steels (AHSS) are summarized in the following sections.

4.1 Stiffness

Stiffness is defined as the resistance to deflection or elastic deformation. It is measured by the modulus of elasticity, which is equal to the slope of the elastic portion of the stress-strain curve. The stiffness of a panel depends on the modulus of elasticity of the material and its geometry. For a rectangular cross section, the stiffness, S, of a panel is given by:

$$S = CEwt^3/12L^3$$

where C is the load distribution constant; E is the modulus of elasticity; and L, w, and t are length, width, and thickness of the panel, respectively. Because the modulus of elasticity of all steels is constant at 210 GPa (210,000 MPa, or 30,458 ksi), changing the steel grade for a specific panel design does not affect its stiffness. Therefore, to improve panel stiffness, its thickness or its moment of inertia ($I = t^3/12$) must be increased. For a specific panel, the stiffness, S, the length, L, and the width, w, are specified; only the thickness, t, can be changed. The mass of the panel can be reduced by reducing its thickness, t, but only to the extent that the stiffness constraint is still met (Ref 4.1).

One approach to stiffening sheet metal is to increase its second moment of inertia through corrugating or dimpling of the sheet metal. Corrugation or dimpling allows the reduction of thickness for weight savings without compromising stiffness (Ref 4.2). Advanced high-strength steels offer greater design flexibility to optimize the stiffness of panels because of their high strength and enhanced formability.

4.2 Strength

Strength is defined as the resistance of a metal to plastic flow. It is measured by the magnitude of the flow stress of a deforming material. The yield strength is defined as the stress needed to yield the material and initiate plastic deformation. The ultimate tensile strength is defined as the maximum flow stress on the stress-strain curve.

Strength in metals is related to dislocation glide through crystal lattices. In pure metals strength is determined by an intrinsic lattice resistance to dislocation glide. Lattice imperfections, such as dislocations, twins, solid solutions, impurities, precipitates, and grain boundaries provide obstacles to dislocation glide and, hence, strengthen the metal alloys. Strength in AHSS is determined by a number of strengthening mechanisms, including solid-solution hardening, precipitation hardening, work hardening, mechanical twinning, grain-boundary hardening, and phase transformation. Strength, in most metals, increases with plastic strain because of work hardening. It reaches a maximum at the tensile strength and then starts to decrease due to localized necking that leads to fracture.

Engineering stress-strain curves provide essential information on the deformation properties of materials. The curves reveal basic data on the strength, ductility, hardening capacity, and toughness of engineering materials. Figure 4.1 shows engineering stress-strain curves for three high-strength steel types: ferrite/bainite DP 400/600, TRIP 550/650, and TWIP 300/600.

The engineering stress-strain curves of Fig. 4.1 show that twinning-induced plasticity (TWIP) steel exhibits much higher uniform elongation (e_u = 86%) and total elongation (e_t = 108%) than both ferrite/bainite dual-phase (FBDP) and transformation-induced plasticity (TRIP) steels. High strength is needed in many automotive applications, and strength levels as high as 1800 MPa (261 ksi) have been identified for future requirements of anti-intrusion components.

4.3 Strain Hardening

The rate of increase in strength with plastic strain is defined as the strain-hardening exponent, n. High work-hardening exponent n gives metals the

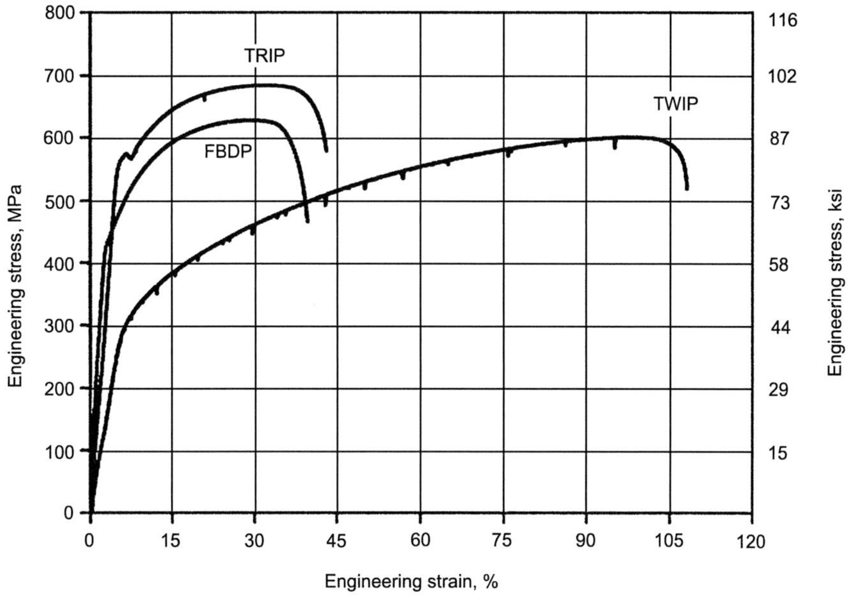

Fig. 4.1 Engineering stress-strain curves for three advanced high-strength steel (AHSS) types. FBDP, ferrite/bainite dual-phase; TRIP, transformation-induced plasticity; TWIP, twinning-induced plasticity. Source: Ref 4.3

ability to stretch by distributing the strains more uniformly and by postponing localized deformation and fracture. Figure 4.2 shows the inverse relationship between the strain-hardening exponent, n, and yield strength for conventional and medium-strength steels.

Advanced high-strength steels are characterized by high n-value at low strains, which delays the onset of plastic instability and the development of localized necking. A high n-value also means a higher potential to increase strength during forming. In dual-phase (DP) steels, the dispersion of hard martensite in soft ferrite produces a high work-hardening rate that diminishes with strain. In TRIP steels, the retained austenite transforms to martensite as deformation proceeds. This means that the initial strain-hardening rate is low but increases progressively with strain. These characteristics make AHSS highly suitable for press forming operations.

Figure 4.3 shows the instantaneous n-values for TRIP 350/600, DP 350/600, and HSLA 350/450 steels. All three steels have the same yield strength of 350 MPa (50 ksi). The differences in their work hardening explain the enhanced formability of DP steel over high-strength, low-alloy (HSLA) steel. It also explains the better formability of TRIP steel over DP steel.

The instantaneous value of the strain-hardening exponent, n, for Fe-18Mn-0.6C-1.5Al TWIP steel is shown in Fig. 4.4. The value of n increased steadily up to a true strain of $\varepsilon = 0.25$, where it assumed a constant

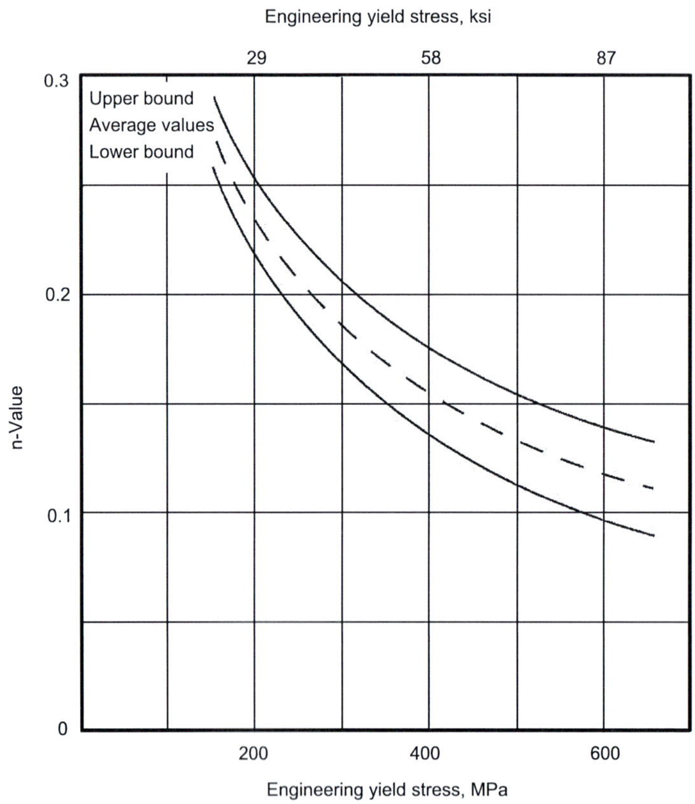

Fig. 4.2 Variation of strain-hardening exponent with yield stress for conventional steels. Source: Ref 4.1

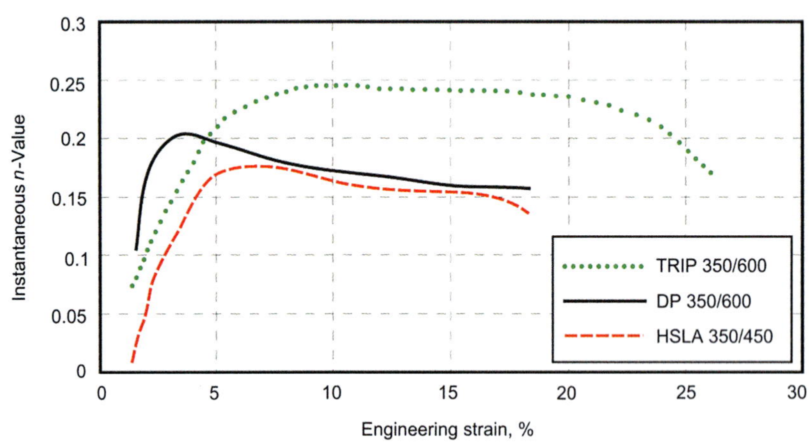

Fig. 4.3 Instantaneous *n*-values for different steel grades. Source: Ref 4.1

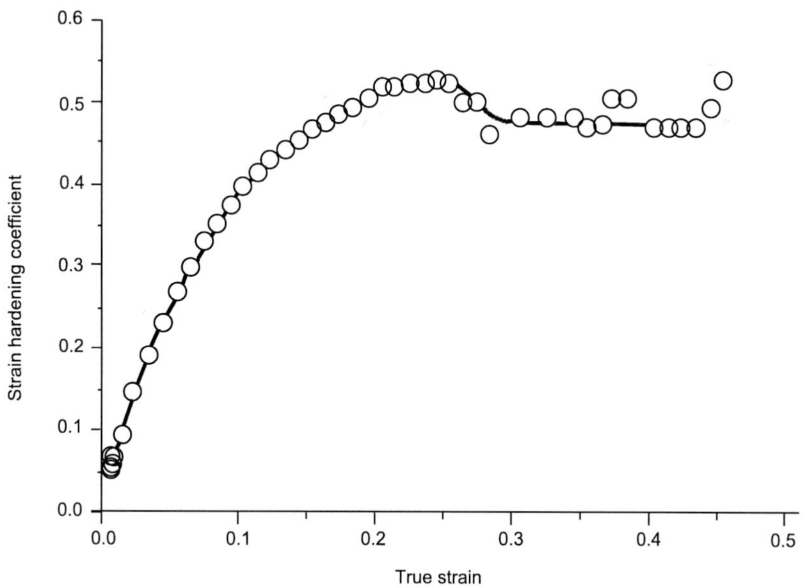

Fig. 4.4 Instantaneous *n*-values of Fe-18Mn-0.6C-1.5Al TWIP steel. Source: Ref 4.4

value of approximately 0.5 until fracture. Model calculations indicate that strain hardening is related to the formation of strain-induced twins and that it reached a constant value of $n = 0.5$, which coincided with the twin volume fraction saturation of $\varepsilon = 0.25$ (Ref 4.4).

4.4 Fatigue

Fatigue in metals is caused by repeated or cyclic stresses that can result in component damage or failure. High-cycle fatigue refers to load cycling below the yield strength of the metal and within the elastic deformation range. In this case, the component survives for a large number of cycles before fatigue failure. Low-cycle fatigue refers to load cycling above the yield strength but below the tensile strength of the metal and within the plastic deformation range. In this case, the component survives for a small number of cycles before fatigue failure. Some components may be free of initial cracks while others may contain or are assumed to have small cracks.

Fatigue characteristics are measured, plotted, and presented in the form of an *S-N* curve where *S* is the stress amplitude and *N* is the number of cycles to failure. Figure 4.5 shows a typical *S-N* curve on which the location of the endurance limit, σ_e, of the metal is superimposed. The endurance limit is the most important property characterizing fatigue strength.

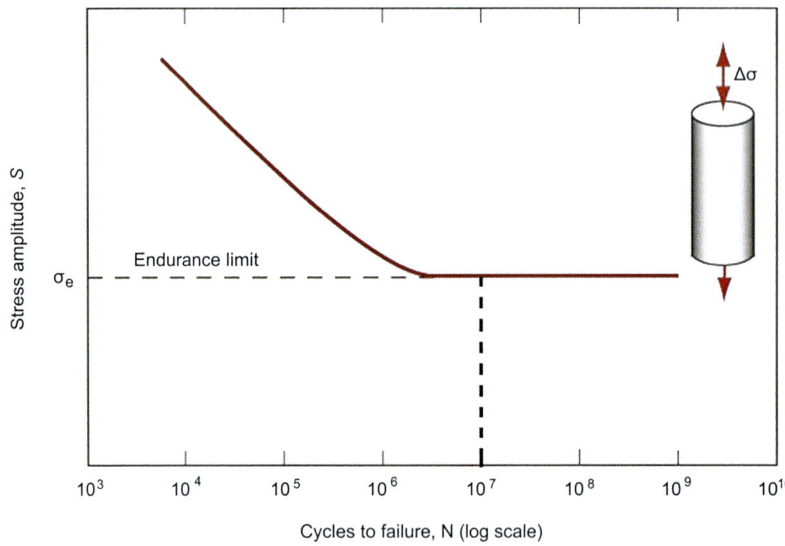

Fig. 4.5 *S-N* curve showing fatigue characteristics and endurance limit of a metal displaying a fatigue limit. Source: Adapted from Ref 4.5

It is defined as the stress amplitude below which fatigue failure will not occur. Fatigue failure is characterized by three steps:

- Crack initiation at points of high stress concentrations inside or on the surface of a component
- Crack propagation with each stress cycle
- Crack reaching a critical size leading to fracture and failure of the component

The endurance limit, σ_e, is directly related to tensile strength, σ_{ts}, by the approximate relationship:

$$\sigma_e \sim 0.33\ \sigma_{ts}$$

For many steels, endurance limits range between 35 and 60% of the tensile strength (Ref 4.5).

Because AHSS have very large tensile strength ranges, their fatigue strengths will be considerably higher than those of lower-strength grade steels. Designing for high-cycle fatigue life is similar to elastic-limited design to avoid yield. This means that the maximum stress in a loaded component should be less than its endurance limit, σ_e, rather than its yield stress, σ_y. Fatigue life can be extended by:

- Choosing strong materials
- Lowering the mean stress level

- Reducing material defects
- Eliminating surface irregularities and sharp features

The fatigue characteristics and endurance limit of TRIP 450/780 steel are compared with two HSLA steels of similar yield strength in Fig. 4.6. The endurance limit for TRIP steel is much higher than those of the two HSLA steels.

The variation of fatigue limit (endurance limit) with cyclic yield stress for TRIP, DP, and HSLA steels is shown in Fig. 4.7. The high endurance limit for TRIP steel results from the progressive transformation of retained austenite to martensite during plastic deformation. This process relaxes stress fields and introduces compressive stresses in the metal crystals, hence improving the fatigue strength of TRIP steel (Ref. 4.1).

The fatigue performance of Fe-22Mn-0.52C TWIP steel is found to improve with monotonic pre-deformation. The improvement is attributed to a higher volume fraction of twins and nucleation of new twins that results in more obstacles to dislocation glide, thus increasing the strength and fatigue performance of the pre-deformed TWIP steel (Ref 4.6).

4.5 Crashworthiness

Crashworthiness is an important automotive attribute that is becoming increasingly important. It measures the structural ability of the vehicle to plastically deform while maintaining a sufficient survival space for its occupants in crashes involving deceleration loads. One of the mandated re-

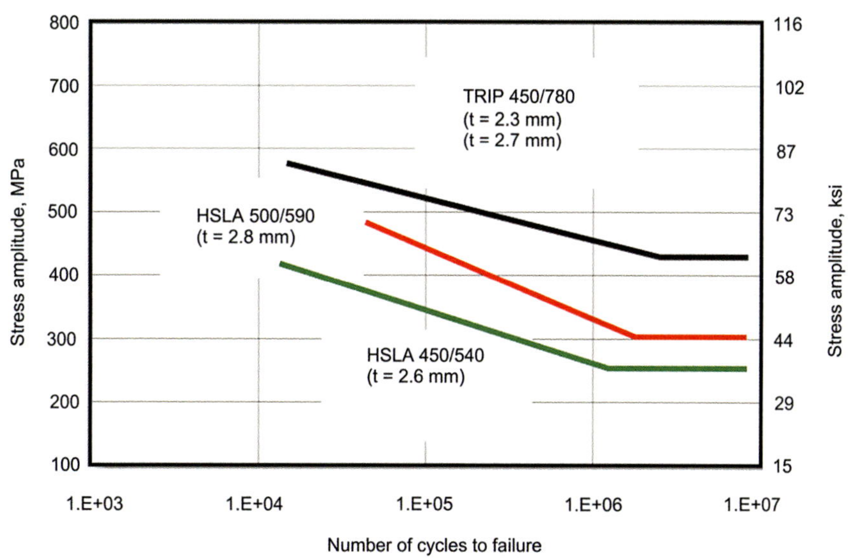

Fig. 4.6 *S-N* curve for transformation-induced plasticity (TRIP) and high-strength, low-alloy (HSLA) steels. Source: Ref 4.1

quirements for new vehicles is to improve occupant safety during frontal and side collisions.

From the point of view of crashworthiness, a car is divided into two distinct zones: safety zone and crushing zone. The safety zone includes the occupant compartment, which must be rigid to resist plastic deformation and to prevent intrusion into the occupant compartment (Fig. 4.8a). The crushing zone consists of energy management structures in the engine compartment and trunk designed to surround the occupant compartment to absorb energy and protect occupants (Fig. 4.8b). Material requirement for the safety zone is high strength and that for the crushing zone is high toughness. Body structures evolved to include progressive crush zones to

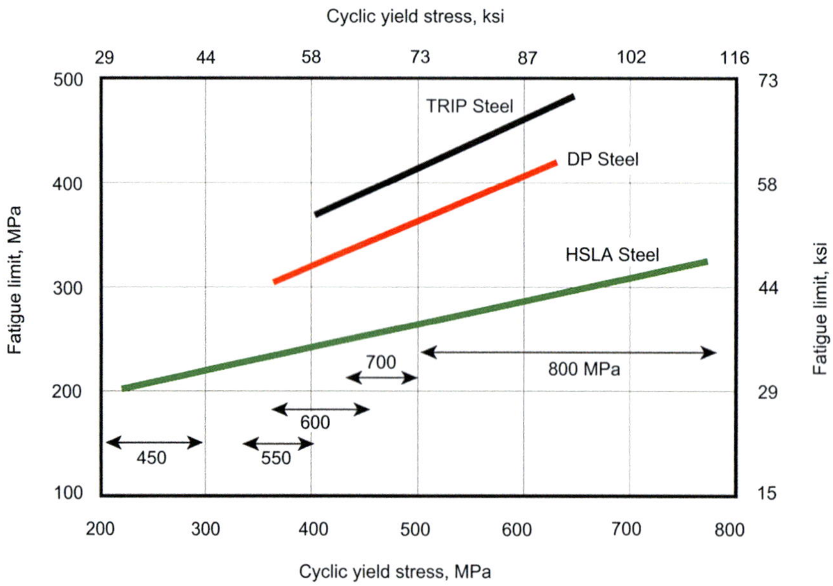

Fig. 4.7 Variation of fatigue limit and yield stress for TRIP, DP, and HSLA steels. Source: Ref 4.1

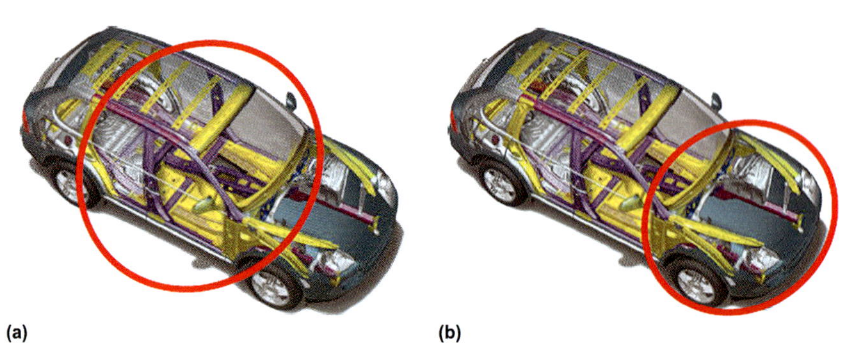

Fig. 4.8 (a) Safety zone to protect occupant. (b) Crushing zone to absorb crash impact energy. Source: Ref 4.7

absorb most of the kinetic energy resulting from a crash. The kinetic energy is spent in the plastic deformation of the crush zones.

Crashworthiness requirements include:

- *Front structure* that is stiff and deformable with crush zones to absorb the kinetic energy resulting from frontal collisions and prevent intrusion into the occupant compartment
- *Rear structure* that is deformable to maintain integrity of the rear occupant compartment and protect the fuel tank from rear impact
- *Side structure* to minimize intrusion and prevent doors from opening during side impact
- *Roof structure* that is strong enough to protect occupants during vehicle rollover
- *Restraint systems* (seat belts and air bags) that work in harmony with the vehicle

The ability of a structure to absorb energy in a crash depends on its geometry and the toughness of its material, which is directly related to the area under the stress-strain curve. In other words, a tough material must have both high strength and high ductility. Figure 4.9 shows two steels, HSLA 450 and DP 600, with equal yield strength but different energy-absorbing capabilities. The area under the curve for DP 600 is greater than that for HSLA 450 and this makes DP 600 a better material for crash applications.

Advanced high-strength steel types suitable for the safety zone would be martensitic (MS) grades due to their high strength, and those suitable for the crushing zone would be DP, TRIP, complex-phase (CP), TWIP, and austenitic stainless steel (AUST SS) grades due to their high toughness.

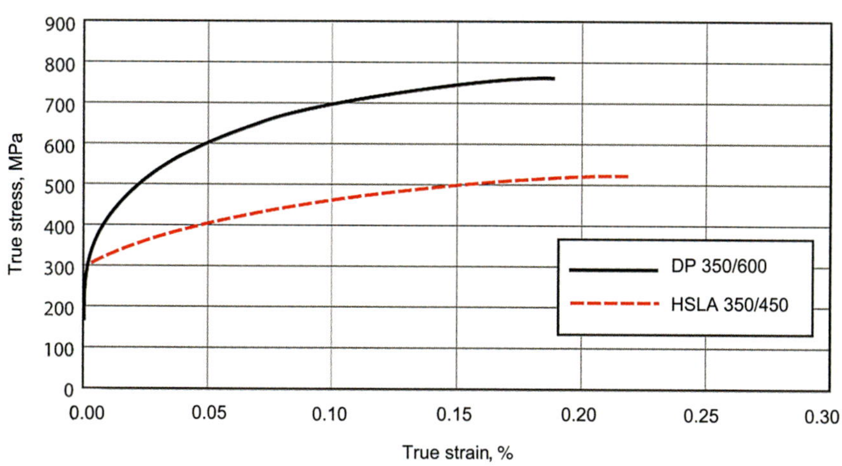

Fig. 4.9 Energy-absorbing capabilities of an HSLA steel and an AHSS. Source: Ref 4.1

Crashworthiness is characterized by the total energy that a structure can absorb. One way to assess crashworthiness is to monitor the folding pattern of a thin-walled structure subjected to a crushing force. Maximum energy absorption is obtained if progressive collapse of the structure is maintained through the entire deformation process. Figure 4.10(a) shows the formation of regular buckling folds in a loaded structure. Energy absorption is compromised if the folding process is disrupted by total buckling caused by improper structure design or material imperfection. Figure 4.10(b) shows the formation of total buckling, which disrupts the formation of regular buckling folds. Many research studies were conducted on structures with different cross sections made from various materials and subjected to several crushing mechanisms to trigger the formation of regular buckling folds (Ref 4.8).

The requirements for a good energy-absorbing material are high specific strength and good ductility. Such requirements are possessed by AHSS, especially the TRIP, TWIP, and AUST SS grades. Transformation-induced plasticity steels have large amounts of retained austenite (5 to 20%) that transforms into martensite during plastic deformation causing a substantial increase in their strength and work-hardening behavior. Twinning-induced plasticity steels deform by twinning and show an even greater work-hardening rate at high levels of deformation. The TWIP

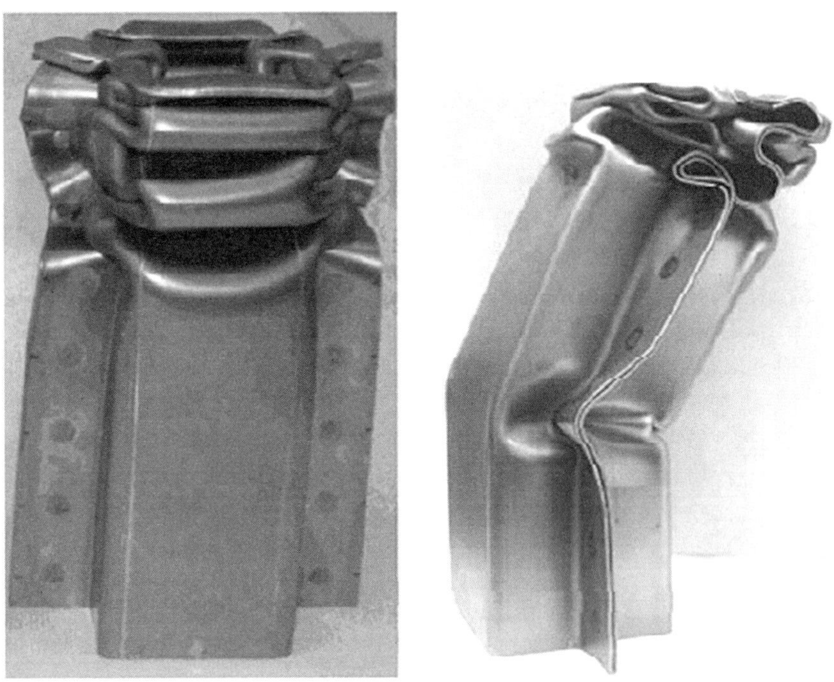

(a) (b)

Fig. 4.10 Formation of (a) regular buckling folds and (b) total buckling structures subjected to crushing loads. Source: Ref 4.8

steels are also characterized by very high ductility. Increase in the flow stress due to bake hardening improves the energy-absorption characteristic of these materials.

4.6 Formability

Formability of a metal is defined as its ability to be formed into simple and complex shapes by various manufacturing operations known as deformation processes. Forming of ductile metals is limited by plastic instability that leads to flow localization and fracture. In sheet metal forming, flow localization and component thinning occurs under complex conditions of loading. A number of tests that measure the tendency of metals for flow localization can be used. These include uniaxial tensile testing, hemispherical punch forming, deep drawing, and hole expansion. Such tests are usually performed at room temperature because most sheet forming operations are conducted at ambient temperatures.

Uniaxial Tension. The uniaxial tensile test provides a number of parameters that characterize formability of a metal. High work-hardening exponent, n, accounts for the ability of a metal to stretch and to distribute the strains more uniformly in the presence of an applied stress. Total elongation determines the extent to which a metal can be stretched before failure. Percent total elongation measures the formability of the material.

Hemispherical punch forming tests provide two measurements of formability known as limiting dome height (LDH) and forming limit diagram (FLD). The two tests use identical tooling and sample configurations but differ in data collection. Both tests use a hemispherical punch and a die with drawbeads to stretch locked sheet metal samples of varying widths (Fig. 4.11). The LDH test records the height of the formed dome at the moment of fracture.

Limiting dome heights for TRIP 600 and HSLA 550 steels are shown in Fig. 4.12. The superior formability of TRIP steel compared to HSLA of similar strength is evident.

The relationship between formability and strength is shown in Fig. 4.13. Formability, as measured by a normalized dome height (H/d), decreases as strength increases from the soft mild steel to the hard MS 1400 steel.

An example of an equibiaxial hemispherical stretch forming of AHSS is shown in Fig. 4.14. The dome height decreases with increase in the strength of the steel from mild (left) to DP 600 (middle) and DP 1000 (right).

Unlike the LDH test, which measures the limiting dome height of samples, the FLD test uses similar but gridded samples for measuring major and minor strains of deformed grids. The grids are imprinted on the surface of the sheet metal using chemical or photo etching. Plastic deformation during punch forming causes the etched circles to deform into el-

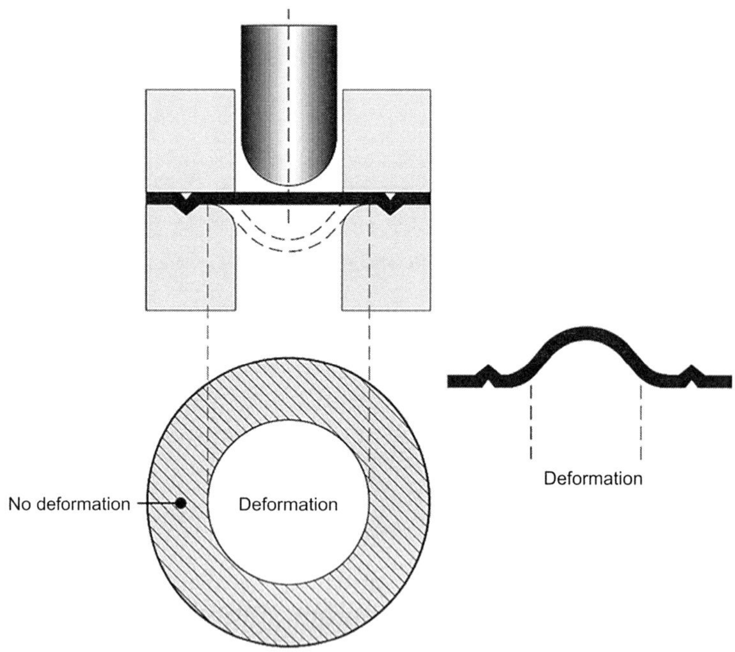

Fig. 4.11 Tooling used for the forming limit diagram and limiting dome height tests. Source: Ref 4.1

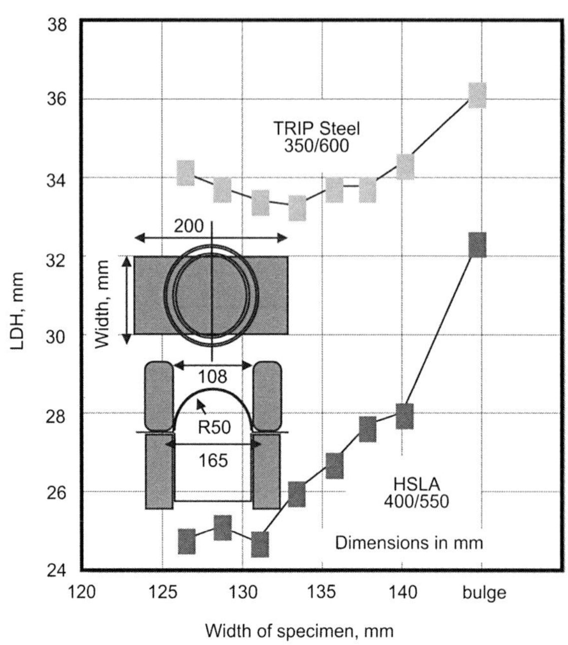

Fig. 4.12 Limiting dome height tooling and measurements for TRIP 600 and HSLA 550 steels. Source: Ref 4.1

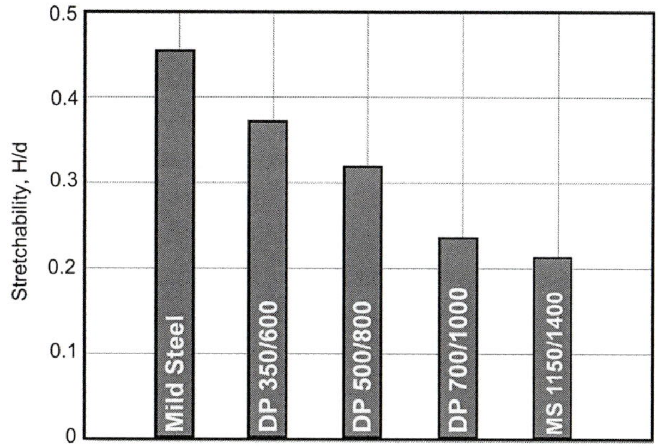

Fig. 4.13 Stretchability of various grades of steel. Source: Ref 4.1

Fig. 4.14 Effect of steel strength on dome height. Source: Ref 4.9

lipses. The amount of plastic strain at each grid can be calculated from the major and minor diameters of the ellipses.

Grid deformation occurs in samples subjected to various deformation paths simulating a number of sheet metal forming conditions. Forming paths, such as uniaxial, biaxial, and plane strain conditions, are simulated by varying the sizes of the tested samples. A hemispherical punch is used to stretch locked samples until fracture. Grid measurements are then taken in the failed location and plotted as major and minor strains separating grids that failed or necked from safe grids. The demarcation line between the safe and failed grids constitutes the FLD for the material. Figure 4.15 shows FLDs for TRIP 600 and MS 1400 steels. The FLD levels drop down drastically with increase in tensile strength of the steel.

Deep drawing is a sheet metal forming process that measures the ability of a material to be formed into a shape where the depth of the drawn part is significant compared to the dimensions of the blank. Good examples of deep-drawn parts are beverage cans and kitchen sinks. A test simulating deep drawing consists of a flat-bottomed punch pushing a circular sheet metal blank into a die cavity to produce flat-bottomed circular cups (Fig. 4.16): The sheet metal experiences compressive forces in zones 1 and 2 due to the blankholder force of the pressure plates. The sheet metal

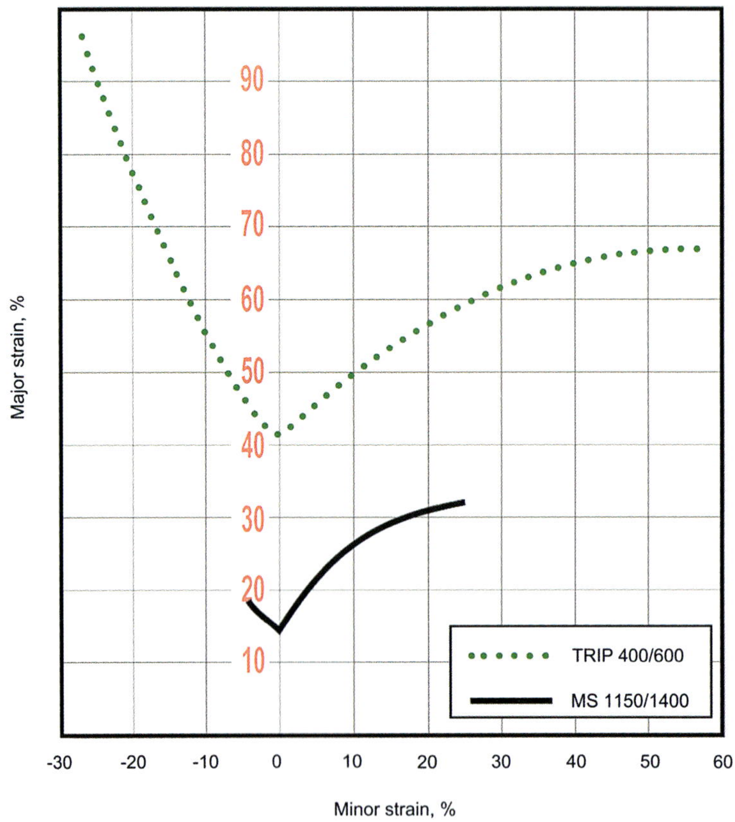

Fig. 4.15 Forming limit diagrams of two AHSS grades. Source: Ref 4.1

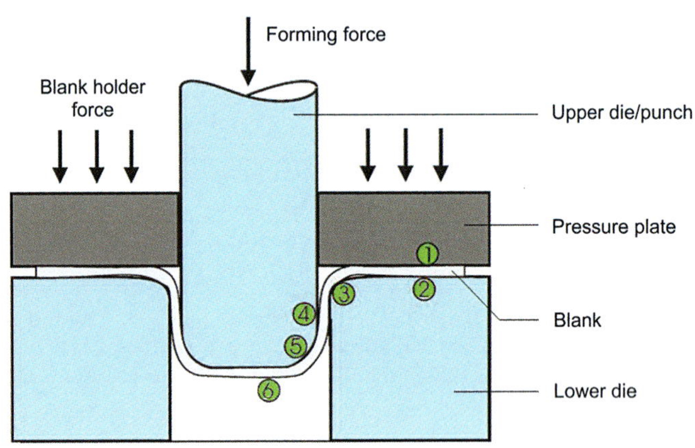

Fig. 4.16 Tooling for deep drawing of flat-bottomed cups

bends in zone 3, stretches in zone 4, and stretch-bends in zone 5. In zone 6, the sheet metal experiences biaxial stretching and friction with the bottom of the punch.

Drawability is expressed by the limiting drawing ratio (LDR), which is defined as the largest ratio of the blank-to-cup diameters that can be drawn successfully into a circular cup. If materials are compared with the same blank size and cup diameter, then their cup height can serve as a measure of the drawability of these materials. Because ductility decreases with increase in strength, LDR is also expected to decrease with increase in the strength of the material. Figure 4.17 shows that drawability, in general, decreases with increase in strength from mild to martensitic steel.

A demonstration of the effect of material strength on cup height is shown in Fig. 4.18. Cup height is largest for mild steel and smallest for martensitic M1400 steel.

Hole Expansion. Stretch flanging or edge stretchability is a common feature in automotive stampings because it corresponds to cut-outs for windows and other openings. In stretch flanging, a tensile strain is imposed on the sheared edge causing splitting of the flange (Fig. 4.19). The sheared edge forming limit is determined by stretching a punched hole

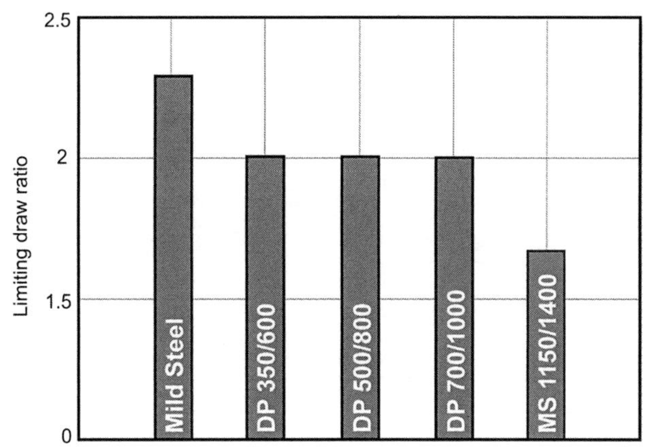

Fig. 4.17 Variation of limiting draw ratio with different steel types and grades. Source: Ref 4.1

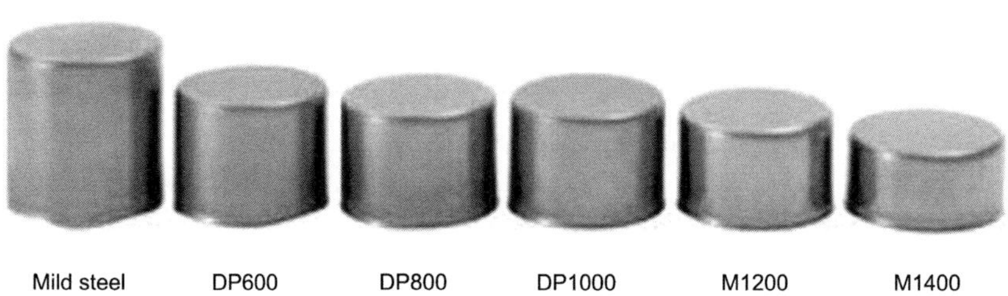

Fig. 4.18 Variation of cup height with strength for different types and grades of steel. Source: Ref 4.10

until a sheared edge crack develops. The hole expansion test measures the ability of a material to undergo local or edge stretching without failure.

The hole expansion test includes a tool set that is composed of a circular cavity and a conical punch. A sketch of the hole expansion tooling showing the conical punch, die, and sheet blank is illustrated in Fig. 4.20.

The conical punch is pushed into a square sample with punched hole of an initial diameter, d_0. The punch moves into the hole, expanding it until a crack develops at the edge of the expanding hole. The diameter of the cracked hole, d_f, is measured and the percent hole expansion is calculated from the ratio of the final hole diameter, d_f, to its initial diameter, d_0:

$$\text{Hole expansion (\%)} = [(d_f - d_0)/d_0]\ 100$$

Figure 4.21 shows a sample before and after stretch flanging using the hole expansion test. Stretch flangeability or sheared edge stretching limit of a metal is determined by the diameter of the hole at the onset of cracking.

Hole expansion test results for DP 980 and three types of CP 1000 steels are shown in Fig. 4.22. Although the four steels have similar ultimate tensile strengths of 1000 MPa (145 ksi), the CP family shows higher hole

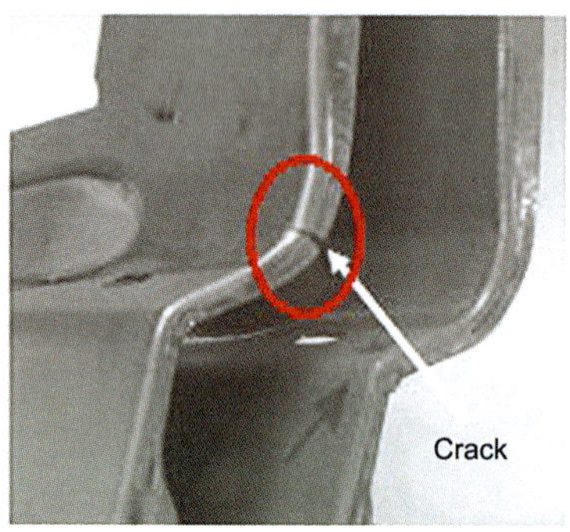

Fig. 4.19 Edge cracking in automotive component. Source: Ref 4.11

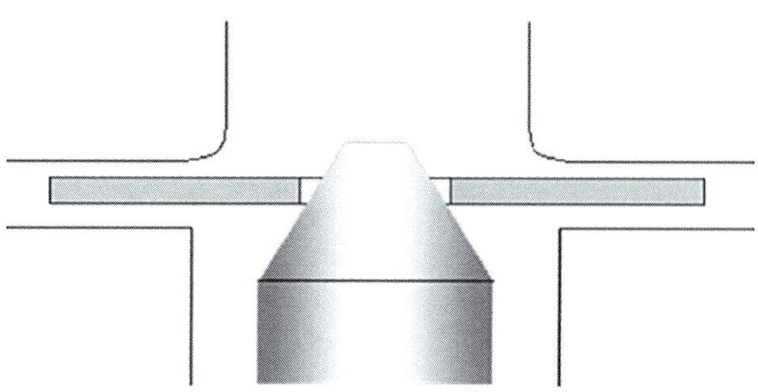

Fig. 4.20 Hole expansion test tool

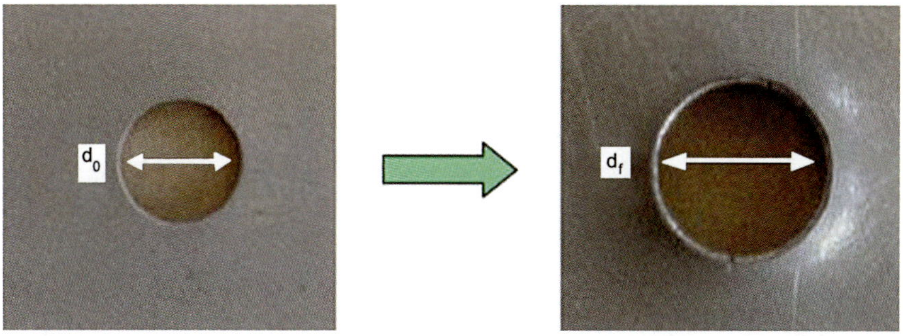

Fig. 4.21 Initial blank and final expanded hole sample. Source: Ref 4.11

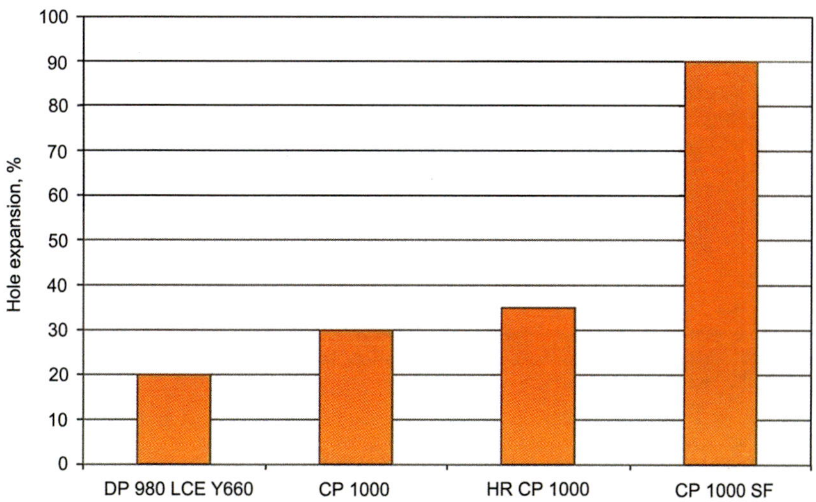

Fig. 4.22 Hole expansion behavior for dual-phase (DP) and three complex-phase (CP) steel types. Source: Ref 4.12

expansion values compared to that of DP 980 steel. The CP 1000 SF steel is specially designed for stretch flangeability; it shows the highest hole expansion value.

Sheared edge stretching limits, determined by the percent hole expansion, are plotted against the tensile strength of various steels (Fig. 4.23). Percent hole expansions are found to decrease strongly with increase in tensile strength of steel. The largest drop in hole expansion values occurs for steels with tensile strength below 700 MPa (100 ksi). Increasing strength beyond 700 MPa appears to have no effect on the sheared edge limit of steels.

Initial holes in the blanks for the hole expansion test are produced by punching, machining, or laser cutting. A clean edge cut is important for blanked, punched, or machined holes in AHSS because the quality of the blank hole affects the level of hole expansion. Machining usually produces better finished holes than blanking or punching.

Figure 4.24 compares hole expansion test results for punched and machined blank holes of an HSLA steel and various grades and types of AHSS. Machined holes produced higher hole expansion values than punched ones for all tested steels.

Cutting tools must also be sharp to produce clean edge cuts. Figure 4.25 compares hole expansion values for sharp and worn tools. Worn tools re-

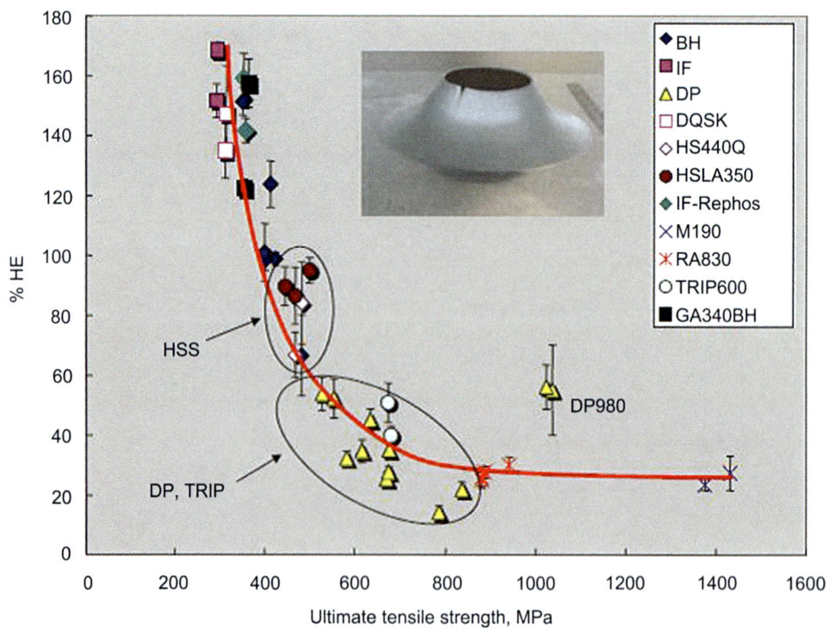

Fig. 4.23 Sheared edge stretching limits for various steels. HE, hole expansion. Source: Ref 4.11

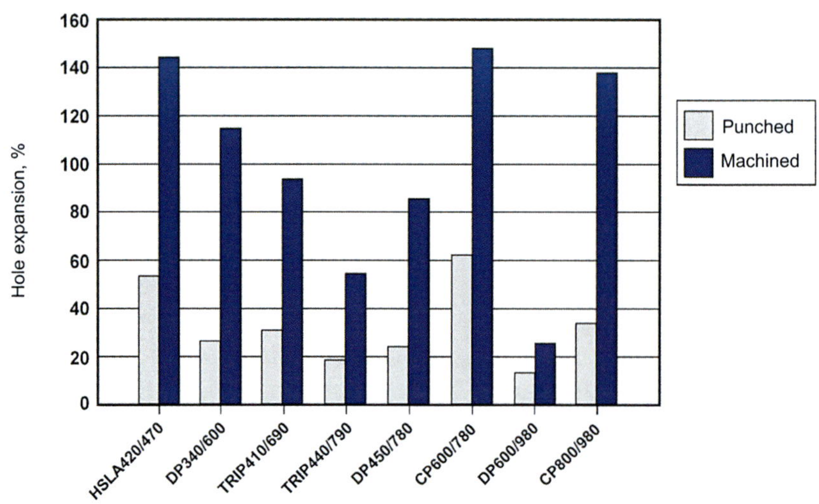

Fig. 4.24 Effect of cutting tool on hole expansion. Source: Ref 4.1

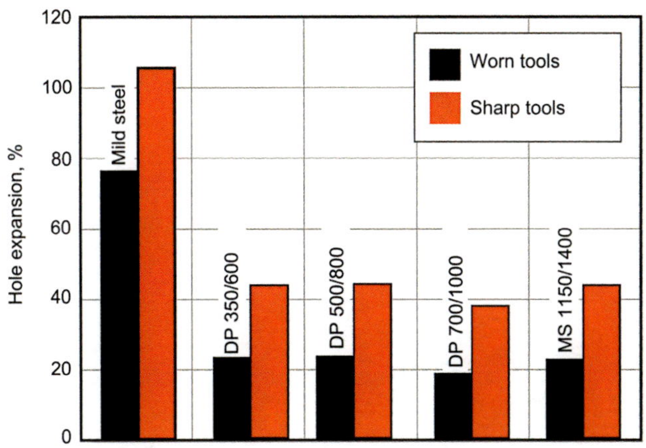

Fig. 4.25 Variation of percent hole expansion with some types of AHSS. Source: Ref 4.1

sult in 25% reduction in stretch flangeability in mild steel, but the reduction increases to 50% in AHSS. Laser cutting produces the cleanest edge cut, producing higher stretch flangeability than other cutting methods.

4.7 Toughness

Toughness is defined as the resistance of a metal to fracture or its ability to absorb energy and deform plastically before fracturing. A rough measure of toughness is the area under the stress-strain curve up to fracture. This area determines the amount of plastic work per unit volume at fracture and it is important in energy-absorbing applications. For a metal to be tough, both strength and ductility must be high. The larger the area under the stress-strain curve, the more the energy that can be absorbed before fracture.

Figure 4.26 compares the dynamic energy absorption of different grades of automotive steels tested at a high strain rate of 10^3 s^{-1}. The superior energy-absorption capabilities of the three AHSS (TWIP, TRIP, and DP) are evident compared to low-carbon rephosphorized (Rephos), bake-hardenable (BH), interstitial-free (IF), and high-strength, low-alloy (HSLA) steels. The outstanding energy absorption of TWIP steel is the result of its very high strength and excellent ductility.

Toughness can also be defined as the resistance of a metal to propagate cracks. Cracks in metals result from the coalescence of tiny holes produced by inclusions (particles of carbides, oxides, nitrides) during plastic deformation. The energy absorbed by an advancing crack is used to enlarge the plastic zone around its tip. The size of the plastic zone and, hence, the toughness of the metal decreases rapidly as the yield stress, σ_y, increases. This means that increasing the yield strength causes the plastic

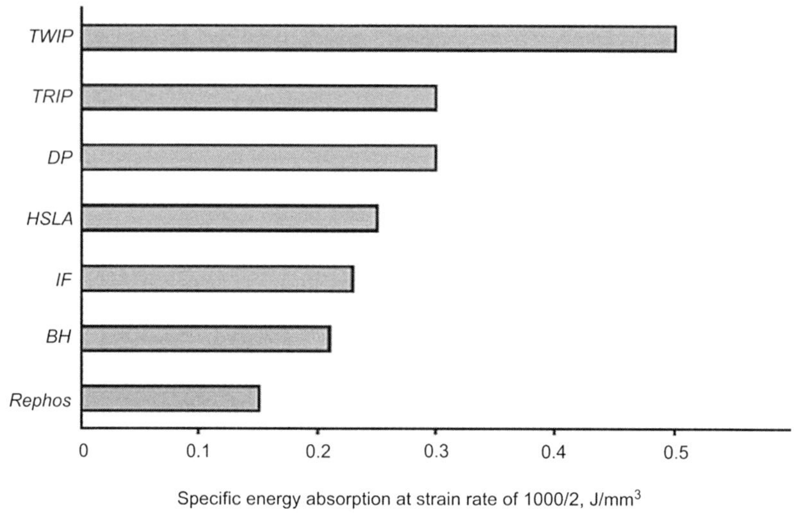

Fig. 4.26 Energy-absorption capabilities for common types of automotive steels. Source: Ref 4.4

zone around the tip of the crack to shrink and this causes the toughness to decrease.

Figure 4.27 shows the variation of fracture toughness with yield strength for different grades of steel. The inverse relationship between fracture toughness and yield strength is obvious. Steels toward the bottom right have high strength and low toughness; they fracture before they yield. Steels toward the top left have low strength and high toughness; they yield before they fracture. For strength-limited designs, structures rely on yielding before they fracture. Materials selection for structures involves a compromise between strength and fracture toughness.

Composition and heat treatment can produce many combinations of yield strength and toughness. Figure 4.27 shows clearly the outstanding fracture toughness of stainless steels and the extremely low fracture toughness of quenched alloy and carbon steels. The figure also shows the benefits of quench and temper heat treatments in raising the fracture toughness of alloy and carbon steels. One way to increase toughness in alloys with no loss of strength is to reduce the number of inclusions that nucleate voids and coalesce into cracks that lead to ductile fracture.

The first-generation AHSS (DP, TRIP, CP, and MS) have high strength but low ductility; hence, they have lower toughness than those of the second generation (TWIP, L-IP, and AUST SS) that have high strength as well as high ductility. Toughness of steel can be increased by increasing its strength using heat treatment, which produces fine grains or fine precipitates, and also by cold working and alloying.

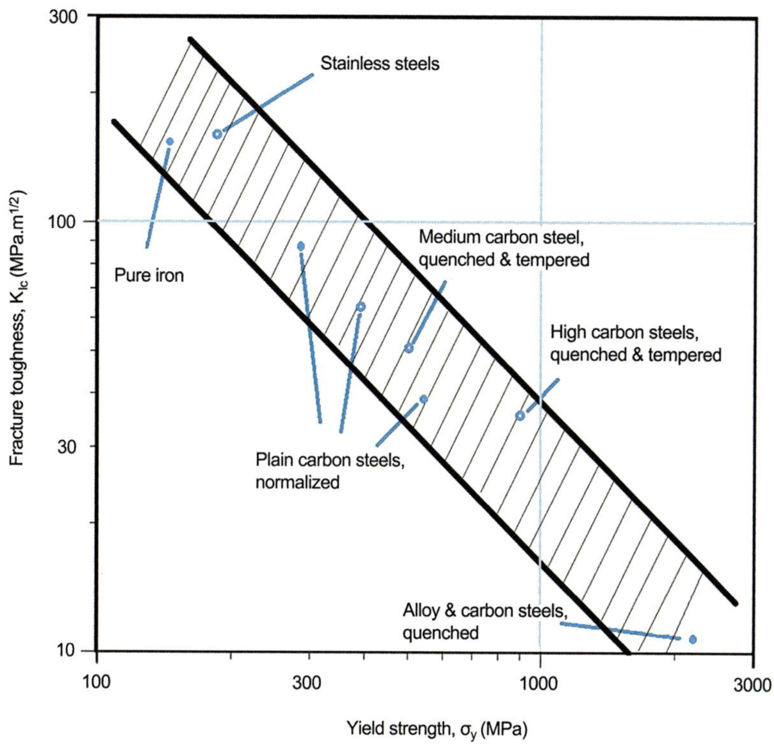

Fig. 4.27 Variation of fracture toughness with yield strength for different steel grades. Source: Adapted from Ref 4.13

4.8 Bake Hardening

Bake hardening is the increase in strength of prestrained steel resulting from a combination of work hardening during part forming and strain aging during the subsequent paint/bake cycle (177 °C, or 350 °F, for 30 min) of automobiles. Figure 4.28 shows the deformation characteristics of bake hardening and the contributions of work hardening and bake hardening to the total strength of bake-hardened steel.

Bake hardening is a controlled aging process that is related to the presence of carbon and/or nitrogen in solid solution in the steel. The bake-hardening mechanism involves the diffusion of carbon and nitrogen atoms to form an atmosphere around dislocations generated by prestraining or forming. Blocking and pinning dislocations lead to a lower dislocation-free-mean path and result in strengthening the steel. Bake hardening makes a formed steel panel stronger after baking than after forming. This implies that the panel will form better because of its lower initial strength but will gain strength and hardness for dent resistance when assembled in the automobile body.

Figure 4.29 shows an illustration of the bake-hardening process, which includes forming and the paint/bake cycle. Superimposed on the load elongation chart is the process of pinning of dislocations by diffused carbon or nitrogen solutes during the paint curing temperature.

Bake hardening exhibits strong dependence on the type of steel, the degree of prestrain, the baking temperature, and the baking time. In Fig. 4.30, bake hardening of 2% strained and baked samples of DP, TRIP, and HSLA steels were determined and their strain hardening properties were evaluated. Figure 4.30 compares their work-hardening and bake-hardening contributions to yield strength. High-strength, low-alloy steels showed no bake-hardening effect. The increase in yield strength of TRIP steel in a typical paint/bake cycle is approximately 60 MPa (10 ksi) for the

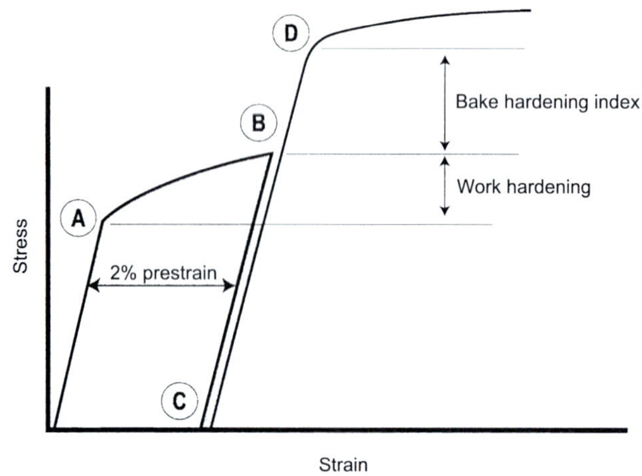

Fig. 4.28 Deformation characteristics in bake-hardened steel. Source: Ref 4.1

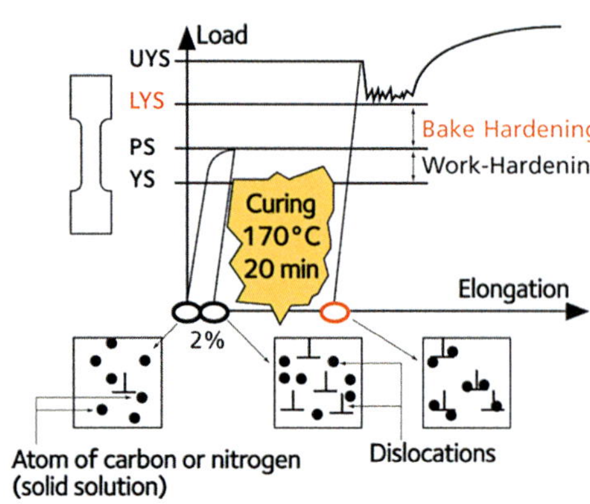

Fig. 4.29 Illustration of the bake-hardening process. UYS, upper yield strength. LYS, lower yield strength. PS, pre-strain. YS, yield strength. Source: Ref 4:12

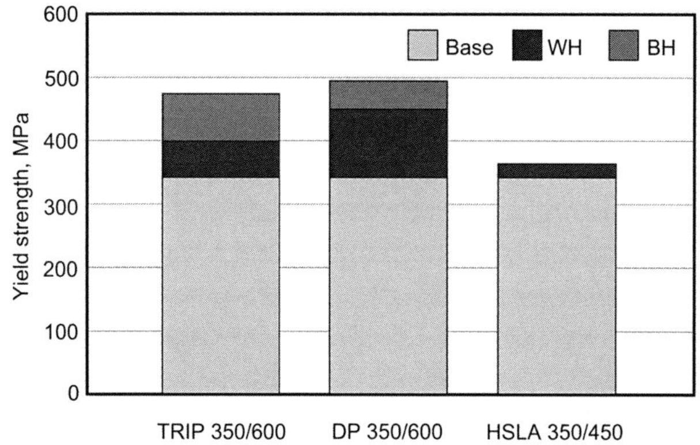

Fig. 4.30 Contribution of work hardening and bake hardening to yield strength of three steels. Source: Ref 4.1

2% prestrain and 96 MPa (15 ksi) for the 20% prestrain. Bake hardening of DP and TRIP steels can, therefore, contribute significantly to the strength level of formed and paint baked parts.

REFERENCES

4.1 "Advanced High Strength Steels Applications Guidelines," Version 4.1, WorldAutoSteel, www.worldautosteel.org, June 2009

4.2 M.F. Ashby, *Materials Selection in Mechanical Design,* 3rd ed., Elsevier, 2005

4.3 M.-H. Cai et al., Strain Hardening Behavior of High Performance FBDP, TRIP and TWIP Steels, *Steel Res. Int.,* Vol 82 (No. 3), 2011, p 242–248

4.4 B.C. De Cooman, K. Chin, and J. Kim, High Mn TWIP Steels for Automotive Applications, *New Trends and Developments in Automotive System Engineering,* M. Chiaberge, Ed., InTech, Shanghai, 2011, p 101–128. http://www.intechopen.com/books/new-trends-and-developments-in-automotive-system-engineering/high-mn-twip-steels-for-automotive-applications

4.5 W.D. Callister, Jr. and D.G. Rethwisch, *Materials Science and Engineering: An Introduction,* 8th ed., Wiley, 2010

4.6 T. Niendorf, C. Lotze, D. Canadinc, A. Frehn, and H.J. Maier, The Role of Monotonic Pre-deformation on the Fatigue Performance of a High-Manganese Austenitic TWIP Steel, *Mat. Sci. Eng. A,* Vol 499 (No. 1–2), Jan 15, 2009, p 518–524

4.7 J. Hall, "Evolution of AHSS in Automotive Applications," General Motors Company presentation at Great Designs in Steel Seminar, May 18, 2011 (Livonia, MI), AUTOSTEEL

4.8 Z. Gronostjski, A. Niechajowicz, and S. Polak, Prospects for the Use of New-Generation Steels of the AHSS Type for Collision Energy Absorbing Components, *Arch. Metall. Mater.,* Vol 55 (No. 1), 2010

4.9 Uddeholm Tooling and SSAB Swedish Steel, "Tooling Solutions for Advanced High Strength Steels," presented at Uddeholm Swedish Rally, 2005

4.10 Uddeholm Tooling and SSAB Swedish Steels, "Tooling Solutions for Advanced High Strength Steels—Selection Guidelines," 2009, www.ssab.com and www.uddeholm.com

4.11 S. Sadogopan, "Formability Characterization of Advanced High Strength Steels," Ispat Inland Inc. presentation at Great Designs in Steel Seminar, Feb 18, 2004 (Livonia, MI), AUTOSTEEL

4.12 ArcelorMittal Web site, www.arcelormittal.com

4.13 M. Ashby, H. Shercliff, and D. Cebon, *Materials: Engineering, Science, Processing and Design,* 2nd ed., Elsevier, 2010

CHAPTER 5

Dual-Phase Steels

DUAL-PHASE (DP) STEELS have the widest usage in the automotive industry because of their excellent combination of strength and ductility. They have higher tensile strength than conventional steels of similar yield strength. Dual-phase steels also have a higher initial work-hardening rate and lower yield strength/tensile strength ratio than high-strength, low-alloy (HSLA) steels of similar strength.

5.1 Composition and Microstructure of Dual-Phase Steels

Dual-phase steels are a group of steels with a duplex microstructure consisting of a soft ferrite matrix and 10 to 40 volume percent of hard martensite islands. This type of microstructure allows achieving ultimate tensile strength in the range of 500 to 1200 MPa (70 to 175 ksi). The mi-

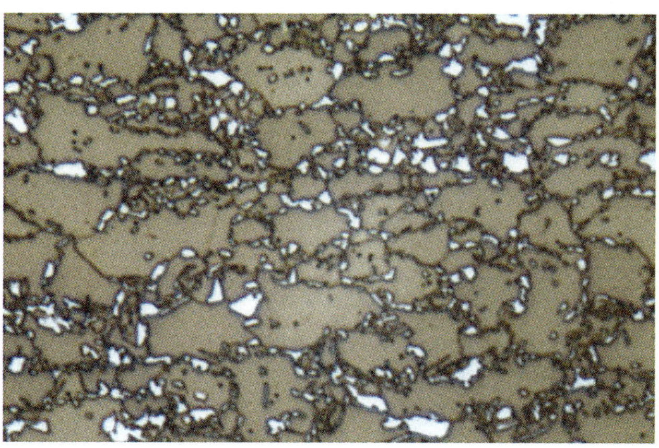

Fig. 5.1 Microstructure of DP 600 showing ferrite matrix and martensite islands. Source: Ref 5.1

crostructure of DP 600 steel is shown in Fig. 5.1. A schematic of the microstructure of DP steel is shown in Fig. 5.2. Dual-phase steels derive their strength from the martensite phase and their ductility from the ferrite phase. The volume fraction of martensite determines the strength level of this steel.

The effect of carbon and alloying elements on DP steels is summarized in Table 5.1.

Special heat treating practices that involve quenching and tempering are used to generate the martensite phase. The dual-phase structure is produced by quenching low-carbon steels from the $\alpha + \gamma$ phase region to form a microstructure of martensite islands in a ferrite matrix. The austenite-to-martensite phase transformation determines the ending microstructure and

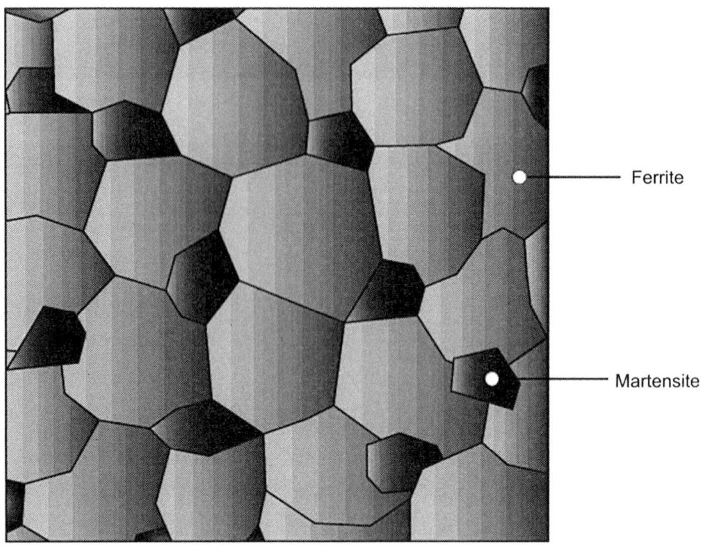

Fig. 5.2 Schematic microstructure of dual-phase (DP) steel. Source: Ref 5.2

Table 5.1 Effect of alloying elements in dual-phase steels

Alloying element	Effect of and reason for adding
C, 0.06–15%	Austenite stabilizer Strengthens martensite Determines the phase distribution
Mn, 1.5–2.5%	Austenite stabilizer Solid-solution strengthener of ferrite Retards ferrite formation
Si	Promotes ferritic transformation
Cr, Mo, up to 0.4%	Austenite stabilizer Retards pearlite and bainite formation
V, up to 0.06%	Austenite stabilizer Precipitation strengthener Refines microstructure
Nb, up to 0.04%	Austenite stabilizer Reduces M_s temperature Refines microstructure and promotes ferrite transformation from nonrecrystallized austenite

Source: Ref 5.3

the resulting mechanical properties of the steels. This microstructure gives DP steels their high strain-hardening capability and better formability compared to HSLA grades. Other unique properties of DP steels include low elastic limit, high tensile strength, high uniform strain, and high total elongation. Dual-phase steels can also exhibit a bake-hardening effect, which is the increase in yield strength resulting from prestraining and elevated-temperature aging. This process occurs during the paint/bake cycle that leads to increase in the yield strength and results in improved dent and crush resistance. The extent of the bake-hardening effect in AHSS depends on the specific chemistry and thermal histories of the steels.

Dual-phase steels were developed four decades ago and they were designed to provide ultimate tensile strengths of 600 to 1000 MPa (90 to 145 ksi). Research efforts to increase the strength of DP steels without suffering a loss in ductility have taken the route of grain refinement. In recent years, new processing methods were employed to produce ultrafine ferrite grains of 1.2 μm (0.00004 in.). In one study, three low-carbon DP steels with a martensite volume fraction of 30% and different ferrite grain sizes of 1.2, 2.4, and 12.4 μm were produced by thermomechanical treatment (Ref 5.4). Their microstructures were observed and their mechanical properties were evaluated. Results of the investigation showed that both yield and tensile strength increased, whereas uniform elongation and total elongation were slightly reduced by grain refinement. Figure 5.3 shows scanning electron microscope (SEM) micrographs of the three grain sizes that were produced: (a) coarse grains of 12.4 μm, (b) fine grains of 2.4 μm, and (c) ultrafine grains of 1.2 μm. The martensite fraction was determined to be approximately 30% by volume. Electron diffraction measurements showed that the amount of retained austenite in the samples was between 1 to 3% by volume.

Table 5.2 lists grain size information and tensile data obtained by testing the three grain sizes of the DP steel. The data show clearly that reducing the grain size appreciably increases the strength and slightly reduces the ductility and the strain hardening of the DP steel.

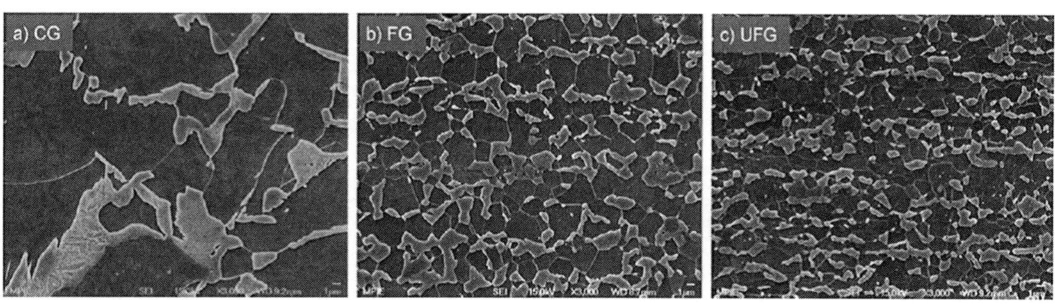

Fig. 5.3 Micrographs showing three grain sizes for DP steel. (a) CG, coarse grains. (b) FG, fine grains. (c) UFG, ultrafine grains. Source: Ref 5.4

Dual-phase steels have a wide range of chemical compositions. They are more highly alloyed than HSLA steels because they contain more than 1% of alloying elements. The chemical composition of DP steels varies from grade to grade but the major alloying elements are carbon, manganese, and silicon. Smaller amounts of other elements, such as aluminum, nitrogen, phosphorus, and sulfur may also be present. The spectrum of chemical compositions of some commercially available DP steels is given in Table 5.3. The table shows the maximum percentage of the major alloying elements.

5.2 Processing of Dual-Phase Steels

The microstructure of DP steels consists of a ferrite matrix containing islands of martensite. The dual microstructure of DP steels is produced by controlled cooling from the austenite phase for hot rolled products or from the two-phase austenite and ferrite for continuously annealed cold rolled and hot dip coated products. In cold rolled and hot dipped sheets, this microstructure develops by intercritical annealing of the cold rolled sheets followed by quenching to room temperature. Quenching transforms the austenite in the ferrite-austenite microstructure to martensite. This produces the dual ferrite-martensite microstructure of DP steels.

Table 5.2 Tensile data for three grain sizes of dual-phase steel, martensite volume fraction ~30%

Microstructure DP Steel	Grain size, μm	Yield strength, MPa (ksi)	Ultimate tensile strength, MPa (ksi)	Uniform elongation, %	Total elongation, %	Strain hardening, *n*
Coarse grain	12.4	445 (65)	870 (126)	7.2	7.7	0.21
Fine grain	2.4	483 (70)	964 (140)	7.4	8.9	0.18
Ultrafine grain	1.2	525 (76)	1037 (150)	7.1	7.3	0.18

Source: Adapted from Ref 5.4

Table 5.3 Chemical composition of dual-phase steels

	Composition, % (max)		
Steel grade	C	Mn	Si
FF 280 DP(a)	0.14	1.6	0.40
DP 450(a)	0.08	1.6	0.40
DP 500(a)	0.14	1.6	0.40
DP 600(a)	0.14	2.1	0.40
DP 780 Y450(a)	0.17	2.2	0.60
DP 780 LCE Y450(a)	0.10	2.0	0.40
DP 780 Y500(a)	0.17	2.2	0.60
DP 780 LCE Y500(a)	0.10	2.0	0.40
DP 980 LCE Y600(a)	0.11	2.9	0.70
DP 980 LCE Y660(a)	0.11	2.9	0.70
DP 980 Y700(a)	0.18	2.4	0.60
DP 980 LCE Y700(a)	0.11	2.9	0.70
DP 1180(a)	0.18	2.4	0.60
DP 600(b)	0.09	1.0	0.25
DP 780(b)	0.09	1.0	0.30

(a) Cold rolled. (b) Hot rolled. Source: Ref 5.1

The hard martensite islands dispersed in the soft ferrite matrix are responsible for the high strength of DP steels. The amount of volume fraction of the hard martensite determines the level of strength of the steel.

Dual-phase steels are produced by various processing routes. The simplest way to obtain a dual-phase ferrite-martensite steel is by intercritical annealing of a ferrite-pearlite microstructure in the two-phase ($\alpha + \gamma$) region, followed by rapid cooling to transform austenite to martensite. Figure 5.4 shows a simplified sketch describing the steps for producing DP steel by the intercritical annealing process.

For any given carbon content, C_0, the amount of austenite, γ, increases with an increase in the intercritical annealing temperature, becoming equal to 100% at the Ac_3 temperature. However, the carbon content of austenite decreases with an increase in the intercritical annealing temperature, reaching its minimum value of C_0 at the Ac_3 temperature line. For a given intercritical temperature, the amount of austenite increases with the increase in the carbon content of steel.

The conclusion is that high intercritical annealing temperature results in high austenite fractions of decreased carbon content, while low annealing temperature results in low austenite fraction with high carbon content. This means that the intercritical annealing temperature determines the maximum amount of austenite that can be available for transformation to martensite. It also determines the austenite carbon content and, therefore, the hardenability of the steel. The final critical step in the transformation of austenite to martensite depends decisively on the cooling rate following intercritical annealing.

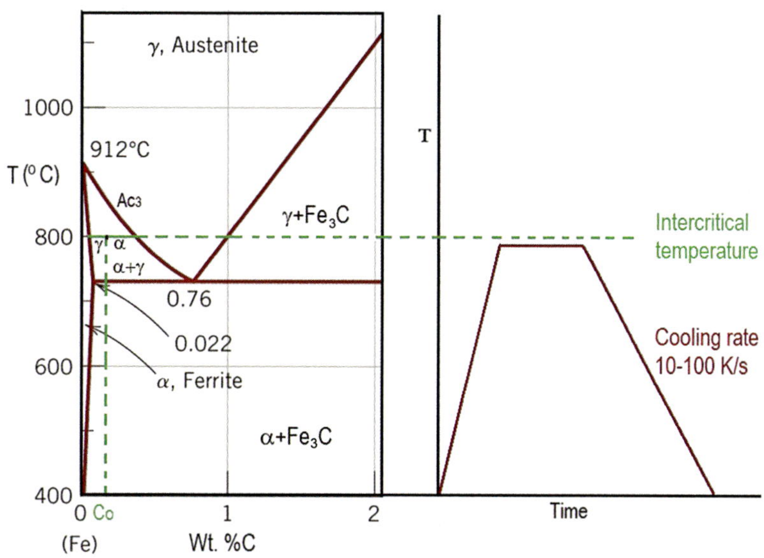

Fig. 5.4 Production of DP steel by intercritical annealing

There are three basic approaches for the commercial production of DP steels:

- Hot rolling where DP microstructure is produced during the conventional hot rolling cycle
- Continuous annealing where hot or cold rolled steel strips are uncoiled and annealed at the intercritical temperature
- Batch annealing where hot or cold steel is annealed in the coiled condition

The dual-phase microstructure can be produced after hot rolling of sheet products by first slow cooling the sheet to allow the austenite to transform into ferrite. After that, the sheet is fast cooled to the coiling temperature, which must be below the martensite start temperature M_s (Fig. 5.5). Alloying elements shift the lobes of the continuous cooling curves to help achieve the desired microstructure.

Dual-phase microstructure can also be produced in cold rolled sheet products by the application of continuous annealing followed by hot dip galvanizing. A schematic diagram of the thermal history during the annealing process is shown in Fig. 5.6. During the annealing process, sheet metal is heated to the temperature region where the structure is part austenite and part ferrite. A temperature slightly above Ac_1 in the range of 730 to 760 °C (1345 to 1400 °F) is typical. During this stage, approximately 15% of the ferrite and pearlite microstructure transforms to austenite. The steel is then quenched to cause the austenite to transform to martensite. The final microstructure will, therefore, contain mostly soft ferrite with

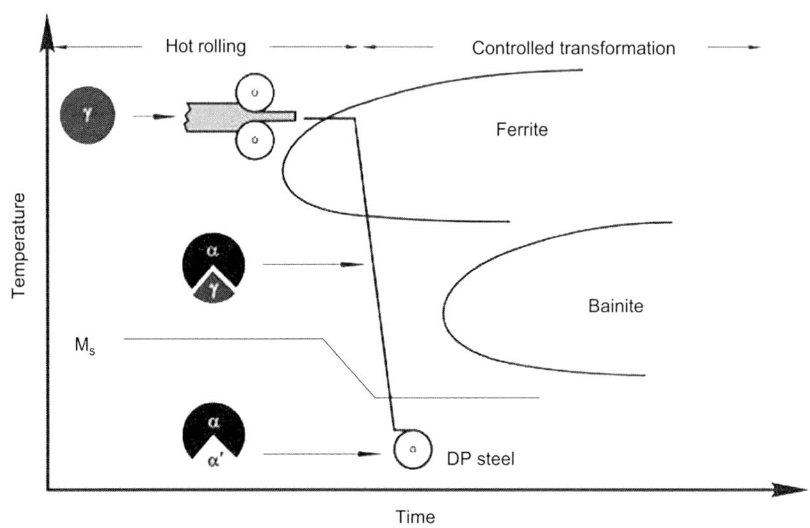

Fig. 5.5 Thermal history for producing hot rolled DP steel strips. Source: Ref 5.3

islands of hard martensite. The strengthening effect in DP steel is almost directly proportional to the volume fraction of hard martensite.

Galvannealed DP steel is produced by the continuous annealing line (CAL) process (Fig. 5.7). This annealing/coating process is comprised of six stages: heating, soaking, slow cooling, quenching, galvannealing, and air cooling. Each step produces a different microstructure and in the end the dual microstructure of ferrite and martensite is produced.

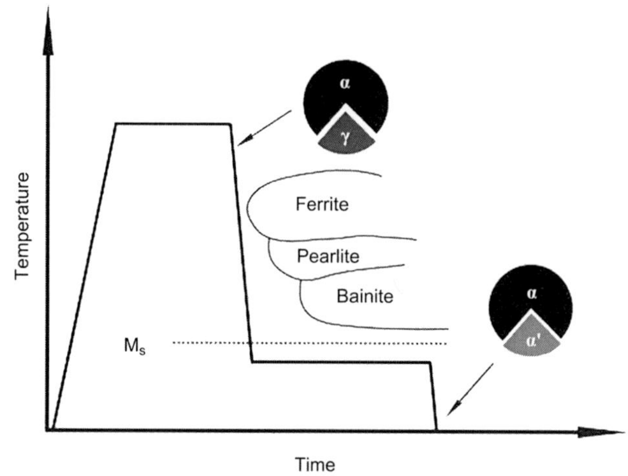

Fig. 5.6 Temperature changes during continuous annealing of DP steel sheets. Source: Ref 5.3

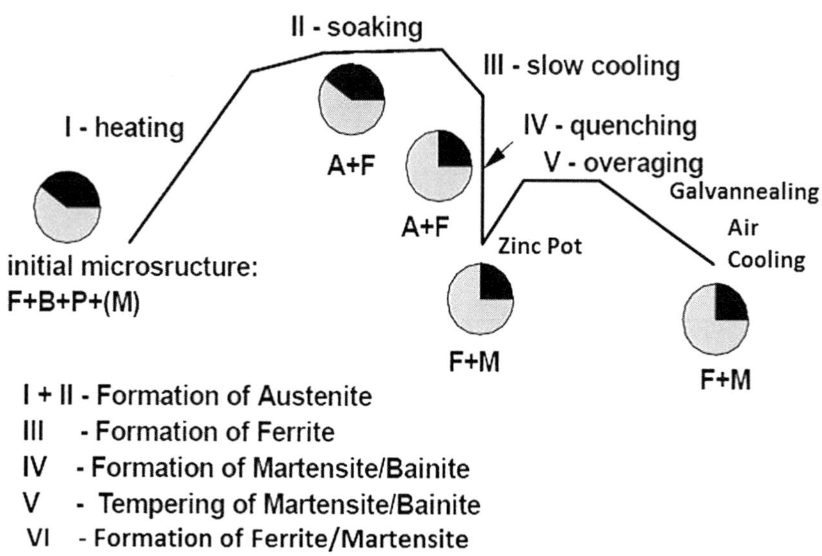

Fig. 5.7 The continuous annealing line (CAL) process for producing coated DP Steel. Source: Adapted from Ref 5.5

5.3 Deformation Mechanism of Dual-Phase Steels

Dual-phase steels deform solely by dislocation glide, which occurs mainly in the soft ferrite matrix. The hard martensite islands obstruct dislocation glide and cause an increase in the stress level. The soft ferrite phase is generally continuous, giving DP steels excellent ductility. When deformation occurs, strain is concentrated in the soft ferrite matrix and this is responsible for the high strain-hardening rate of DP steels. Strengthening DP steels is a function of both the volume fraction of martensite and the strength of the martensite phase.

5.4 Mechanical Properties of Dual-Phase Steels

The key mechanical properties of AHSS are strength to enable downgaging and ductility to allow forming a component. Figure 5.8 shows the location of DP steels in the tensile strength-total elongation space. Tensile strength ranges from 420 to 1030 MPa (60 to 150 ksi) and total elongation ranges from 5 to 28%.

Dual-phase steels contain ferrite and martensite phases. Higher strength results from higher volume fractions of martensite. Typical mechanical properties for DP steels are listed in Table 5.4.

Engineering stress-strain curves for various DP steel types compared to that of mild steel are shown in Fig. 5.9. The familiar deformation pattern of decreasing ductility with increasing strength is evident. Depending on the strength of the DP steel type, ductility can vary from 8 to 32%.

True stress-strain curves for the same group of DP steel types compared to that of mild steel are shown in Fig. 5.10. Increasing tensile strength

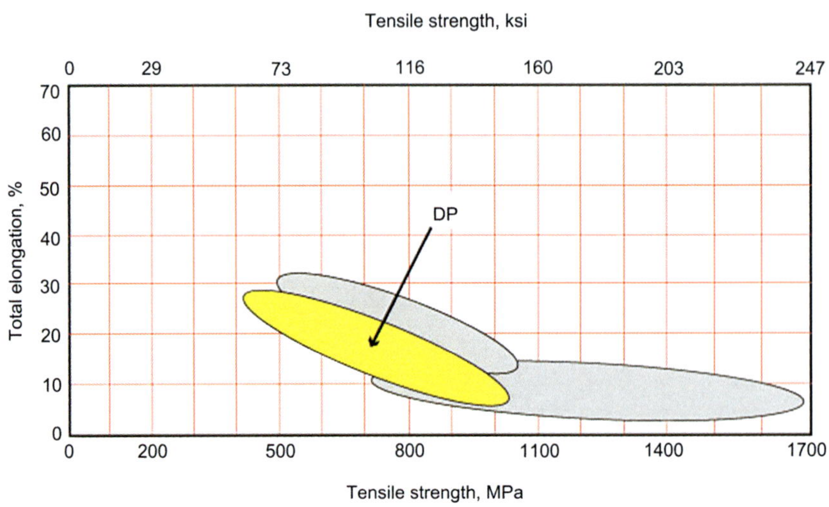

Fig. 5.8 Location of DP steels in the tensile strength-elongation space. Source: Ref 5.2

Table 5.4 Mechanical properties of various dual-phase steel types

Product	Yield strength, MPa (ksi)	Tensile strength, MPa (ksi)	Total elongation, %
EG DP500	315 (46)	547 (79)	31
GA DP590	355 (51)	620 (90)	26
CR DP590	370 (54)	625 (91)	26
CR DP780	470 (68)	820 (119)	18
CR DP980	675 (98)	1030 (149)	13

Source: Adapted from Ref 5.5, 5.6

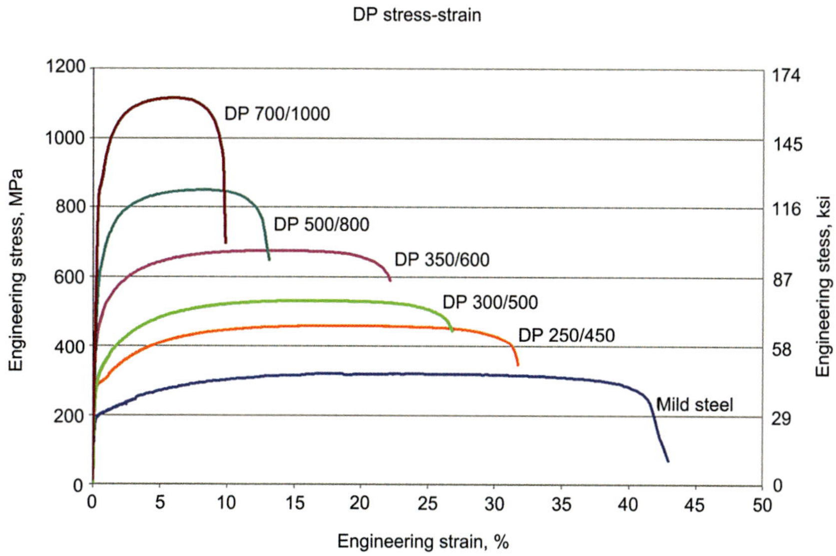

Fig. 5.9 Engineering stress-strain curves for various DP steel types compared to mild steel. Source: Ref 5.2

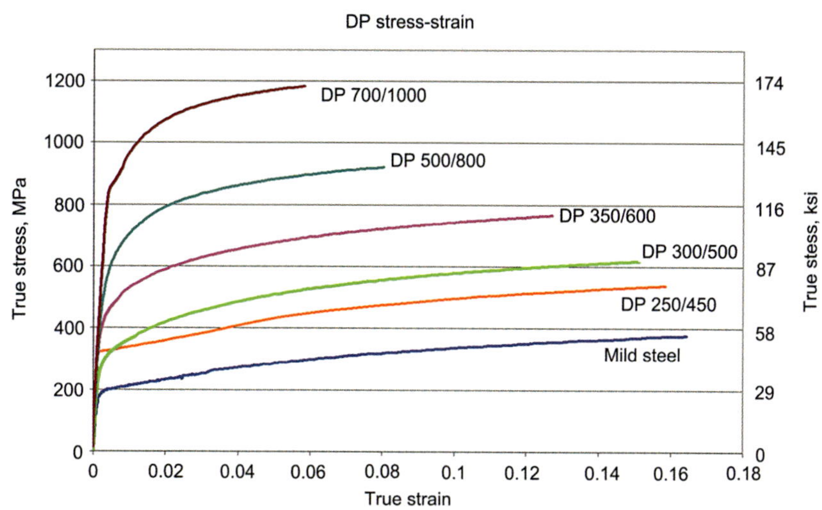

Fig. 5.10 True stress-strain curves for a series of DP steel types compared to mild steel. Source: Ref 5.2

from 450 to 1000 MPa (64 to 143 ksi) reduced true strain from 0.16 to 0.06.

Engineering stress-strain curves for DP 600 steel tested in the three directions to rolling—longitudinal, transverse, and diagonal—are shown in Fig. 5.11. Rolling direction has a small effect on ductility and a smaller effect on the strength level of DP 600 steel.

Endurance limit (fatigue strength) is directly related to tensile strength of the steel. As a result of their high strength, DP steels have good fatigue properties. The *S-N* curves for different types of DP steels are shown in Fig. 5.12. Endurance limit increases with increase in the tensile strength of DP steels. The curves plot maximum stress versus number of cycles to failure for a tension-tension loading ratio $R = 0.1$.

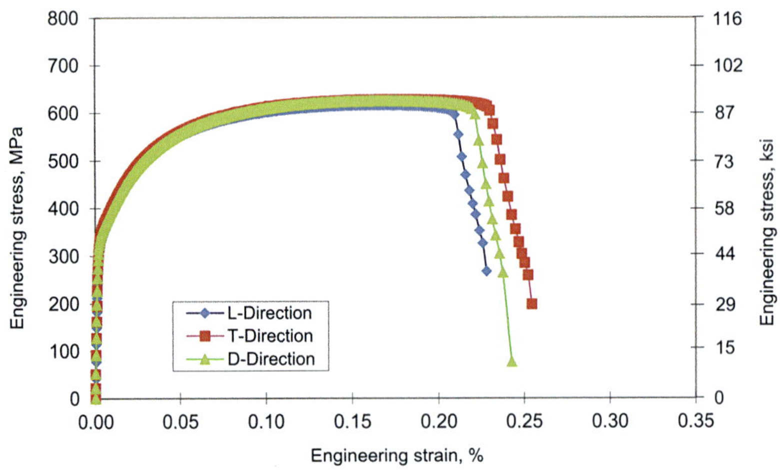

Fig. 5.11 Engineering stress-strain curves for DP 600 steel tested in three directions to rolling. Source: Ref 5.7

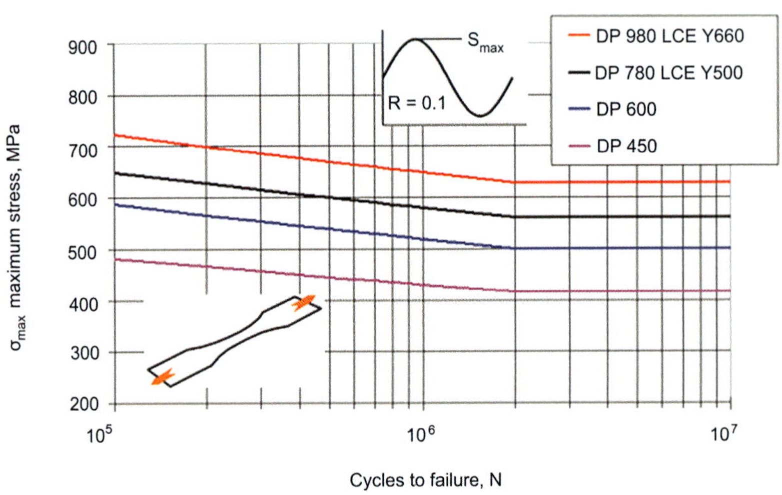

Fig. 5.12 Fatigue curves for various types of DP steel. Source: Ref 5.1

5.5 Formability of Dual-Phase Steels

General formability is related to the ductility of metals and alloys. Forming limit diagrams (FLDs) are commonly used to define maximum allowable strains without necking for different out-of-plane deformation paths, such as uniaxial, biaxial, and plane strain. Figure 5.13 shows FLDs for various types of DP steels. Forming limit diagram levels decrease with increase in the tensile strength of DP steel types. This behavior is expected because FLD levels are directly related to ductility, which decreases with increase in the strength of the steel.

5.6 Special Attributes of Dual-Phase Steels

Dual-phase steels have excellent combinations of strength, ductility, and strain-hardening capacity compared to conventional high-strength steels. Their yield strength can be increased further through bake hardening.

Dual-phase steels have a high initial work-hardening rate. This characteristic allows strain to be distributed more uniformly across greater volumes of material and delays the onset of localized necking and local thinning. As a result, DP steels exhibit high values of tensile strength and ductility. As a result of their high strength, DP steels have high energy-

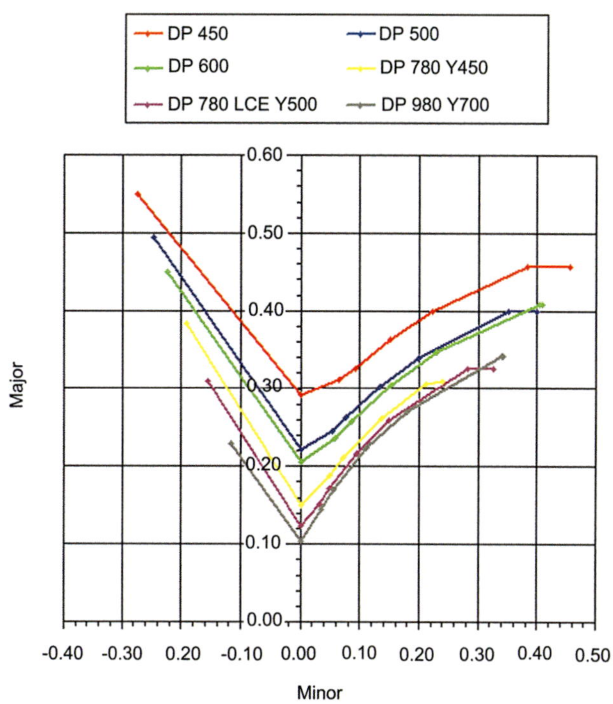

Fig. 5.13 Forming limit diagrams for various cold rolled DP steel grades. Source: Ref 5.1

absorption capacity and fatigue strength and, therefore, are suited for automotive structural and safety parts such as cross members, reinforcements, wheel webs, shock towers, and longitudinal rails.

REFERENCES

5.1 ArcelorMittal Web site, www.arcelormittal.com

5.2 "Advanced High Strength Steels Applications Guidelines," Version 4.1, WorldAutoSteel, www.worldautosteel.org, June 2009

5.3 W. Bleck and K. Phiu-On, Microalloying of Cold-Formable Multi Phase Steel Grades, *Mater. Sci. Forum,* Vol 500–501, 2005, p 97–112

5.4 M. Calcagnotto, D. Ponge, and D. Raabe, Effect of Grain Refinement to 1 mm on Strength and Toughness of Dual-Phase Steels, *Mat. Sci. Eng. A,* Vol 527 (No. 29–30), Nov 2010, p 7832–7840

5.5 R.H. Wagoner, Advanced High Strength Steel Workshop, Oct 22–23, 2006 (Arlington, VA), sponsored by the National Science Foundation, Department of Energy, and the Auto/Steel Partnership

5.6 D. Bhattacharya, *Developments in Advanced High Strength Steels,* Research and Development Center, Mittal Steel, East Chicago, IN, 2005

5.7 M.Y. Demeri, Forming of Advanced High Strength Steels, *Metal Working: Sheet Forming,* Vol 14B, *ASM Handbook,* ASM International, 2006, p 530–538

CHAPTER 6

Complex-Phase Steels

6.1 Compositions and Microstructures of Complex-Phase Steels

Complex-phase (CP) steels consist of a very fine microstructure of ferrite-bainite matrix with small volume fractions of hard phases of martensite, retained austenite, and pearlite. They contain <0.15 wt% C and use many of the same alloy elements found in dual-phase (DP) and transformation-induced plasticity (TRIP) steels but additionally have small quantities of niobium, titanium, and/or vanadium to form fine strengthening precipitates. Figure 6.1 shows the microstructure of CP 800, which contains 42% ferrite, 40% bainite, 13% martensite, and 5% retained austenite. The phase percentages vary with the cooling rate from austenite. High cooling rates suppress the formation of ferrite and favor the transformation to martensite.

Complex-phase steels are strengthened by solid-solution, precipitation, grain refinement, and phase transformation mechanisms. They provide ultimate tensile strengths of 800 MPa (116 ksi) and greater. They are characterized by high deformability and high energy absorption, which makes them ideal candidates for automotive crash applications, such as bumper and B-pillar reinforcements.

The spectrum of chemical compositions of some commercially available CP steels is given in Table 6.1. The table shows the maximum percentage of the major alloying elements for a particular steel producer.

6.2 Processing of Complex-Phase Steels

The microstructure of CP steels contains small amounts of martensite, retained austenite, and pearlite in a ferrite-bainite matrix. Processing CP steels is essentially similar to that of TRIP steels, shown in Fig. 7.3 in

Fig. 6.1 Microstructure of hot rolled CP 800. Source: Ref 6.1

Table 6.1 Chemical composition of complex-phase steels

	Composition, % (max)		
Steel grade	C	Mn	Si
CP 600(a)	0.10	1.60	0.40
CP 800 Y500(a)	0.17	2.20	0.60
CP 800 Y600(a)	0.17	2.20	0.60
CP 1000(a)	0.18	2.40	0.60
CP 1000 SF(a)	0.18	2.40	0.60
CP 750(b)	0.25	1.40	0.40
CP 800(b)	0.10	2.00	0.25
CP 1000(b)	0.14	1.70	0.25
MS 1200(b)	0.15	1.50	0.25

(a) Hot rolled. (b) Cold rolled. Source: Ref 6.2

Chapter 7, "Transformation-Induced Plasticity (TRIP) Steels." However, a less stringent cooling practice can be imposed during the last stage of processing because the microstructure of CP steels does not require the presence of retained austenite.

6.3 Deformation Mechanism of Complex-Phase Steels

The chemical composition and microstructure of CP steels is very similar to that of TRIP steels, but CP steels contain some quantities of niobium, titanium, and/or vanadium that cause the precipitation-strengthening effect.

As mentioned earlier, complex-phase steels are strengthened by solid-solution, precipitation, grain refinement, and phase transformation mechanisms. They deform by dislocation glide that occurs mainly in the ferrite-bainite matrix. The hard phases of martensite, retained austenite, and pearlite obstruct dislocation glide and cause an increase in the stress level.

6.4 Mechanical Properties of Complex-Phase Steels

The key mechanical properties of AHSS are strength to enable down-gaging and ductility to allow forming a component. Figure 6.2 shows the location of CP steels in the tensile strength-total elongation space. Tensile strength ranges from 420 to 1030 MPa (60 to 150 ksi) and total elongation ranges from 5 to 28%.

Ranges of mechanical properties of various types of CP steels are listed in Table 6.2. The tensile strength ranges from 590 to 1005 MPa (85 to 145 ksi) and the total elongation ranges from 15 to 28%.

Engineering stress-strain curves for various CP steel types compared to that of mild steel are shown in Fig. 6.3. The familiar deformation pattern of decreasing ductility with increasing strength is evident. Depending on the strength of the CP steel type, ductility varies from 3 to 13%.

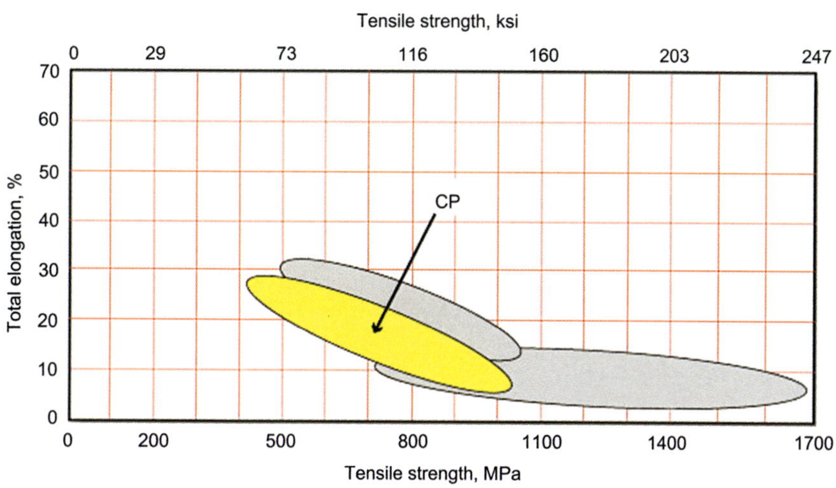

Fig. 6.2 Location of complex-phase (CP) steels in the tensile strength-elongation space. Source: Ref 6.3

Table 6.2 Mechanical properties of complex-phase steels

Product	Yield strength, MPa (ksi)	Tensile strength, MPa (ksi)	Total elongation, %
HR CP590 SF	510 (74)	590 (86)	28
CR CP590 HY	515 (75)	690 (100)	23
CR CP980 HY	795 (115)	1005 (146)	15
GA CP590 HY	505 (73)	620 (90)	26

Source: Ref 6.4, 6.5

True stress-strain curves for the same group of CP steel types are compared to that of mild steel in Fig. 6.4. Increasing tensile strength from 850 to 1200 MPa (123 to 174 ksi) reduced true strains from 0.07 to 0.015.

Endurance limits (fatigue strengths) of CP steels are high because of their high tensile strength. The *S-N* curves for different types of CP steels are shown in Fig. 6.5. Endurance limit increases with increase in the tensile strength of CP steels. The curves plot maximum stress versus number of cycles to failure for a tension-tension loading ratio $R = 0.1$.

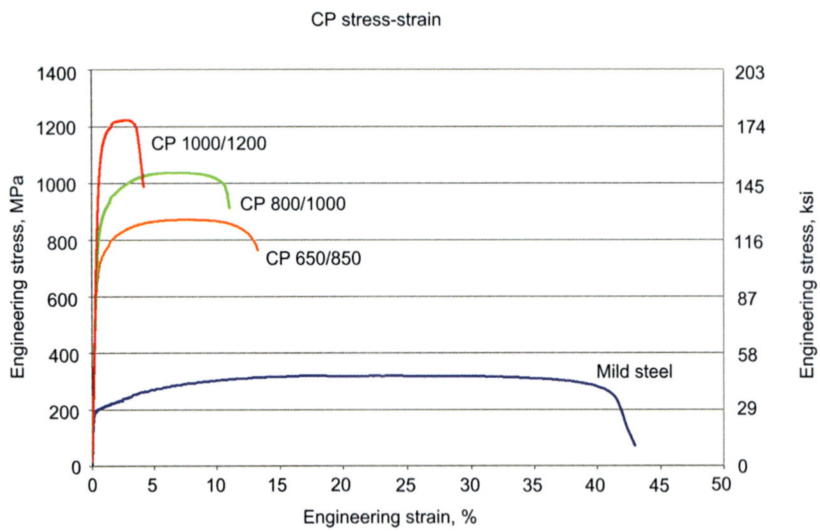

Fig. 6.3 Engineering stress-strain curves for a series of CP steel types. Source: Ref 6.3

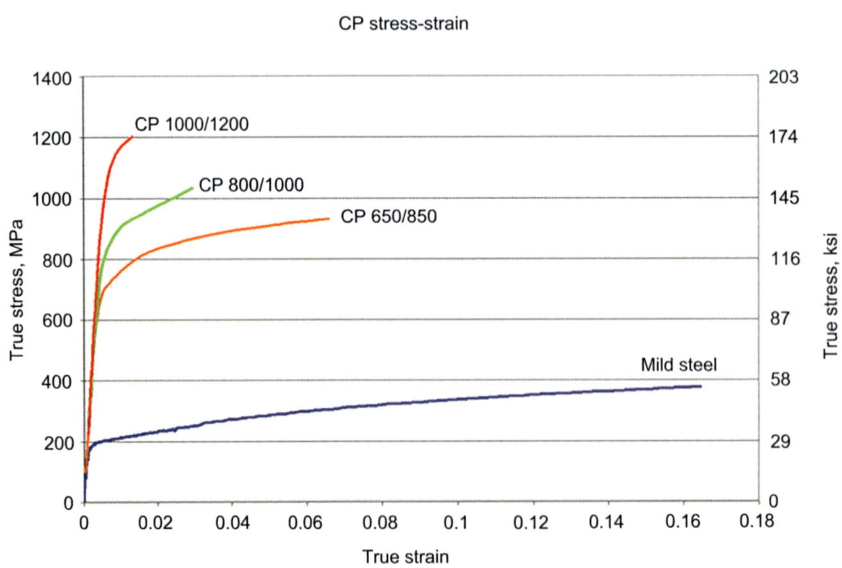

Fig. 6.4 True stress-strain curves for a series of CP steel types. Source: Ref 6.3

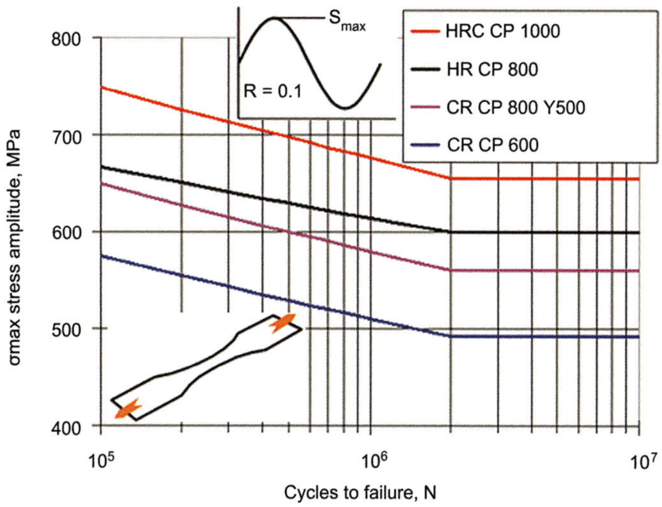

Fig. 6.5 Fatigue curves for various types of CP steels. Source: Ref 6.2

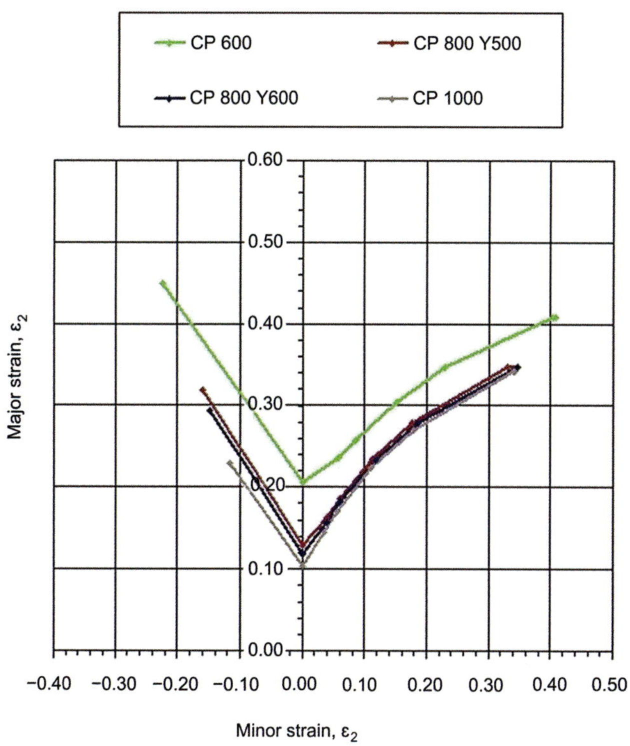

Fig. 6.6 Forming limit diagrams for various CP steel types. Source: Ref 6.2

6.5 Formability of Complex-Phase Steels

Complex-phase steels have lower ductility than DP and TRIP steels but they still have good ductility for their strength level. Figure 6.6 shows forming limit diagrams (FLDs) for various types of CP steels. Forming limit diagram levels decrease with increase in the tensile strength of CP steel types. This behavior is expected because FLD levels are directly related to ductility, which decreases with increase in the strength of the steel.

Comparison between FLDs for mild steel, DP 600, and CP 800 is shown in Fig. 6.7. The FLD for DP 600 is higher than that of CP 800.

6.6 Special Attributes of Complex-Phase Steels

Complex-phase steels have very high yield and tensile strengths and are suitable for impact strength and high-energy-absorption applications such as bumpers, B-pillar reinforcements, and anti-intrusion parts. They also have good fatigue properties that make them appropriate for suspension system components. Complex-phase steels have good roll forming and bending. Specially formulated CP steel, such as CP590SF and CP1000SF, has excellent stretch flangeability, as demonstrated by the high level of the hole expansion test.

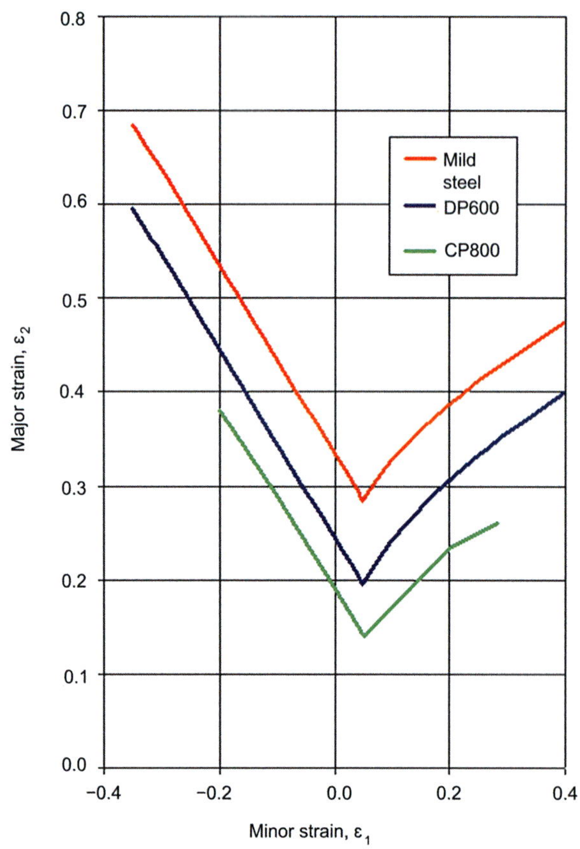

Fig. 6.7 Comparison of forming limit diagrams for mild steel, DP 600, and CP 800 steels. Source: Ref 6.3

REFERENCES

6.1 F. Hairer et al., Influence of Heat Treatment on the Microstructure and Hardness of a Low Alloyed Complex Phase Steel, *MS&T 2009: Proceedings from the Materials Science & Technology Conference,* Oct 25–29, 2009 (Pittsburgh, PA), MS&T Partner Societies, 2009, p 78–83

6.2 ArcelorMittal Web site, www.arcelormittal.com

6.3 "Advanced High Strength Steels Applications Guidelines," Version 4.1, WorldAutoSteel, www.worldautosteel.org, June 2009

6.4 R.H. Wagoner, Advanced High Strength Steel Workshop, Oct 22–23, 2006 (Arlington, VA), sponsored by the National Science Foundation, Department of Energy, and the Auto/Steel Partnership

6.5 D. Bhattacharya, *Developments in Advanced High Strength Steels,* Research and Development Center, Mittal Steel, East Chicago, IN, 2005

CHAPTER 7

Transformation-Induced Plasticity Steels

TRANSFORMATION-INDUCED PLASTICITY (TRIP) steels are one of the newest and interesting materials developed by the steel industry. They are characterized by their excellent strength and high ductility. *Transformation-induced plasticity* refers to a deformation mechanism where soft retained austenite is transformed into hard martensite during plastic deformation. The excellent strength and high ductility of TRIP steels allow the production of more complicated parts for lightweight automotive applications.

7.1 Composition and Microstructure of Transformation-Induced Plasticity Steels

Transformation-induced plasticity steels are iron-carbon alloys that contain 0.1 to 0.4 wt% C and other alloying elements such as silicon, aluminum, titanium, nickel, and vanadium. Silicon and aluminum are used mainly to stabilize the austenite phase at room temperature, while titanium, nickel, and vanadium are added to improve the strength of the steel. Transformation-induced plasticity steels consist of two phases: a ferrite/bainite matrix and a 5 to 20% volume fraction of metastable retained austenite, which progressively transforms to martensite during plastic deformation. The transformation provides significant dispersion hardening and resistance to localized deformation. This combination of phases gives TRIP steels the high formability of austenite during the initial stages of the stamping process followed by the high strength of martensite at the end of the forming process.

Transformation-induced plasticity steels are characterized by high tensile strengths and high strain-hardening rates. The excellent formability of

TRIP steels combined with their high strength makes them attractive candidate materials for lightweight applications. Because the martensite formation occurs at extremely fast rates, the TRIP phenomenon occurs at high deformation rates, as encountered during vehicle crash conditions. Therefore, TRIP steels are also being considered for applications requiring maximization of crash energy absorption.

The microstructure of TRIP steel consists of varying amounts of ferrite, bainite, and retained austenite, depending on the strength level desired. Figure 7.1 shows the microstructure of plastically deformed TRIP steel where some of the retained austenite transformed into martensite. Figure 7.2 shows a schematic drawing of Fig. 7.1 detailing the various phases of the microstructure. The main differences between dual-phase (DP) and TRIP steels are:

- Presence of retained austenite in the TRIP steels that transforms to martensite during deformation
- Presence of bainite in the TRIP steel

The spectrum of chemical compositions of some commercially available TRIP steels is given in Table 7.1. The table shows the maximum percentage of the major alloying elements for a particular steel producer (Ref 7.3).

Other references give different compositions of TRIP 700, as shown in Table 7.2.

Transformation-induced plasticity steels contain higher quantities of carbon than DP steels to stabilize the retained austenite at room tempera-

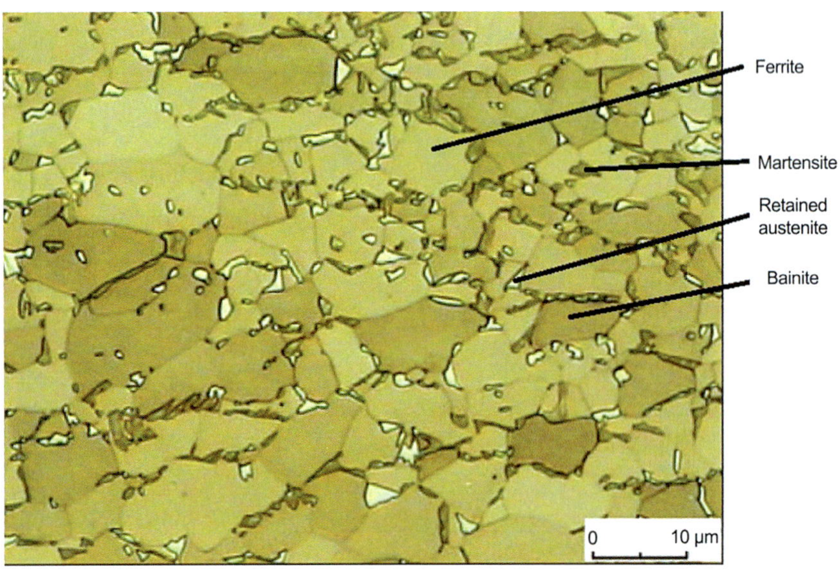

Fig. 7.1 Microstructure of transformation-induced plasticity (TRIP) steel. Light areas are retained austenite. Source: Ref 7.1

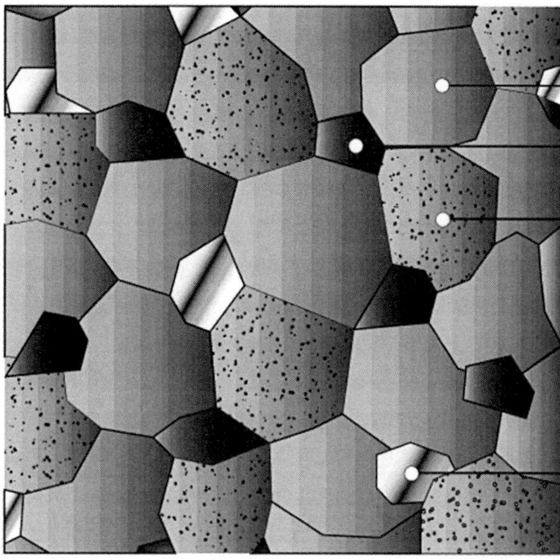

Fig. 7.2 Schematic of TRIP steel microstructure. Source: Ref 7.2

Table 7.1 Chemical composition of cold rolled transformation-induced plasticity (TRIP) steels

Steel grade	Composition, % (max)		
	C	Mn	Al+Si
TRIP 590	0.175	2.0	2.0
TRIP 690	0.200	2.0	2.0
TRIP 780	0.250	2.0	2.0

Source: Ref 7.3

Table 7.2 Chemical composition of TRIP 700 type steel

Steel grade	Composition, wt%			
	C	Si+Cr+Mo	Al+Mn	Fe
TRIP 700	0.2	0.4	2.9	bal

Source: Ref 7.4

ture. Silicon and/or aluminum accelerate the ferrite/bainite formation and suppress carbide precipitation by maintaining the carbon content within the retained austenite.

7.2 Processing of Transformation-Induced Plasticity Steels

Transformation-induced plasticity steels possess a complex microstructure of bainite, martensite, and retained austenite all embedded in a soft ferrite matrix. This microstructure is achieved via a cold rolling route by subjecting the sheet steel to the two-stage heat treatment shown in Fig. 7.3.

The first step is annealing the steel in the austenite-ferrite region at 800 °C (1472 °F) for 60 s, then rapid cooling to 450 °C, or 842 °F (below bainite start temperature), followed by an isothermal hold at 450 °C for 90 s, and final cooling to room temperature. Transformation-induced plasticity steels require the use of an isothermal hold at an intermediate temperature to produce bainite. The developed microstructure contains 11.8% volume fraction of retained austenite. The retained austenite transforms to martensite during plastic deformation, which also produces mechanical twins. Twin boundaries act as effective barriers to dislocation motion.

Figure 7.3 shows that at the beginning of annealing, the carbon content in austenite is 0.2%. After slow cooling in the ferrite region, carbon content in austenite increases to approximately 0.4%. Another cooling produces bainite at the coiling temperature where the carbon content in the remaining austenite increases to 1.2%.

Another schematic representation of the processing schedule for the production of hot rolled TRIP steel, with microstructure evolution, is shown in Fig. 7.4. The figure also shows the differences between processing of TRIP and DP steels. The steel is intercritically annealed in the temperature range of 780 to 880 °C (1436 to 1616 °F) to produce a microstructure of almost equal fractions of ferrite (α) and austenite (γ). This is followed by cooling and isothermal annealing of the steel in the temperature range of 350 to 450 °C (662 to 842 °F) to transform a considerable portion of the austenite to bainite. Finally, the steel is cooled to ambient temperature.

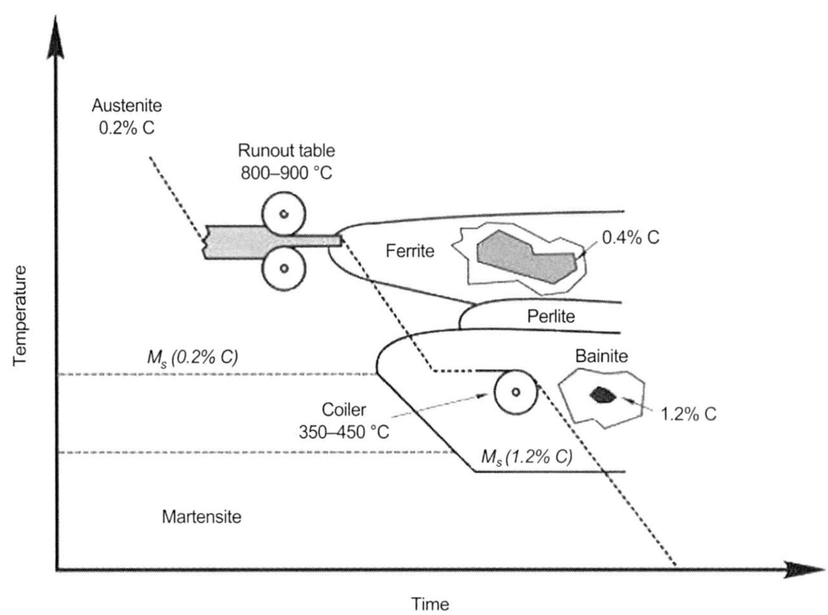

Fig. 7.3 Cooling schedule in the production of the TRIP sheet. Source: Ref 7.5

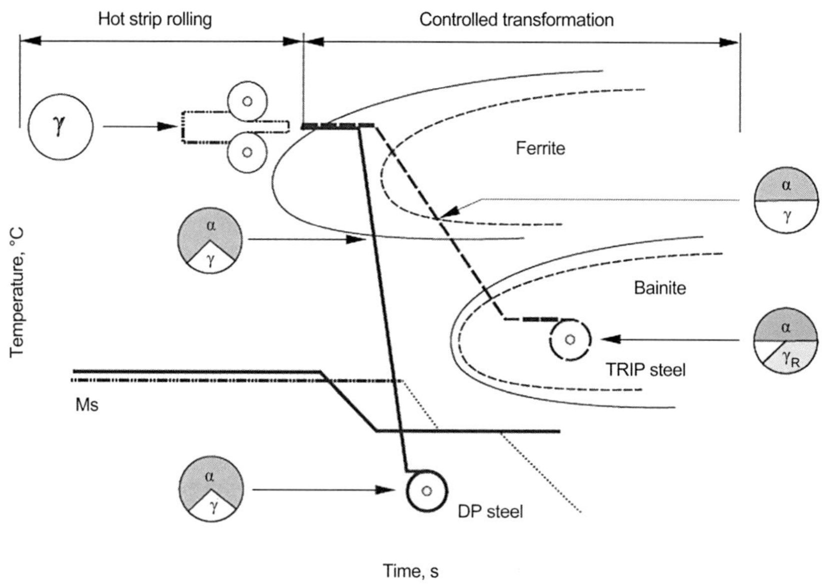

Fig. 7.4 Time-temperature schedule for the production of hot rolled TRIP and dual-phase (DP) steels. Source: Ref 7.6

The heat treatment schedule ensures (Ref 7.7):

- Sufficient carbon enrichment of the retained austenite
- Lowering of the martensite start temperature
- Maintaining metastable austenite at room temperature

7.3 Deformation Mechanism of Transformation-Induced Plasticity Steels

Transformation-induced plasticity steels contain soft ferrite matrix, hard bainite, and retained austenite. The TRIP phenomenon occurs when the retained austenite transforms to hard and brittle martensite during plastic deformation. The transformation of austenite into martensite is instantaneous and diffusionless. A certain amount of stress is required to initiate phase transformation in crystal planes.

The strain level at which retained austenite begins to transform to martensite is controlled by the carbon content of the steel. At low carbon levels, the retained austenite begins to transform almost immediately on deformation. This increases the work-hardening rate and formability during the stamping process. At high carbon levels, the retained austenite is more stable and persists into the final part. It begins to transform to martensite at strain levels beyond those produced during forming. The subsequent deformation usually occurs as a result of a crash event.

The TRIP effect, where austenite to martensite transformation occurs during plastic deformation, is demonstrated in the drawing of a flanged cylindrical cup shown in Fig.7.5. The fraction of transformed austenite to martensite depends not only on the amount of strain but also on the deformation mode (plane strain, compressive strain). The amount of martensite generated in the blank flanged area is less than that in the cup wall area because the wall area undergoes tensile deformation while the flanged area experiences compressive deformation. Tensile deformation is more favorable to the TRIP effect than the compressive one.

7.4 Mechanical Properties of Transformation-Induced Plasticity Steels

The key mechanical properties of AHSS are strength to enable downgaging and ductility to allow forming of a component. Figure 7.6 shows the location of TRIP steels in the tensile strength-total elongation space. Tensile strength ranges from 500 to 1050 MPa (70 to 150 ksi) and total elongation ranges from 12 to 32%.

Ranges of yield strength, tensile strength, total elongation, and strain-hardening exponent for three types of TRIP steels are listed in Table 7.3. Also listed is the increase in yield strength due to bake hardening.

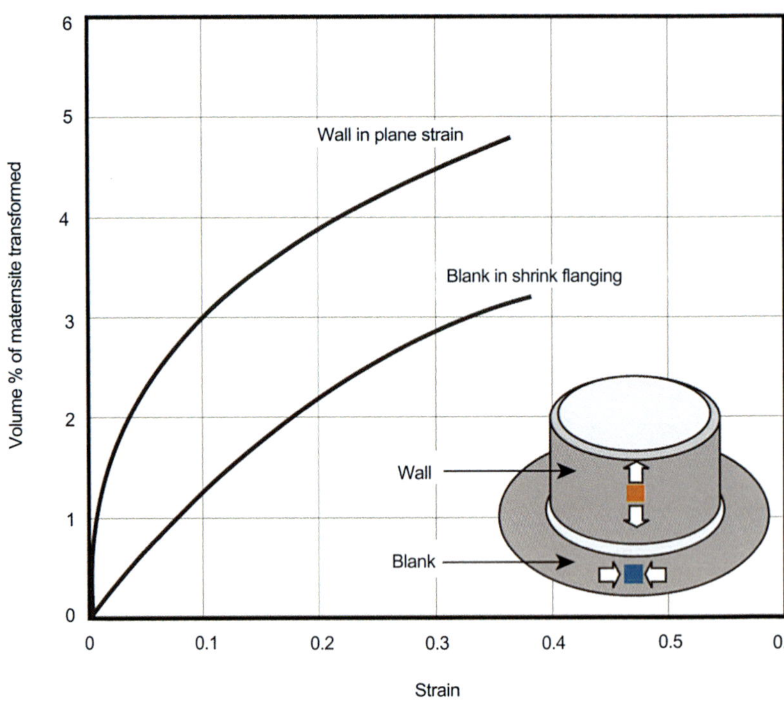

Fig. 7.5 Variation of volume fraction of transformed martensite with strain in flanged cup. Source: Ref 7.2

The mechanical properties of two types (TRIP 590 and TRIP 780) and two coatings (galvannealed GA and galvanized G) of TRIP steels as well as cold rolled (CR) TRIP steel are listed in Table 7.4. Steel coating appears to affect the mechanical properties through its effect on processing and microstructure.

Engineering and true stress-strain curves for various types of TRIP grade steels are shown in Fig. 7.7 and 7.8, consecutively. Mild steel is included for comparison. Transformation-induced plasticity steels have high strength and also good formability that can be attributed to their high work-hardening rate and the fact that their microstructures consist of two

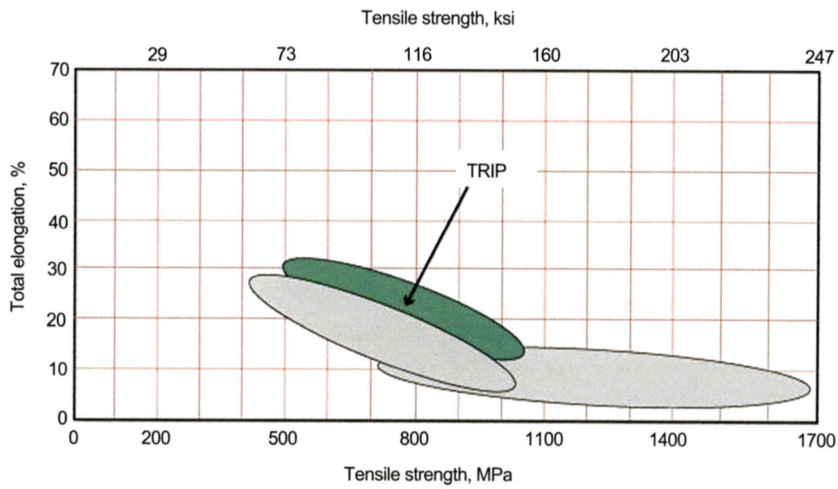

Fig. 7.6 Location of TRIP steels in the tensile strength-elongation space. Source: Ref 7.2

Table 7.3 Mechanical properties of various types of TRIP steels

Steel grade	Yield strength, MPa (ksi)	Ultimate tensile strength, MPa (ksi)	Elongation, %	n	BH_2, MPa (ksi)
TRIP 590	380–480 (55–70)	590–700 (86–102)	≥26	≥0.20	40 (6)
TRIP 690	410–510 (59–74)	690–800 (100–116)	≥25	≥0.19	40 (6)
TRIP 780(a)	450–550 (65–80)	780–900 (113–131)	≥23	≥0.18	40 (6)

(a) Cold rolled. Source: Ref 7.3

Table 7.4 Mechanical properties of various types and coatings of TRIP steels

Product	Yield strength, MPa (ksi)	Minimum tensile strength, MPa (ksi)	Total elongation, %
CR TRIP 590	350–495 (51–72)	590 (86)	31
GA TRIP 590	360–510 (52–74)	590 (86)	26
G TRIP 590	380–480 (55–70)	590 (86)	27
CR TRIP 780	410–500 (59–73)	780 (113)	21
GA TRIP 780	410–560 (59–81)	780 (113)	19
G TRIP 780	440–500 (64–73)	780 (113)	21

Source: Ref 7.8, 7.9

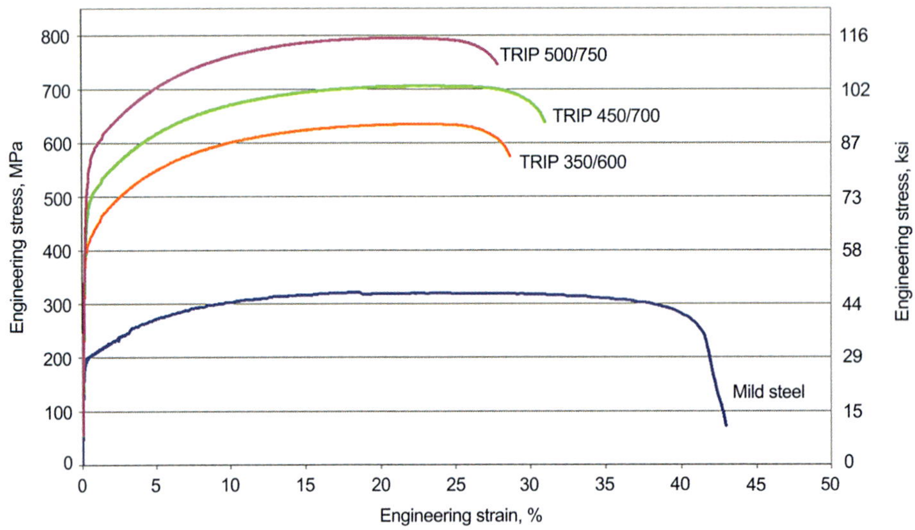

Fig. 7.7 Engineering stress-strain curves for a series of TRIP steel grades. Source: Ref 7.2

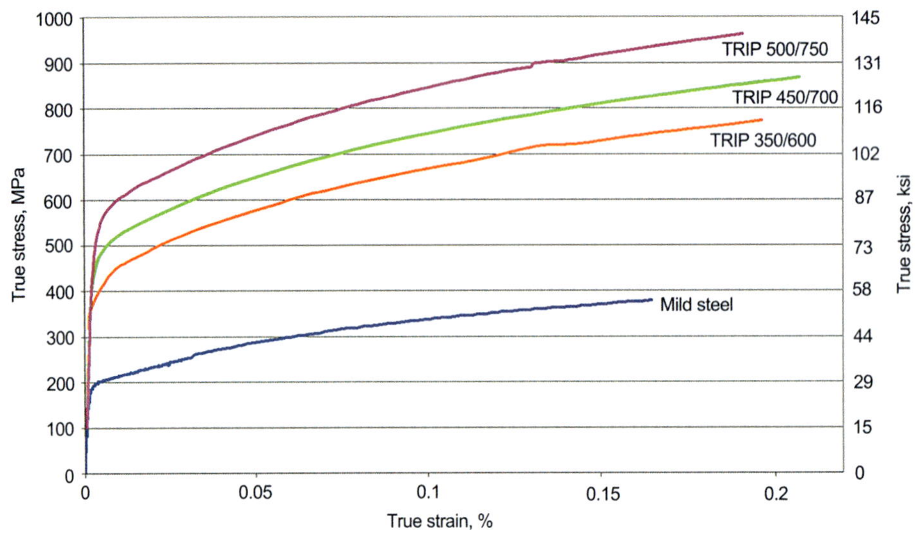

Fig. 7.8 True stress-strain curves for a series of TRIP steel grades. Source: Ref 7.2

phases: a ferrite/bainite matrix and a 5 to 20% volume fraction of retained austenite. This metastable retained austenite transforms progressively to martensite during forming.

Transformation-induced plasticity steels display high work-hardening rates in the entire range of tensile deformation. They have excellent energy absorption properties because of their high toughness (combination of strength and ductility). The TRIP effect can also improve vehicle safety during a crash.

Transformation-induced plasticity steels have significantly better fatigue properties than conventional high-strength, low-alloy (HSLA) steels of equivalent strength. As a result of their high toughness and the post-forming TRIP effect initiated by crash deformation, TRIP steels are suitable for parts designed to absorb impact energy.

7.5 Formability of Transformation-Induced Plasticity Steels

Transformation-induced plasticity steels exhibit better ductility at a given strength level than the other first-generation AHSS members (dual-phase, complex-phase, and martensitic). This enhanced formability comes from the transformation of retained austenite to martensite during plastic deformation. The increased formability of TRIP steels allows the production of more complicated parts than with other high-strength steels (HSS), giving automotive engineers more freedom in part design to optimize weight and structural performance.

Forming limit diagrams (FLDs) are used to assess the formability of metals during punch forming. They define the maximum allowable strains for different deformation paths. Figure 7.9 shows FLDs for TRIP 690, TRIP 780, and DP 600 steels. The better formability of TRIP steels is evident.

7.6 Special Attributes of Transformation-Induced Plasticity Steels

Transformation-induced plasticity steels are characterized by a comparatively low initial strain-hardening rate that increases and maintains itself into high strain ranges of 5 to 20% strain. This characteristic delays the onset of plastic instability and provides excellent ductility for successful stamping of complex automotive components. The advantage of TRIP steels is that they have much higher ductility than other steels with similar strength. Automotive parts that cannot be made with DP steel can often be made with the highly ductile TRIP steels.

Transformation-induced plasticity steels have an excellent bake-hardening capacity that produces an increase in the yield strength in a typical paint baking cycle of approximately 70 MPa (10 ksi) for a 2% prestrain and 96 MPa (14 ksi) for a 20% prestrain. The excellent tensile strength of TRIP steels makes them ideal for downgaging and energy absorption applications. Because the endurance limit for metals is directly proportional to tensile strength ($\sigma_e = 0.33\ \sigma_{ts}$), TRIP steels have high endurance limits and hence have greater resistance to fatigue failure compared to conven-

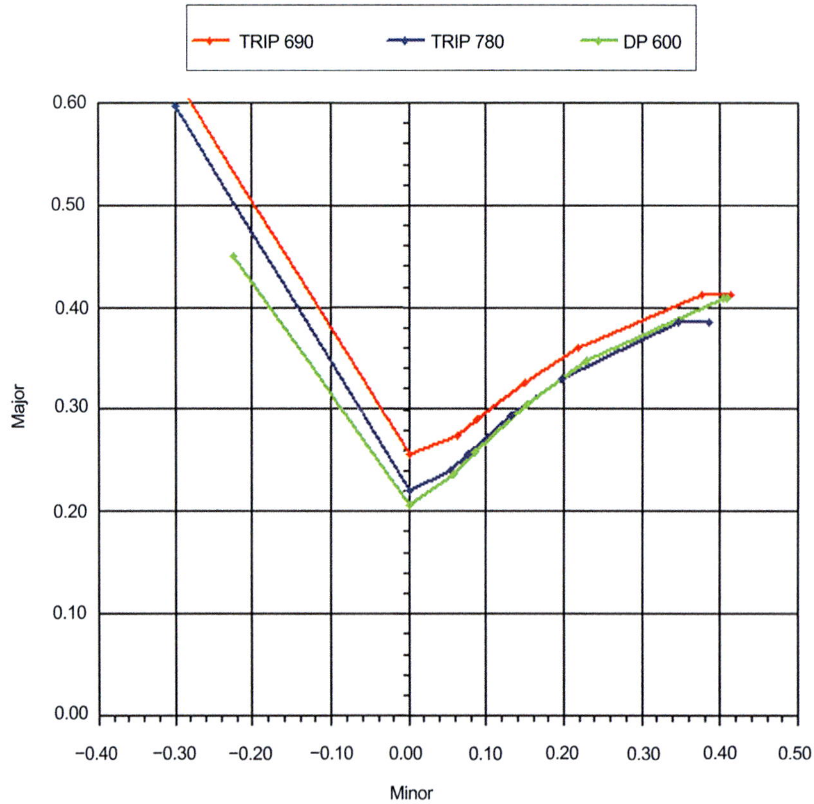

Fig. 7.9 Forming limit diagrams for TRIP and DP steels. Source: Ref 7.2

tional HSLA steels. Applications for TRIP steels include automotive structural and safety parts such as cross members, beams, B-pillar reinforcements, and bumper reinforcements.

Transformation-induced plasticity steels are low-alloy steels that achieve their excellent strength and ductility by engineering of their microstructure without using large amounts of expensive alloying elements. Processing TRIP steels relies on an intercritical annealing step, which is not difficult to implement. This means that TRIP steels can be produced for about the same price as other HSS.

REFERENCES

7.1 R. Kuziak, R. Kawalla, and S. Waengler, Advanced High Strength Steels for the Automotive Industry, *Arch. Civ. Mech. Eng.*, Vol 8 (No. 2), 2008, p 103–118

7.2 "Advanced High Strength Steels Applications Guidelines," Version 4.1, WorldAutoSteel, www.worldautosteel.org, June 2009

7.3 ArcelorMittal Web site, www.arcelormittal.com

7.4 S. Kurtze, "Characterization of the Dynamic Behavior and Microstructural Evolution of High Strength Sheet Steels," Doctor of Technology thesis, Tampere University of Technology, Publication 833, Tampere, Finland, Oct 2009
7.5 W. Bleck and K. Phiu-On, Microalloying of Cold-Formable Multi Phase Steel Grades, *Mater. Sci. Forum,* Vol 500–501, 2005, p 97–112
7.6 J. Zrnik et al., Recent Progress in High Strength Low Carbon Steels, *METABK,* Vol 45 (No. 4), 2006, p 323–331
7.7 C.-S. Oh et al., Dilatometric Analysis on Phase Transformations of Intercritical Annealing of Fe-Mn-Si and Fe-Mn-Si-Cu Low Carbon TRIP Steels, *Met. Mater.-Int.,* Vol 10 (No. 5), 2004, p 399–406
7.8 R.H. Wagoner, Advanced High Strength Steel Workshop, Oct 22–23, 2006 (Arlington, VA), sponsored by the National Science Foundation, Department of Energy, and the Auto/Steel Partnership
7.9 D. Bhattacharya, *Developments in Advanced High Strength Steels,* Research and Development Center, Mittal Steel, East Chicago, IN, 2005

CHAPTER 8

Martensitic Steels

MARTENSITIC (MS) STEELS are produced by quenching carbon steel from the austenite phase into martensite. Martensite is formed when rapid cooling of austenite traps carbon atoms that do not have time to diffuse out of the transforming face-centered cubic (fcc) to body-centered cubic (bcc) crystal structure. This martensitic transformation begins during cooling when the austenite reaches the martensite start temperature (M_s). As the temperature is decreased, more of the austenite transforms to martensite and finally, when the martensite finish temperature (M_f) is reached, the transformation is complete.

Martensite can also be produced by the application of stress, as in transformation-induced plasticity (TRIP) steels where plastic deformation induces martensitic transformation. Martensite is not an equilibrium phase because quenching does not give sufficient time for carbon to diffuse out of the austenite structure. Martensitic transformation can be thermally induced by quenching or stress induced by plastic deformation. The diffusion of carbon out of martensite can be accelerated by the application of heat in a process called tempering. Tempering produces small carbide particles in the tempered martensite microstructure. Tempered martensite is often the objective of heat treatments aimed at obtaining optimal strength/ductility property combination.

8.1 Compositions and Microstructures of Martensitic Steels

Martensitic steels microstructure consists of a martensitic matrix containing small amounts of ferrite and/or bainite. They contain a high fraction of martensite with the carbon content determining the strength level of the material. The austenite that exists during hot rolling or annealing of the material is transformed to martensite during quenching and/or post-forming heat treatment. Martensitic steels are often subjected to postquench tempering to improve ductility and can provide good formability even at extremely high ultimate tensile strengths of up to 1500 MPa (217 ksi). Carbon is added to martensitic steels to increase hardenability and also to strengthen the martensite. Elements such as manganese, silicon, chromium, molybdenum, boron, vanadium, and nickel are also used in various combinations to increase hardenability. Figure 8.1 shows the needlelike structure of martensite.

Martensite is important because it is the hardest structure obtained from the transformation of austenite. As formed, martensite is of little use and must be softened for metalworking. Softening is accomplished by reheating or tempering martensite at temperatures below the eutectoid temperature of 727 °C (1341 °F). Properties of tempered martensite depend on the tempering temperature and the hold time at that temperature.

8.2 Processing of Martensitic Steels

Martensitic steels are produced from the austenite phase by rapid quenching the hot rolling or annealing products to transform most of their austenite to martensite. These steels are often subjected to postquench tempering to improve ductility.

8.3 Deformation Mechanism of Martensitic Steels

Trapped carbon atoms in martensite cause lattice distortions in its body-centered tetragonal (bct) structure. The distorted lattice resists dislocation glide and increases the strength and hardness of martensite. The strength-

Fig. 8.1 Microstructure of martensite. Source: Ref 8.1

ening mechanism in tempered martensite shifts from resistance to dislocation glide in a distorted lattice to resistance to dislocation glide by Fe_3C precipitates.

8.4 Mechanical Properties of Martensitic Steels

The key mechanical properties of advanced high-strength steels (AHSS) are strength to enable downgaging and ductility to allow forming a component. Figure 8.2 shows the location of MS steels in the tensile strength-total elongation space. Martensitic steels have the highest tensile strength level of all multiphase steels. Tensile strength ranges from 720 to 1680 MPa (105 to 245 ksi) and total elongation ranges from 3 to 15%.

Typical mechanical properties of various types of MS steels are listed in Table 8.1. Their yield and tensile strengths are extremely high but their ductility is very low. In most cases, heat treatment is used to improve the steel ductility to acceptable levels. In other situations, hot forming of MS steels is used to overcome their formability issues.

Engineering and true stress-strain curves for two types of MS steels are compared to those of mild steel in Fig. 8.3 and 8.4, respectively. The MS

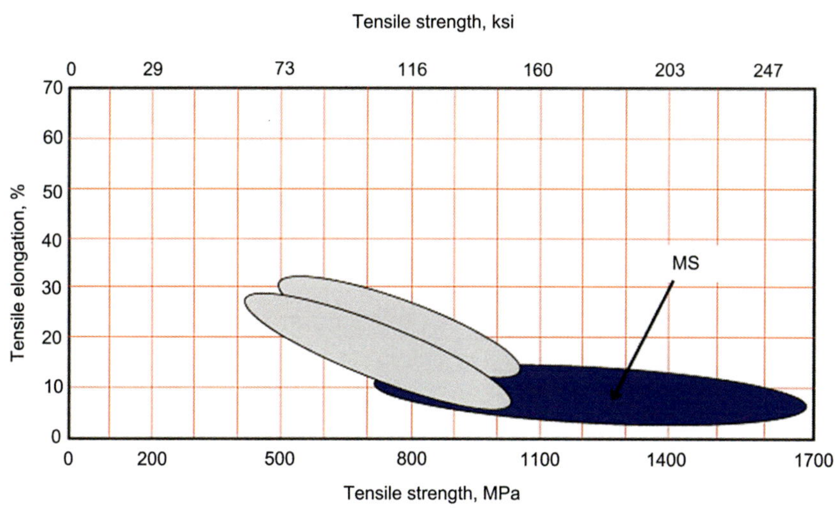

Fig. 8.2 Location of martensitic (MS) steels in the tensile strength-elongation space. Source: Ref 8.2

Table 8.1 Typical mechanical properties of various types of martensitic steels

Product	Yield strength, MPa (ksi)	Tensile strength, MPa (ksi)	Total elongation, %
MS 130	923 (134)	1054 (153)	5.4
MS 160	1020 (148)	1178 (171)	5.1
MS 190	1213 (176)	1420 (206)	5.1
MS 220	1350 (196)	1585 (230)	4.7

Source: Ref 8.3, 8.4

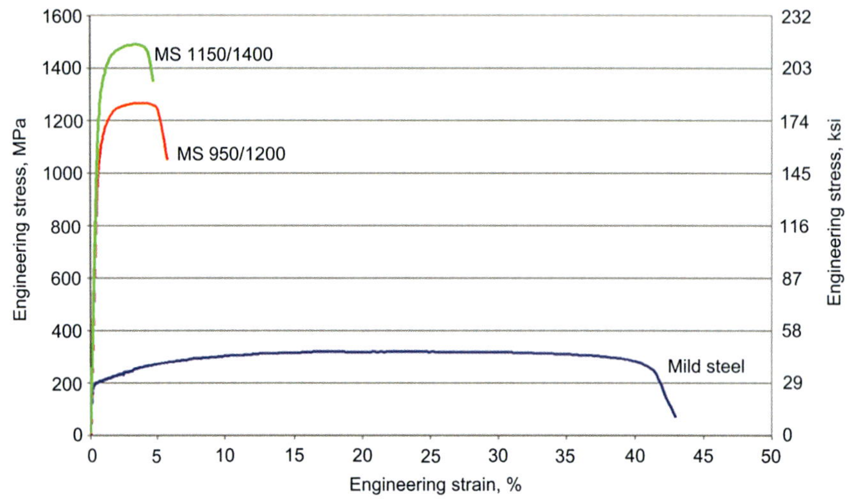

Fig. 8.3 Engineering stress-strain curves for MS steels compared to mild steel. Source: Ref 8.2

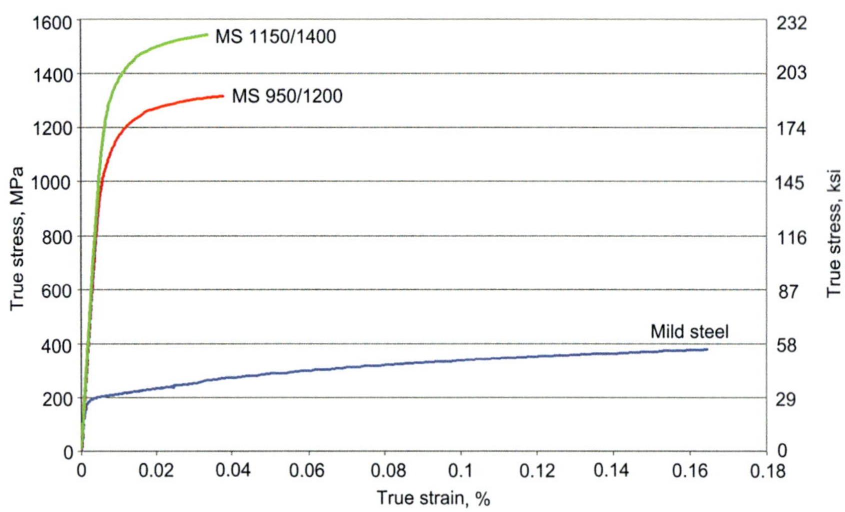

Fig. 8.4 True stress-strain curves for MS steels compared to mild steel. Source: Ref 8.2

steels have extremely high yield strength (1150 MPa, or 165 ksi) and tensile strength (1400 MPa, or 200 ksi) but very low ductility (<5%).

8.5 Hot Forming of Martensitic Steels

Martensitic steels have very high strength and very low ductility that restrict their utilization to simple parts. To overcome the low formability of MS steels, the hot forming process was developed. The idea behind the process is to first austenitize the steel blank, then form it while it is in the ductile austenite phase, and finally water quench or die quench the formed part to transform its soft austenite microstructure to the strong and hard martensite phase. A schematic of the process is shown in Fig. 8.5 where the steel blank is heated into the austenite region above 850 °C (1562 °F) in a continuous feed furnace, formed at that temperature, and then quenched in the die to form martensite. Following heat treatment, forming, and quenching, the microstructure is 100% martensite.

Figure 8.6 shows a schematic of the hot forming process and its effect on the tensile strength of the steel. The starting sheet metal has a tensile strength of 600 MPa (87 ksi) and has a ferrite-pearlite microstructure. Heating the steel blank transforms the microstructure into austenite. The heated blank is then transferred to a press for forming. After forming, the part is quenched in the stamping tool. Quenching transforms austenite into martensite microstructure with a tensile strength of 1600 MPa (232 ksi).

The mechanical properties of the final hot formed parts are characterized by high yield and tensile strength, high hardness, high impact and fatigue strength, and the absence of springback. Hot formed steels are

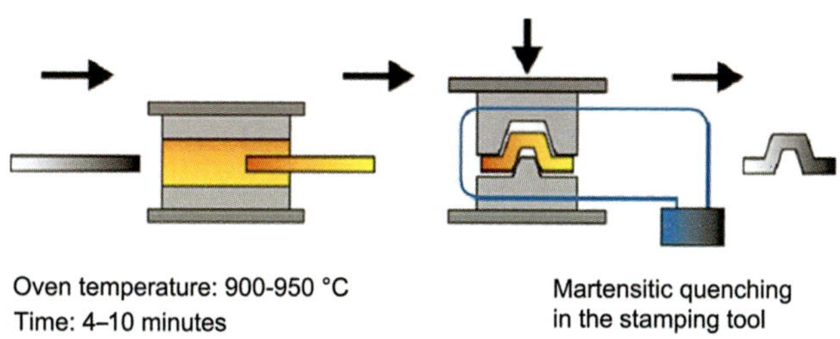

Fig. 8.5 Hot forming of MS steel. Source: Ref 8.5

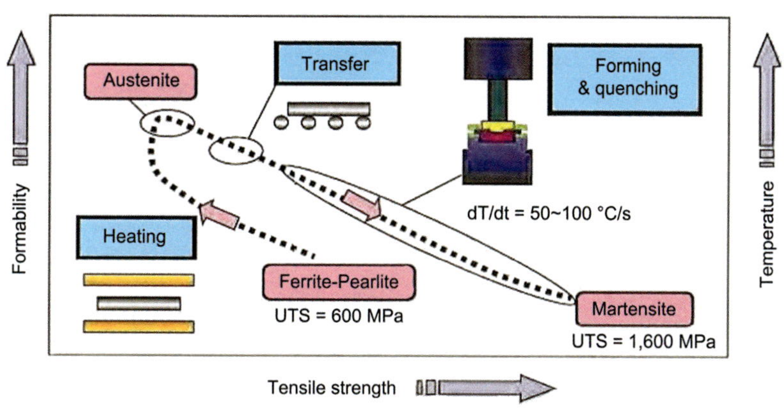

Fig. 8.6 Tensile strength and formability during hot forming. UTS, ultimate tensile strength. Source: Ref 8.6

suitable for structural and safety component applications such as bumper beams, fender beams, B-pillars, and door, floor, and roof reinforcements.

8.6 Tempering Martensite

The transformation from austenite to martensite produces a brittle microstructure with low fracture toughness and diminished ability for deformation processing. Ductility is regained by a reheating process called tempering. On tempering, the excess carbon in the deformed bct lattice combines with iron atoms to form Fe_3C precipitates. The strengthening mechanism in tempered martensite shifts from resistance to dislocation glide in a distorted lattice to resistance to dislocation glide by precipitates. The resulting yield stress evolves with tempering time because the process of nucleation and growth of the Fe_3C precipitates is time dependent. Figure 8.7 shows evolution of yield stress during tempering of quenched carbon steel (martensite). Tempering reduces the yield strength to levels between quenched martensite (2500 MPa, or 360 ksi) and normalized ferrite-pearlite microstructure (350 MPa, or 50 ksi). Manipulating properties of steel by using quench and temper treatment is a widespread technique for controlling the toughness of steel.

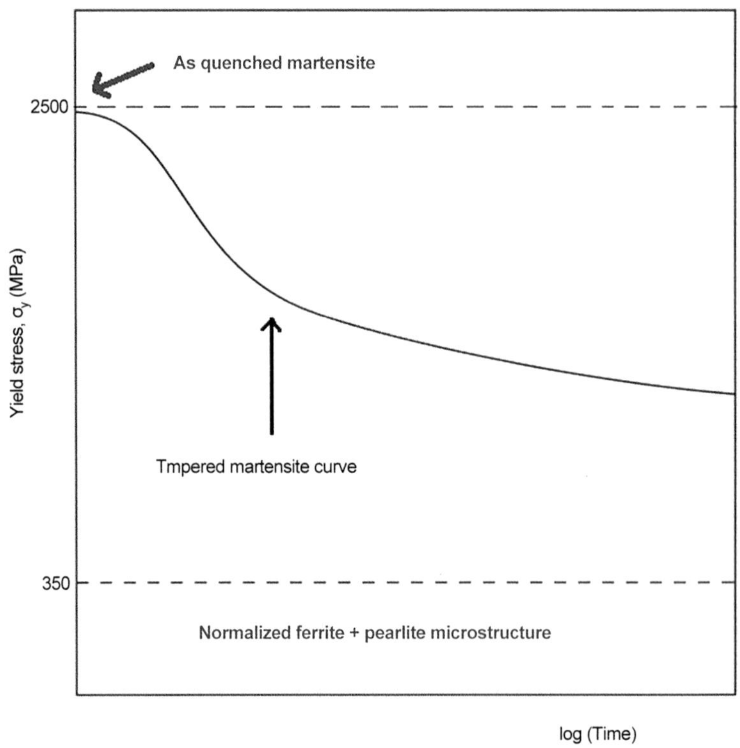

Fig. 8.7 Evolution of yield stress during tempering of martensite. Source: Adapted from Ref 8.7

8.7 Special Attributes of Martensitic Steels

Martensitic steels have the highest tensile strength level, up to 1700 MPa (245 ksi), of any AHSS. Typical applications for MS steels are those that require high strength and good fatigue resistance, with relatively simple shapes for traditionally stamped parts and more complex shapes for hot formed parts. Typical applications of MS steels include bumper reinforcements, door intrusion beams, side sill reinforcements, and more recently, roll formed rocker panel reinforcements.

REFERENCES

8.1 W.D. Callister, Jr. and D.G. Rethwisch, Materials Science and Engineering: An Introduction, 8th ed., Wiley, 2010

8.2 "Advanced High Strength Steels Applications Guidelines," Version 4.1, WorldAutoSteel, www.worldautosteel.org, June 2009

8.3 R.H. Wagoner, Advanced High Strength Steel Workshop, Oct 22–23, 2006 (Arlington, VA), sponsored by the National Science Foundation, Department of Energy, and the Auto/Steel Partnership

8.4 D. Bhattacharya, *Developments in Advanced High Strength Steels,* Research and Development Center, Mittal Steel, East Chicago, IN, 2005

8.5 ArcelorMittal Web site, www.arcelormittal.com

8.6 WorldAutoSteel, "FutureSteelVehicle (FSV)," Final Engineering Report, Steel Market Development Institute, May 2011, www.worldautosteel.org

8.7 M. Ashby, H. Shercliff, and D. Cebon, *Materials: Engineering, Science, Processing and Design*, 2nd ed., Elsevier, 2010

CHAPTER 9

Twinning-Induced Plasticity Steels

TWINNING-INDUCED PLASTICITY (TWIP) steels are austenitic steels with a high manganese content of 22 to 30% and other alloying elements such as carbon (<1 wt%), silicon (<3 wt%), and aluminum (<10 wt%). The high manganese content causes the steel to be fully austenitic at room temperature. Aluminum and silicon strengthen the steel by solid-solution hardening. Twinning-induced plasticity steels have an excellent combination of high tensile strength exceeding 1000 MPa (145 ksi), a very high ductility of 60 to 90% elongation, and a high strain-hardening exponent, *n*, of 0.48. Twinning-induced plasticity steels have a low stacking fault energy (SFE) of ~30 mJ/m^{-2}, which produces extensive mechanical twinning under plastic deformation. Deformation twinning produces an excellent combination of high strength and ductility and a reorientation of crystallographic texture.

9.1 Twins and Stacking Faults

Twins are highly symmetrical discontinuities in the crystalline structure characterized by two crystalline regions that are, structurally, mirror images of each other. "Mechanical twins" are produced by deformation, in body-centered cubic (bcc) and hexagonal close-packed (hcp) metals; "annealing twins" are produced by annealing following deformation, in face-centered cubic (fcc) metals.

Figure 9.1(a) shows the twin plane in an undeformed crystal, and Fig. 9.1(b) shows how deformation by twinning results from the application of shear stress (τ) to displace atoms such that their positions on one side of the twin plane are a mirror image of atoms on the other side. In twins, atomic planes move over each other a fraction of the interatomic spacing, depending on the atomic plane distance from the twin plane. Slip, on the other hand, occurs in whole number increments of interatomic spacing. Another difference between slip and twinning is that in slip the crystallographic orientation above and below the slip plane does not change with deformation; in twinning, these orientations change and may produce new slip systems that are more favorable to slip.

Twinning is important in hexagonal crystals because dislocation slip can only occur on one plane, the (0001) plane. If this plane is normal to the axial load then no shear component would exist and brittle fracture would occur. In this case, the slip process is restricted and deformation by twinning takes place. Twinning can take place at a faster rate than slip and that explains the existence of mechanically formed twins, rather than slip bands, at high rates of loading.

The amount of plastic deformation from twinning is small compared to that resulting from dislocation slip. In microstructures, twins appear as thin plates inside the austenite matrix.

Stacking Faults. Atoms in metal crystals are arranged in close-packed planes stacked in a periodic sequence. Crystal structures are generated by the stacking of these close-packed planes on top of each other. A stacking fault exists when there is an interruption in the stacking sequence of the close-packed planes. A stacking fault causes a change in the energy field around it, called stacking fault energy (SFE), in units of mJ/m^{-2}. Stacking fault energy changes with alloy composition and deformation temperature and its magnitude controls the ease of dislocation glide and the activation of deformation mechanisms in the metal. Decreasing the SFE widens the stacking fault, making it more difficult for dislocations to glide and easier for mechanical twins to deform.

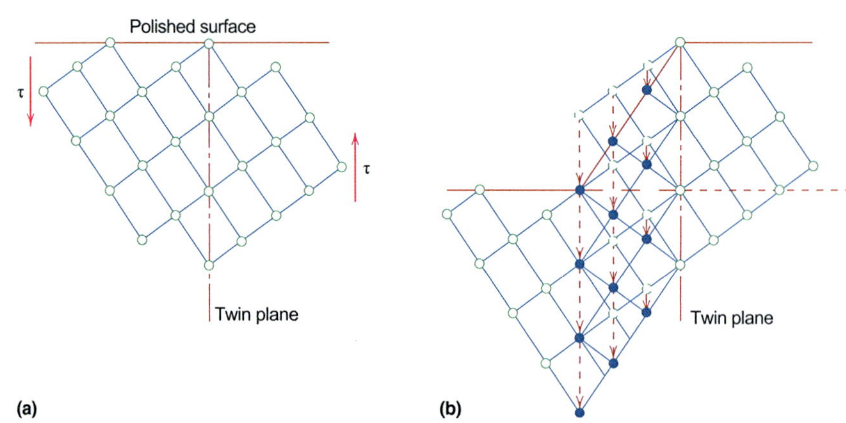

Fig. 9.1 Schematic diagram of deformation by twinning. Source: Ref 9.1

9.2 Compositions and Microstructures of Twinning-Induced Plasticity Steels

Twinning-induced plasticity steels consist of a fully austenitic single-phase microstructure that is retained at room temperature by means of a high percentage of alloying elements, such as manganese, aluminum, and silicon. Manganese is an austenite stabilizer that, in high concentrations, drastically lowers the transformation temperature of austenite to ferrite. Aluminum and silicon strengthen the steel by solid-solution hardening. Some TWIP steel compositions are listed in Table 9.1.

A detailed study of two TWIP alloys with different manganese compositions and different microstructures was conducted to establish the relationship between composition, microstructure, and mechanical properties. Table 9.2 lists the compositions of the two TWIP alloys.

The difference in manganese content of the two alloys resulted in different microstructures due to the effect of manganese on the phase stability of austenite with respect to ε-martensite. Alloy Fe-30Mn possesses an austenite single-phase microstructure after annealing, whereas alloy Fe-24Mn has a mixture of austenite and 50% ε hcp martensite. Figure 9.2 shows the annealed microstructure of (a) TWIP Fe-30Mn and (b) TWIP Fe-24Mn.

Manganese is the major alloying element in TWIP steels. It is responsible for preserving the austenite structure at ambient temperature. The main influence of manganese in TWIP steels is to control the SFE of mechanical twins. Increasing manganese content increases the SFE from low values (<20 mJ/m^2) to moderate values (>20 mJ/m^2). Low SFE favors martensitic transformation-induced plasticity, the TRIP effect; while high SFE favors mechanical twinning-induced plasticity, the TWIP effect.

Table 9.1 Compositions of different types of twinning-induced plasticity (TWIP) steels

TWIP type	Mn	Al	Si	C	Cr+Mo	Nb	Fe
TWIP 1	28	1.6	0.28	0.08	<0.01	<0.001	bal
TWIP 2	25	1.6	0.24	0.08	<0.01	0.05	bal
TWIP 3	27	4.1	0.52	0.08	<0.01	0.05	bal
TWIP 4	28.1	0.9	0.54	0.17	<0.01	<0.001	bal
TWIP 5	23.9	3.5	0.448	0.11	1.0	<0.001	bal

Source: Ref 9.2

Table 9.2 Compositions of two types of TWIP steels

	Composition, wt%			
TWIP type	Fe	Mn	C	S
TWIP Fe-24Mn	75.57	24.30	0.0163	0.0052
TWIP Fe-30Mn	69.08	30.79	0.0163	0.0060

Source: Ref 9.3

9.3 Thermodynamics of Twinning-Induced Plasticity Steels

Microstructures of TWIP steels are obtained under nonequilibrium conditions due to the thermal treatment used in producing them. The metastable Fe-Mn phase diagram, shown in Fig. 9.3, can be used to understand the phases present and the microstructures produced under practical nonequilibrium conditions encountered in processing Fe-Mn steels.

At room temperature, the microstructure of Fe-Mn alloys is dominated by the presence of cubic α-martensite at low manganese content, and hexagonal ε-martensite at high manganese content. Both types of martensite

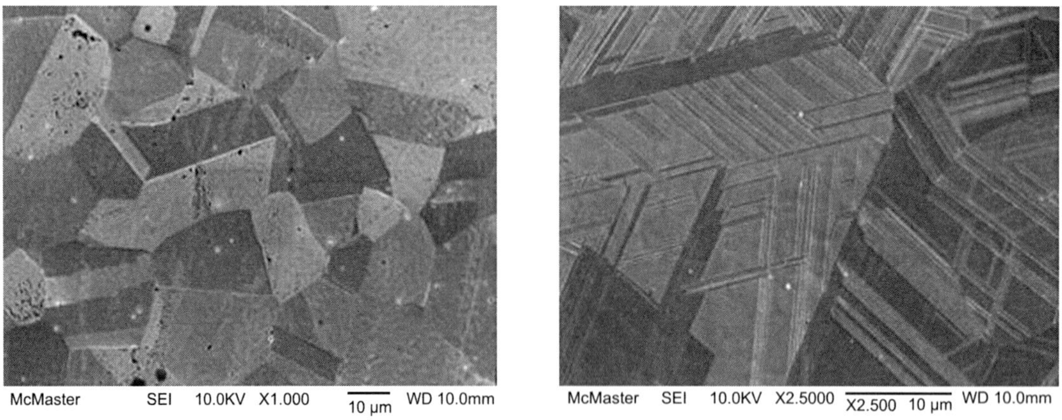

(a) (b)

Fig. 9.2 Microstructure of (a) annealed TWIP Fe-30Mn, and (b) annealed TWIP Fe-24Mn. Source: Ref 9.3

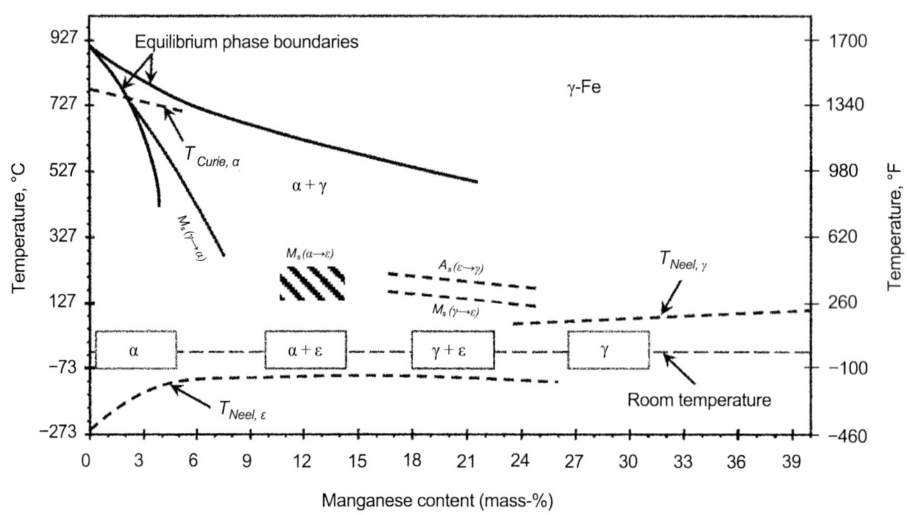

Fig. 9.3 Metastable (nonequilibrium) Fe-Mn phase diagram. Source: Ref 9.4

can also be generated by deformation-induced transformation of retained austenite. Figure 9.3 shows that stabilizing austenite at room temperature requires manganese content in excess of 27 wt%. Stable room temperature austenite can be obtained with manganese content of less than 25 wt% by using small additions of carbon (0.6 wt%) to suppress the formation of any type of martensite. For example, the microstructure of Fe-18Mn-0.6C TWIP steel was found to be a carbide-free, single-phase austenitic structure. Higher carbon content produces Fe_3C particles, which are detrimental in welding steels.

9.4 Processing of Twinning-Induced Plasticity Steels

Processing of TWIP steels includes hot rolling at 900 °C (1650 °F), then annealing at 1150 °C (2100 °F), followed by quenching with nitrogen gas. Thermal treatment of alloys usually affects the room-temperature microstructure and the phase transformation process. However, early research on TWIP Fe-29Mn, TWIP Fe-30Mn, and TWIP Fe-31Mn alloys showed that the steels remained fully austenitic irrespective of the heat treatment used. For the water-quenched TWIP Fe-24Mn alloy, 45% ε-martensite and 55% austenite were reported. To investigate whether the amount of ε-martensite varies with heat treatment of TWIP Fe-24Mn alloy, six different thermal treatments were investigated as shown in Fig. 9.4. The heat treatments included quenching in different media such as water, oil, furnace, dry ice, and liquid nitrogen. A description of the six heat treatments used is listed in Table 9.3.

Vickers microhardness tests of samples obtained from the six heat treatments showed that the hardness values were similar. This implies that the phase fractions of retained austenite and ε-martensite in those samples

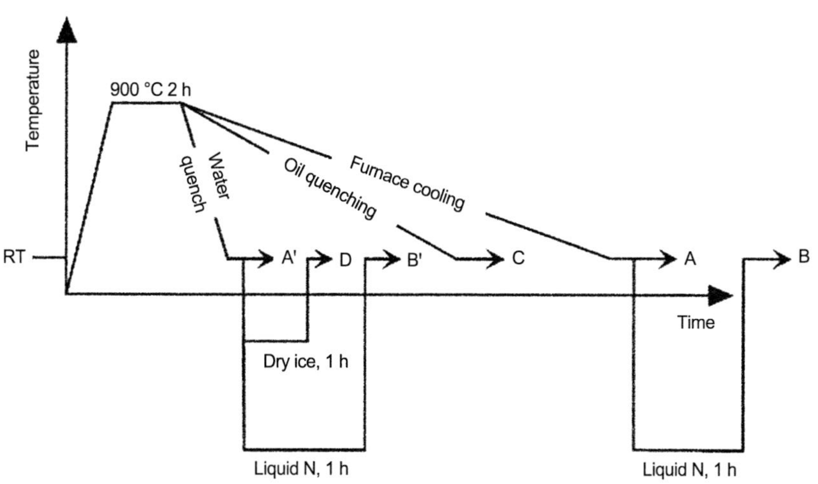

Fig. 9.4 Thermal processes for TWIP Fe-24Mn steel. Source: Ref 9.3

Table 9.3 Thermal treatments used to process Fe-24Mn alloy

Heat treatment	Description
A (FC)	Annealed → furnace cooled to room temperature (293 K)
B (FC + N)	Annealed → furnace cooled to room temperature → soaked in liquid N (77 K) for 1 h → brought to room temperature
C (OQ)	Annealed → oil quenched to room temperature
A' (WC)	Annealed → water quenched to room temperature
D (WC + DI)	Annealed → water quenched to room temperature → soaked in dry ice (195 K) for 1 h → brought to room temperature
B' (WC + N)	Annealed → water quenched to room temperature → soaked in liquid N for 1 h → brought to room temperature

Source: Ref 9.3

were essentially the same. Also, micrographs of the microstructures of the six samples were found to be almost identical. The results led to the conclusion that thermal treatments have no effect on the phase fractions of austenite and ε-martensite.

9.5 Deformation Mechanism of Twinning-Induced Plasticity Steels

In addition to dislocation glide, TWIP steels exhibit two other hardening mechanisms: TRIP and TWIP. A certain amount of stress is required to initiate both mechanisms. In the early stage of straining, twinning is the major deformation mechanism. As straining continues, transformation from austenite to martensite becomes increasingly important. As in TRIP steel, plastic deformation in TWIP steels causes the formation of mechanical twins that act as barriers to dislocation motion, increasing the strength of the steel.

The formation of mechanical twins involves the creation of new crystal orientations that progressively reduce the effective mean free path of dislocations and increase the flow stress, resulting in a high strain-hardening rate for TWIP steel.

The resultant twin boundaries act like grain boundaries in strengthening the steel. The interactions among the twins, martensite, and dislocations also contribute to increase in strength of TWIP steels. Although the dominant deformation mode in TWIP steel is dislocation glide, the number of deformation-induced twins continues to increase with strain, thus reducing the effective glide distance of dislocations. Figure 9.5 shows how deformation twins reduce the free mean path for dislocation glide, causing the high strain-hardening rate observed in TWIP steels.

It is believed that the activation of a specific deformation mechanism (TRIP or TWIP) is strongly dependent on the SFE in the metal crystals. The TRIP effect is observed in very low SFE steels (below 20 mJ/m^{-2}) and is associated with the transformation of austenite into martensite. The TWIP effect is observed in medium SFE steels (20 to 40 mJ/m^{-2}) and is characterized by the formation of deformation twins with nanometer thickness (Ref 9.5). Carbon additions of less than 1 wt% to Fe-Mn alloys reduce the SFE, but the addition of carbon is limited by the formation of Fe_3C particles.

The critical value for SFE to achieve the TWIP effect is not settled yet. Reported values ranged from 18 to 35 mJ/m^2 to activate the TWIP mechanism. Values of SFE below that range result in the TRIP effect and the austenite transformation to ε-martensite. If the SFE is higher, dislocation glide is the only mechanism contributing to deformation.

A schematic showing the relationship between SFE and the operating deformation mechanism is shown in Fig. 9.6. Low SFE is associated with martensitic transformation, while medium SFE is associated with deformation twinning, and high SFE is associated with dislocation glide. A decrease in SFE widens the stacking faults and makes slip across them more difficult. This results in activating twinning as a favorable deformation mode.

It is generally believed that mechanical twins and/or martensite transformation products continue to increase with increase in plastic deforma-

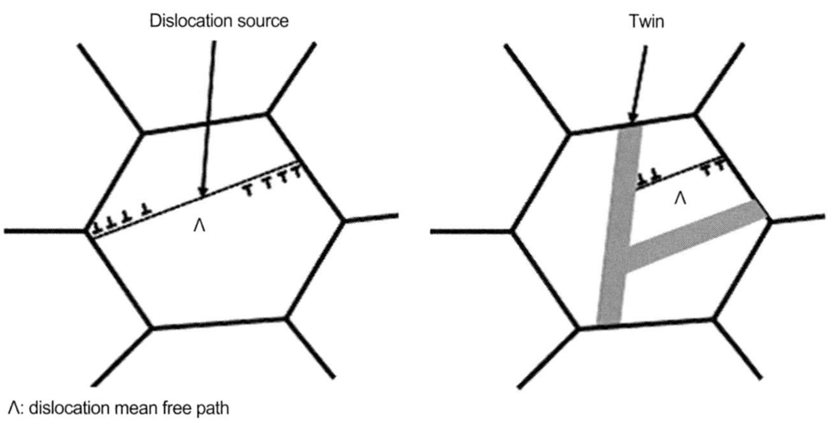

Fig. 9.5 Twins act as obstacles to dislocation glide. Source: Ref 9.4

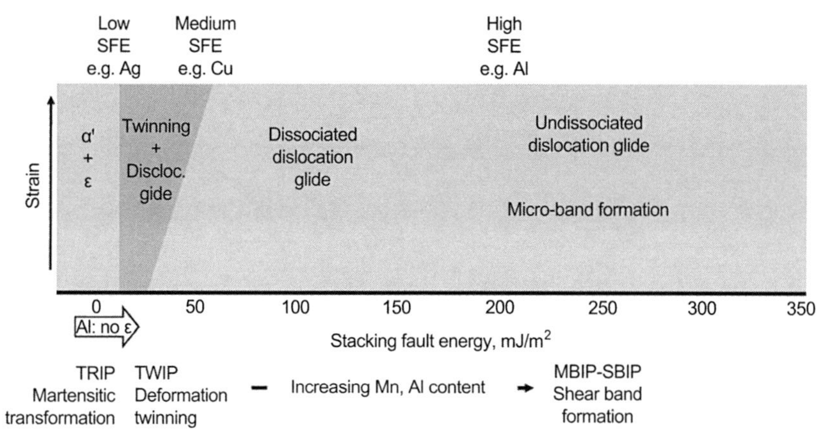

Fig. 9.6 Relationship between stacking fault energy (SFE) and deformation mechanism. Source: Ref 9.4

tion. This means that more obstacles to dislocation glide (reduction in dislocation mean free path) are continuously being added to the material, thus increasing its strength.

The strain-hardening mechanism for TWIP steels can be designed by adjusting the content of alloying elements, such as manganese, aluminum, and carbon. Alloying content affects the SFE and determines the deformation mechanism in the steel.

Mechanical twins appear in TWIP steel with plastic deformation. The volume fraction of twins increases with strain, hence adding more obstacles to dislocation glide and increasing strength, delaying necking and fracture, and maintaining high work-hardening rate. Twins appear as thin plates or thin dark lines inside the austenite matrix. Figure 9.7 shows an electron backscatter diffraction (EBSD) map of mechanical twins forming within the grains of TWIP Fe-20Mn-1.2C steel deformed to 0.30 strain.

Figure 9.8 shows a micrograph of dislocation pileup at a twin boundary. The pileup of dislocations at twin or grain boundaries leads to increase in the strength of the steel.

9.6 Mechanical Properties of Twinning-Induced Plasticity Steels

Twinning-induced plasticity steels combine high strength with high ductility. Their tensile strength exceeds 700 MPa (100 ksi) and their ductility surpasses 50%. Their strain-hardening exponent increases to 0.4 at 30% strain and then remains constant afterward. The high strain-hardening rate

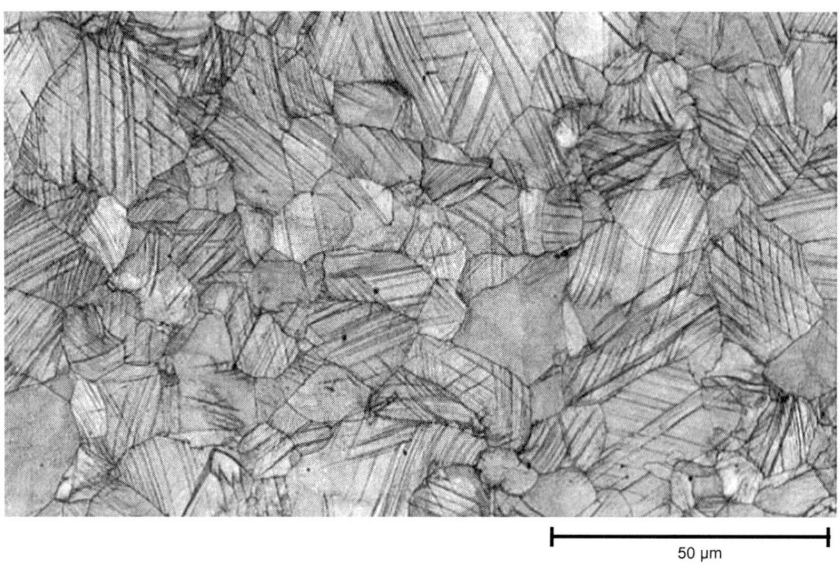

Fig. 9.7 Microstructure showing mechanical twins within TWIP Fe-20Mn-1.2C steel grains deformed to 0.30 strain. Source: Ref 9.6

results from reduction of the dislocation mean free path with mechanical twins acting as obstacles to dislocation glide.

The key mechanical properties of AHSS are strength to enable down-gaging and ductility to allow forming parts. Figure 9.9 shows the location of TWIP steels in the tensile strength-total elongation space. These steels have high tensile strength ranges, from 900 to 1100 MPa (130 to 160 ksi), and total elongation ranges from 55 to 70%.

The exceptional properties of TWIP steels arise from their fully austenitic microstructure and their twinning deformation mode. Figures 9.10 and 9.11 show engineering and true stress-strain curves obtained at room temperature for a TWIP steel whose composition is 22 wt% Mn, 0.6 wt% C, and a balance of Fe. The TWIP steel has high yield strength (600 MPa, or 85 ksi), high tensile strength (1800 MPa, or 260 ksi), and high total elongation (60%).

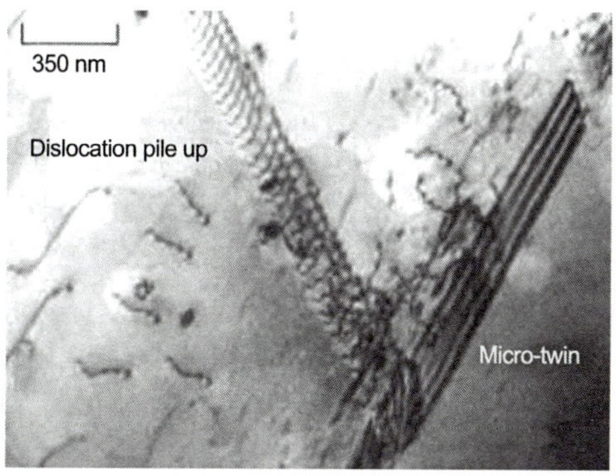

Fig. 9.8 Dislocation pileup at twin boundary. Source: Ref 9.5

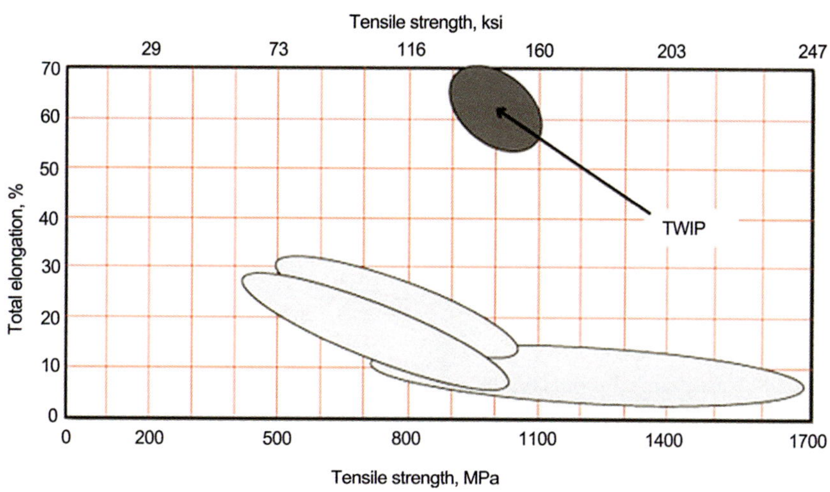

Fig. 9.9 Location of TWIP steels in the tensile strength-elongation space. Source: Ref 9.7

Table 9.4 lists the mechanical properties of three TRIP steel types referenced in Table 9.1.

The effect of high manganese content on the mechanical properties of TWIP steels is listed in Table 9.5.

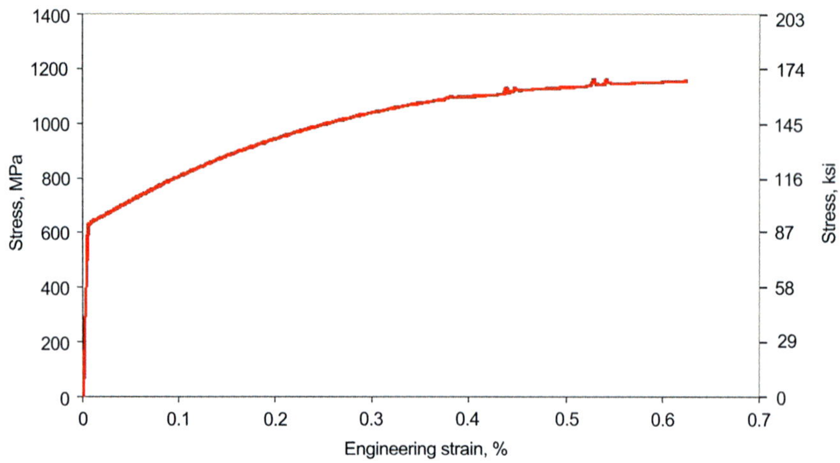

Fig. 9.10 Engineering stress-strain curve for Fe-22Mn-0.6C TWIP steel. Source: Ref 9.8

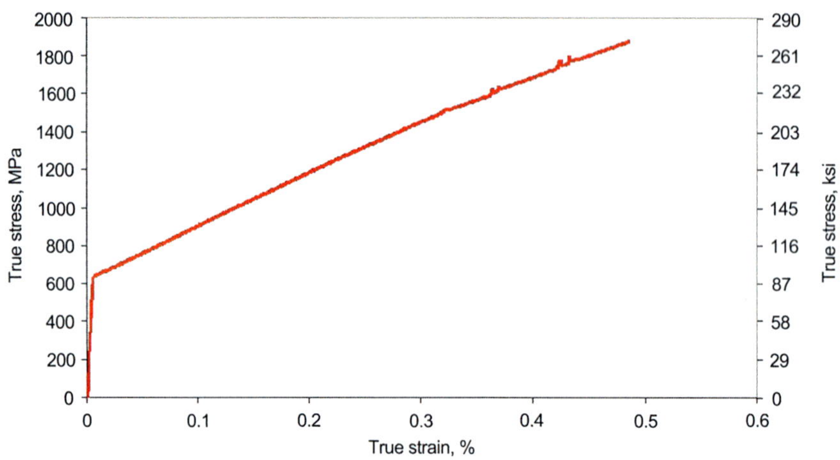

Fig. 9.11 True stress-strain curve for Fe-22Mn-0.6C TWIP steel. Source: Ref 9.8

Table 9.4 Mechanical properties of various TRIP steel types referenced in Table 9.1

TWIP type	Mn, %	Yield stress, MPa (ksi)	Tensile strength, MPa (ksi)	Total elongation, %
TWIP 1	28	325 (47)	495 (72)	64
TWIP 2	25	375 (54)	538 (78)	61
TWIP 3	27	383 (56)	548 (79)	61

Source: Ref 9.2

Deformation behavior of two TWIP steels listed in Table 9.2 is shown in Fig. 9.12 to 9.15. The engineering stress-strain curves for the three samples of Fe-30Mn steel tested at room temperature are shown in Fig. 9.12. The material has yield strength of 150 MPa (20 ksi), an ultimate tensile strength of 490 MPa (70 ksi), a maximum uniform elongation of 37%, and total elongation of 60%.

The true stress-strain curves for the three samples of Fe-30Mn steel are shown in Fig. 9.13. The maximum true strain for the material is approximately 40%.

The engineering stress-strain curves for three samples of Fe-24Mn steel tested at room temperature are shown in Fig. 9.14. The material has yield strength of 150 MPa (20 ksi), an ultimate tensile strength of 820 MPa (120 ksi), a maximum uniform elongation of 35%, and total elongation of 48%.

The true stress-strain curves for three samples of Fe-24Mn steel tested at room temperature are shown in Fig. 9.15. The maximum true strain for the material is approximately 36%.

The stress-strain curves of Fe-24Mn have a higher level of flow stress compared to those of Fe-30Mn. This behavior is due to the more compli-

Table 9.5 Effect of manganese content on the mechanical properties of TWIP steel

TWIP type	Mn, %	Tensile strength, MPa (ksi)	Total elongation, %
TWIP A	23.8	666 (97)	67
TWIP B	33	540 (78)	97

Source: Ref 9.5

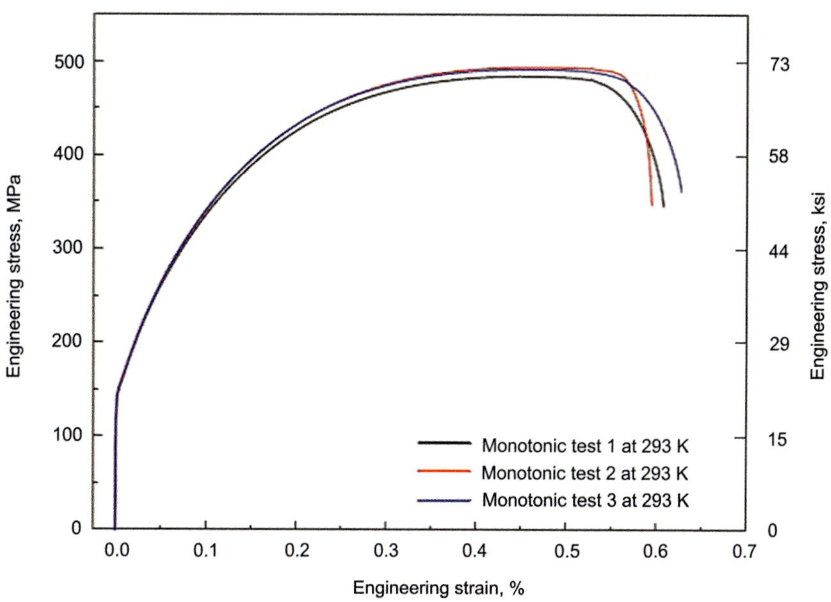

Fig. 9.12 Engineering stress-strain curves of TWIP Fe-30Mn steel. Source: Ref 9.3

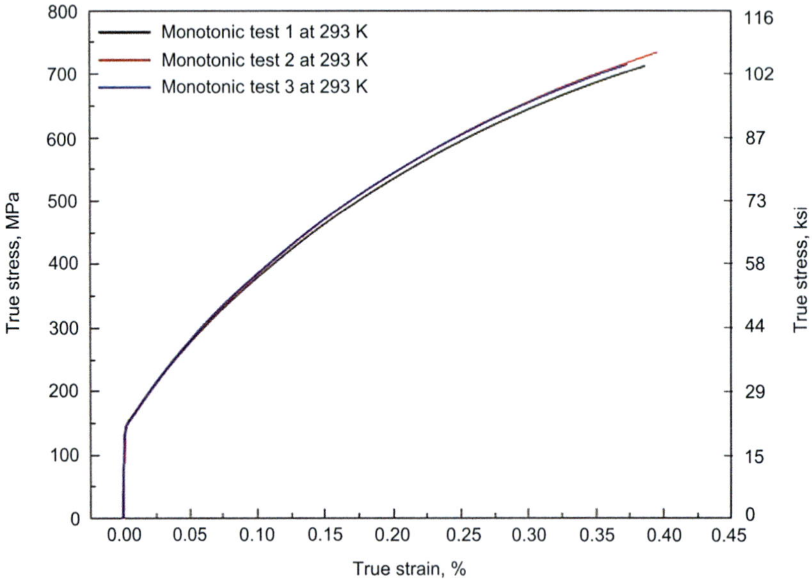

Fig. 9.13 True stress-strain curves of TWIP Fe-30Mn steel. Source: Ref 9.3

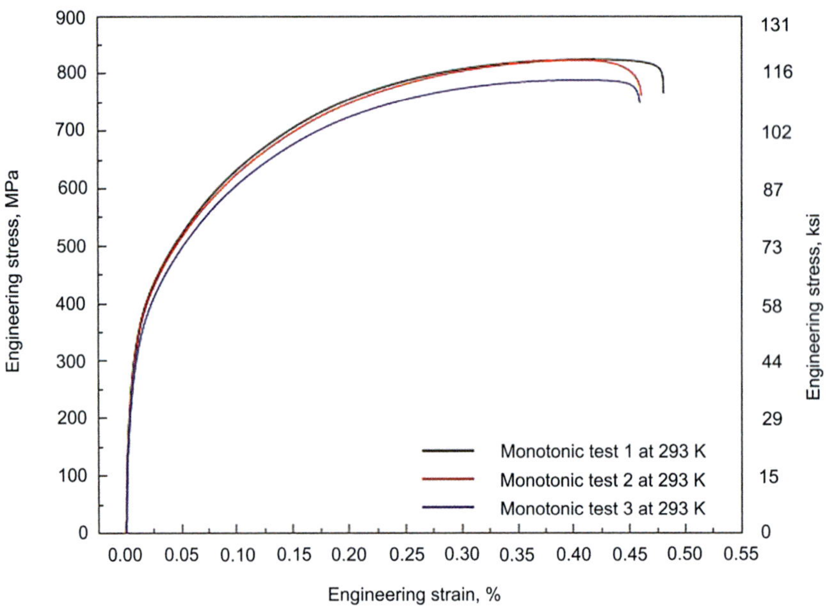

Fig. 9.14 Engineering stress-strain curves for TWIP Fe-24Mn steel. Source: Ref 9.3

cated microstructure of Fe-24Mn compared to the single-phase microstructure of Fe-30Mn.

Room-temperature dynamic strain aging, where dislocations are pinned by carbon atoms, occurs in the most commonly studied Fe-22Mn-0.6C and Fe-18Mn-0.6C TWIP steels. A widely accepted consequence of dynamic strain aging is the observation of negative strain-rate sensitivity, the presence of serrations in the stress-strain curves, and the limited post-uniform elongation of such steels. Figure 9.16 shows stress-strain curves at different strain rates for Fe-18Mn-0.6C-1.5Al TWIP steel.

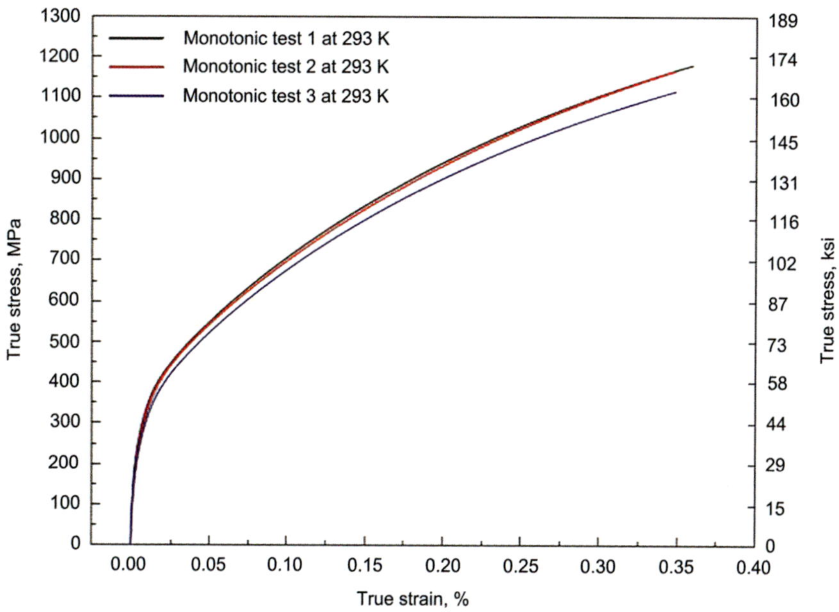

Fig. 9.15 True stress-strain curves for TWIP Fe-24Mn steel. Source: Ref 9.3

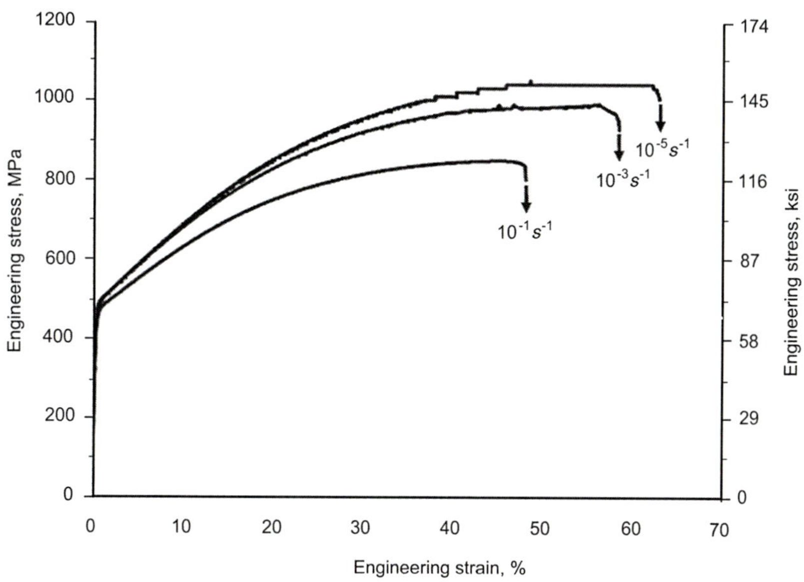

Fig. 9.16 Stress-strain curves at different strain rates for Fe-18Mn-0.6C-1.5Al TWIP steel. Source: Ref 9.4

The negative strain-rate sensitivity of this material is clearly demonstrated by the decrease in the flow stress with increase in the strain rate. In alloys with negative strain-rate sensitivity, the stress needed to deform an alloy at high strain rate is lower than that needed to deform it at a lower strain rate. Figure 9.16 also shows that serrations in the stress-strain curve are suppressed when testing is done at higher strain rate.

9.7 Formability of Twinning-Induced Plasticity Steels

Twinning-induced plasticity steels have outstanding ductility compared to all other steel grades, as shown in Fig. 9.17. This extended ductility is due to the high hardening capacity that delays plastic instability and localized necking. The TWIP steels exhibit gradual increase in instantaneous n values over the whole uniform plastic deformation range. Their instantaneous n values increase from 0.39 to 0.83 with increase in strain from 0.09 to 0.60 (Ref 9.9). The high hardening rate is attributed to the increase in the number of deformation twins with tensile strain, creating more obstacles to dislocation glide and strengthening the steel. Their low post-uniform elongation results from their negative strain-rate sensitivity and this has a negative effect on their stretch formability.

9.8 Special Attributes of Twinning-Induced Plasticity Steels

Twinning-induced plasticity steels have both very high strength (>700 MPa, or 100 ksi), very high ductility (>50%), and a high strain-hardening

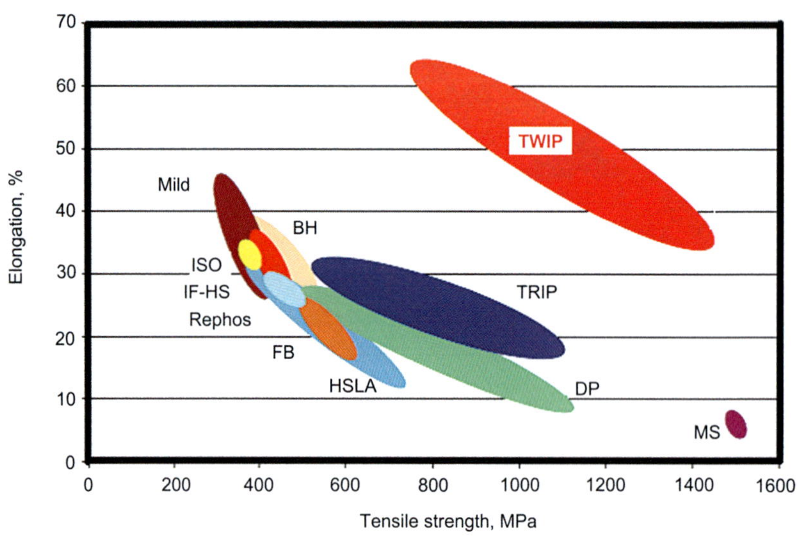

Fig. 9.17 Strength-elongation space showing TWIP steels positions among other steel grades

exponent (n = 0.47). They are ideal for automotive applications, such as crash safety, due to their high energy absorption and high toughness. However, the high alloy content in TWIP steels makes them expensive for large-scale deployment, hence the search for third-generation AHSS.

REFERENCES

9.1 W.D. Callister, Jr. and D.G. Rethwisch, *Materials Science and Engineering: An Introduction,* 8th ed., Wiley, 2010
9.2 S. Kurtze, "Characterization of the Dynamic Behavior and Microstructural Evolution of High Strength Sheet Steels," Doctor of Technology thesis, Tampere University of Technology, Publication 833, Tampere, Finland, Oct 2009
9.3 X. Liang, "Structure and Mechanical Properties of Fe-Mn Alloys," master's thesis, McMaster University, July 2008
9.4 B.C. De Cooman, K. Chin, and J. Kim, High Mn TWIP Steels for Automotive Applications, *New Trends and Developments in Automotive System Engineering,* M. Chiaberge, Ed., InTech, Shanghai, 2011, p 101–128. http://www.intechopen.com/books/new-trends-and-developments-in-automotive-system-engineering/high-mn-twip-steels-for-automotive-applications
9.5 L. Wei et al., Microstructure and Mechanical Properties of High Manganese Austenite TRIP/TWIP Steels, *J. Iron Steel,* Vol 42 (No. 1), 2007
9.6 H. Idrissi et al., On the Mechanism of Twin Formation in Fe-Mn-C TWIP Steels, *Acta Mater.,* Vol 58, 2010, p 2464–2476
9.7 "Advanced High Strength Steels Applications Guidelines," Version 4.1, WorldAutoSteel, www.worldautosteel.org, June 2009
9.8 C. Beal, "Mechanical Behavior of New Automotive High Manganese TWIP Steel in the Presence of Liquid Zinc," Ph.D. thesis, Lyon National Institute of Applied Sciences, Lyon, France, 2011
9.9 M.-H. Cai et al., Strain Hardening Behavior of High Performance FBDP, TRIP and TWIP Steels, *Steel Res. Int.,* Vol 82 (No. 3), 2011, p 242–248

CHAPTER **10**

Austenitic Stainless Steels

AUSTENITIC STAINLESS STEELS (AUST SS) are iron-base alloys containing more than 50% Fe, 16 to 26% Cr, and less than 35% Ni. They have an austenitic, face-centered cubic (fcc) crystal structure as their primary phase. Austenitic stainless steels have exceptional corrosion resistance and impressive mechanical properties. They have excellent strength, toughness, formability, and durability and also have good appearance, recyclability, and low maintenance cost.

If austenite in carbon steel is cooled under equilibrium conditions, it transforms into a mixture of ferrite and cementite. With AUST SS, the high chromium and nickel content suppress this transformation and keep the steel fully austenitic. Stainless steels attain their stainless characteristics because of the formation of an invisible and adherent chromium-rich oxide surface film. This oxide establishes on the surface and heals itself in the presence of oxygen. Some other alloying elements are added to enhance specific characteristics. These include molybdenum, copper, titanium, aluminum, silicon, and niobium.

The first austenitic steel alloy with a composition of Fe, 18% Cr, and 8% Ni was developed in 1912 and became known as stainless steel type 304. The austenitic grades are nonmagnetic and, because of their high chromium and nickel content, are the most corrosion resistant of the stainless group, providing unusually fine mechanical properties. They cannot be hardened by heat treatment but can be hardened considerably by cold working. Corrosion resistance and mechanical properties are commonly the principal factors in selecting a grade of stainless steel for a given application.

Austenitic stainless steels have high ductility, low yield stress, and relatively high ultimate tensile strength when compared to typical carbon steels. They are also ductile and tough at very low subzero temperatures. Type 304 steel is used for chemical processing equipment, food, dairy, and beverage industries, and heat exchangers. Type 316 steel is used in chemical processing, the pulp and paper industry, food and beverage processing and dispensing, and in corrosive environments. Type 321 and type 347 have been developed for corrosive resistance for repeated intermittent exposure to temperatures above 450 °C (840 °F). These grades are primarily used in the aircraft industry.

10.1 Compositions and Microstructures of Austenitic Stainless Steels

The most familiar AUST SS is type 304, sometimes called T304 or simply 304. It is the most widely used alloy of the austenitic group. Type 304 AUST SS is a ferrous alloy containing a maximum of 0.08% C, 18 to 20% Cr, and 8 to 10% Ni. The higher the carbon content, up to 0.1%, the greater the yield strength of the steel. Other alloying elements such as molybdenum, titanium, and niobium may be added to the basic type 304 alloy to obtain other variations of AUST SS.

Austenite microstructure contains parallel bands, called twins, inside the grain. The twins are caused by a certain type of shearing that develops inside the grain during plastic deformation. Austenite is a soft and extremely ductile and tough structure. Figure 10.1 shows a micrograph of austenite grains with twins inside the grains of annealed type 302 steel.

Austenitic stainless steels are solid-solution alloys where each alloying element plays an important role in controlling properties such as corrosion resistance, strengthening, and austenite stabilization. Table 10.1 lists the chemical compositions of the major types of AUST SS. They all share the high content of the alloying elements chromium, nickel, and manganese.

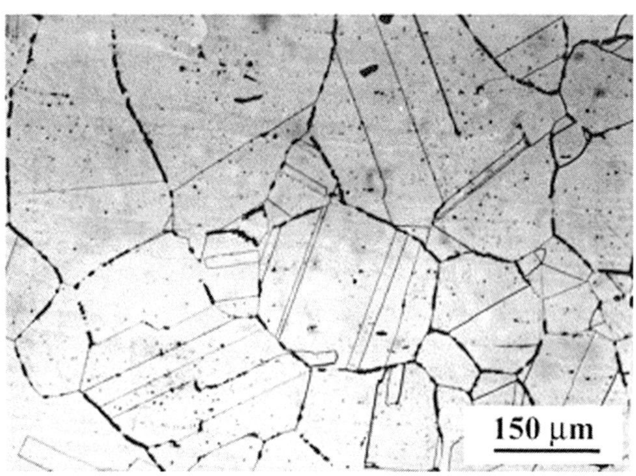

Fig. 10.1 Annealed austenitic stainless steel (AUST SS) type 302. Source: Ref 10.1

Table 10.1 Principal types of austenitic stainless steels (AUST SS)

Type	Composition, wt%						
	C	N	Mn	Si	Cr	Ni	Mo
SS 201	0.15	0.25	5.5–7.5	1.0	16–18	3.5–5.5	...
SS 301	0.15	...	2.0	1.0	16–18	6.0–8.0	...
SS 304	0.08	0.1	2.0	1.0	18–20	8.0–10.5	...
SS 316	0.08	0.1	2.0	1.0	16–18	10.0–14.0	2.0–3.0

Source: Ref 10.1

10.2 Processing of Austenitic Stainless Steels

Continuous casting or hot rolled slabs of AUST SS can be used as hot rolled or they can be further reduced in thickness by cold rolling to work harden the steel. Final annealing of AUST SS to eliminate the effect of cold work is conducted in air or in hydrogen/nitrogen protective atmospheres within the temperature range of 1000 to 1100 °C (1830 to 2010 °F). Annealing causes recrystallization to control grain size and to dissolve precipitated carbides. Cooling from annealing temperatures must be sufficiently rapid to avoid carbide precipitation. No martensite can form from quenching because mechanical deformation is required to transform austenite to martensite.

10.3 Deformation Mechanism of Austenitic Stainless Steels

Austenitic stainless steels have an fcc crystal structure which, at room temperature, provides 12 slip systems for dislocation glide. Combined with the low level of interstitial elements that may act to obstruct dislocation glide, the slip systems give this material good ductility.

10.4 Mechanical Properties of Austenitic Stainless Steels

Austenitic stainless steels cannot be hardened by heat treatment. The steel can be strengthened by cold work to high tensile and yield strengths, while retaining good ductility and toughness, even at cryogenic temperatures. Depending on the composition and amount of cold work, room-temperature yield strength range is 200 to1400 MPa (30 to 200 ksi) and tensile strength range is 900 to1200 MPa (130 to 175 ksi).

The compositions and mechanical properties of three AUST SS are listed in Table 10.2. The annealed SS 304 has the highest tensile strength and highest ductility.

A comparison of the mechanical properties of TRIP 700, DP 750, DP 800, and SS 900 (annealed) is listed in Table 10.3. Austenitic stainless steel has the highest tensile strength and ductility compared to the other listed advanced high-strength steels (AHSS) grades. The high value of the strain-hardening exponent, *n*, is responsible for the excellent ductility of AUST SS.

Uniaxial tensile tests of common types of AUST SS compared to mild and dual-phase (DP) steels are shown in Fig. 10.2. The mechanical properties of those steels are listed in Table 10.4.

It is clear from the stress-strain curves and Table 10.4 that the ductility of stainless steels far exceeds that of mild and DP steels. Their strength is higher than that of mild steel but comparable to that of DP steels.

Toughness is the resistance of a metal to fracture and is related to the total area under the stress-strain curve from yielding up to fracture. This

Table 10.2 Compositions and properties of some AUST SS alloys

Alloy no.	Composition, wt%	Ultimate tensile strength, MPa (ksi)	Elongation, %
SS 302	18 Cr, 9 Ni, 2 Mn	517 (75)	40
SS 304 annealed	18 Cr, 8 Ni	579 (84)	55
SS 316	18 Cr, 12 Ni, 2 Mn	517 (75)	40

Source: Ref 10.1

Table 10.3 Mechanical properties of various AHSS grades

Steel grade	Yield strength, MPa (ksi)	Tensile strength, MPa (ksi)	True stress at tensile strength, MPa (ksi)	Uniform elongation, %	*n*	Total elongation, %
TRIP 700	473 (69)	703 (102)	818 (119)	16.4	0.169	17
DP 750	513 (74)	811 (118)	920 (133)	13.4	0.181	18.8
DP 800	573 (83)	896 (130)	976 (142)	8.9	0.170	9.9
SS 900	306 (44)	937 (136)	1429 (207)	52.5	0.517	59.3

Source: Ref 10.2

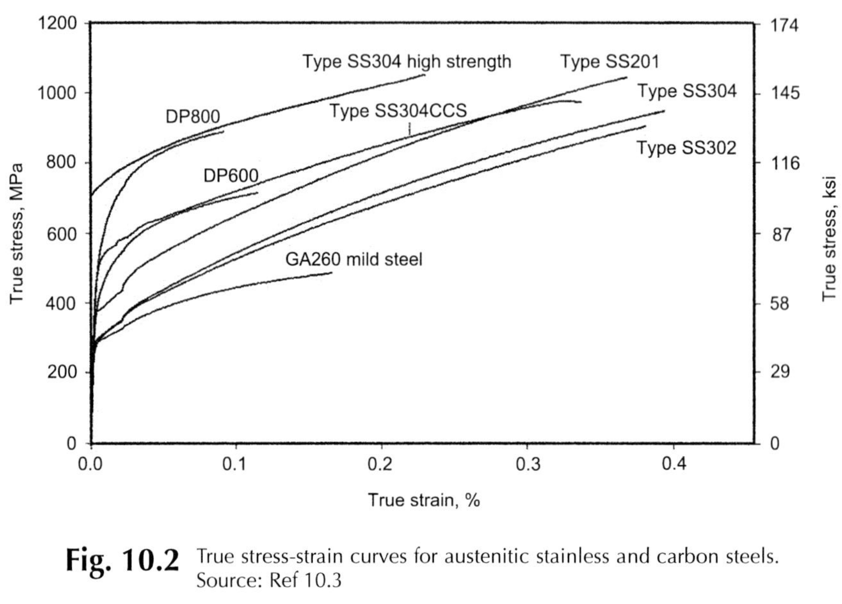

Fig. 10.2 True stress-strain curves for austenitic stainless and carbon steels. Source: Ref 10.3

Table 10.4 Mechanical properties of steels shown in Fig. 10.2

Type	Yield strength, MPa (ksi)	Tensile strength, MPa (ksi)	Uniform elongation, %	Total elongation, %
GA 260	283 (41)	410 (59)	18	32
DP 600	388 (56)	637 (92)	14	21
DP 800	505 (73)	814 (118)	11	16
SS 201	375 (54)	723 (105)	46	55
SS 302	280 (41)	619 (90)	51	56
SS 304	295 (43)	640 (93)	48	56
SS 304 high-strength	726 (105)	836 (121)	24	26
SS 304 CCS	486 (70)	711 (103)	34	40

Source: Ref 10.3

Table 10.5 Toughness of various austenitic stainless and carbon steels

Steel type	Toughness, J/m^3
GA 260	68
DP 600	73
DP 800	71
SS 201	284
SS 302	246
SS 304	268
SS 304 CCS	267
SS 304 high-strength	201

Source: Ref 10.3

Table 10.6 Toughness of various AHSS grades

Steel grade	Toughness, J/m^3
TRIP 700	105
DP 750	101
DP 800	74
SS 900	364

Source: Ref 10.2

area determines the amount of plastic work per unit volume at fracture and it is important in energy-absorbing applications. Table 10.5 lists toughness values for the materials listed in Table 10.4 and shown in Fig.10.2.

Another set of data on toughness of AHSS and AUST SS is listed in Table 10.6. Both tables show that AUST SS have the highest toughness compared to the other listed steels.

10.5 Formability of Austenitic Stainless Steels

Annealed AUST SS are characterized by low yield strengths, high tensile strengths, and very high elongation. This translates into good formability at ambient temperatures. Austenitic stainless steel work hardens considerably and, therefore, can delay localized necking in stretched and deep-drawn applications.

The forming limit curves of various austenitic stainless, DP, and mild steels are shown in Fig. 10.3. Results clearly show the good formability of stainless steels compared to the other steels. The limiting dome height test is used to assess the equi-biaxial stretch forming of metals. This test is believed to be more accurate than the forming limit diagram test in assessing the formability of alloys that undergo stress-induced phase transformation. The phase transformation and the resulting high strain-hardening rate in SS 900 are responsible for the more uniform strain distribution and the resulting higher dome height in the steel. Figure 10.4 shows photos of the limiting dome height for AUST SS 900 and DP 750 steels. Austenitic SS 900 has a higher limiting dome height than DP 750 because it has the capacity to distribute strains more uniformly, thereby delaying fracture and causing more stretching in the dome.

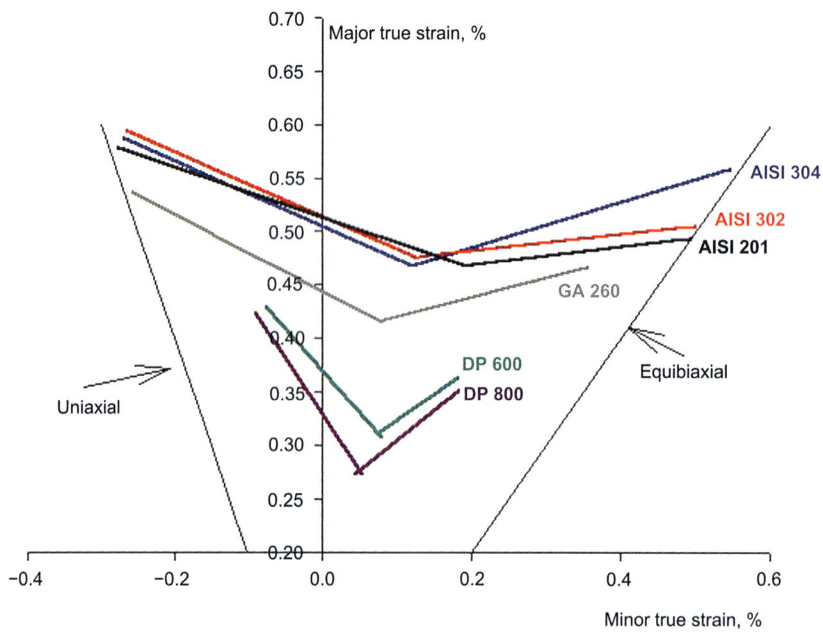

Fig. 10.3 Forming limit diagrams for austenitic stainless and carbon steels. Source: Ref 10.3

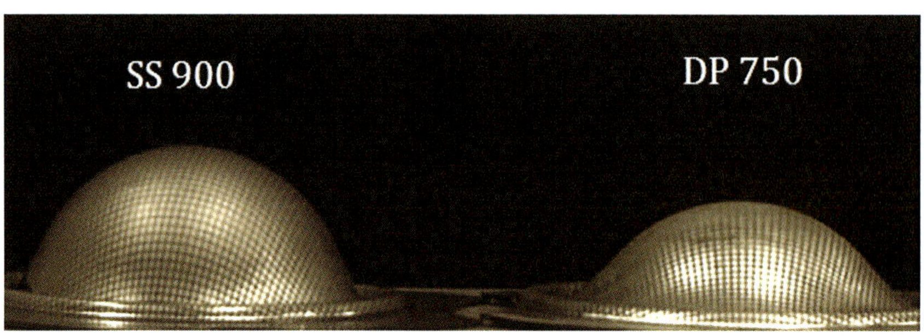

Fig. 10.4 Photos of limiting dome height test for SS 900 and DP 750 steels. Source: Adapted from Ref 10.2

10.6 Special Attributes of Austenitic Stainless Steels

Austenitic stainless steels have exceptional corrosion resistance and impressive mechanical properties including excellent strength, exceptional formability, outstanding durability, and good fatigue strength. They also have excellent toughness down to subzero temperatures, with no steep ductile-to-brittle transition. Such properties make them excellent candidates for structural applications, especially for energy-absorbing components. In addition, stainless steels have pleasing appearance, good recyclability, and low maintenance cost.

Austenitic stainless steels change microstructure from austenite to martensite during plastic deformation. This phenomenon helps critical areas, such as sharp corners and highly stressed locations, by shifting deformation to softer surroundings.

In 1936, Ford Motor Company built a stainless steel Ford Deluxe Sedan prototype car to demonstrate to the public the wonders of stainless steel. The prototype car is shown in Fig. 10.5. It became the centerpiece for advertising but was never commercially produced.

The automotive industry uses approximately 12 kg (26 lb) of stainless steel per car. The major applications include wheel covers, antennas, engine components, trim and decorative items, and wiper blades. The benefits of using AUST SS in the automotive industry are summarized in Table 10.7.

Fig. 10.5 Ford's 1936 stainless steel Deluxe Sedan. Image courtesy of Allegheny Technologies Inc.

Table 10.7 Benefits of using AUST SS in the automotive industry

Benefits of stainless steels	Description
Commercial availability	Currently available everywhere. Austenitic stainless steels are also affordable for some parts.
Weight reduction	Weight reductions of 40–50% are possible by replacing SS for other materials (mild steel, HSS).
Safety	High strain-rate sensitivity and additional strengthening from austenite to martensite transformation during crash result in 10% less intrusions into the occupant cabin. High energy absorption increases safety in collisions.
Material cost	The 40–50% downgaging may keep SS material cost per part affordable in comparison to mild steel. Also, part consolidation is expected to reduce cost further.
Capital investment	Capital investment is low because existing production lines can be used.
Durability	This value will be high for SS because endurance limit is approximately 50% of the tensile strength of metals. High corrosion resistance provides long life for the product.
Formability	Excellent formability allows stamping of complex parts and this also leads to part integration.

Source: Ref 10.4

REFERENCES

10.1 Ford Scientific Research Lab, Archived Microstructure Files, Formability Lab, Dearborn, MI

10.2 C. Magnusson and R. Andersson, "Stainless Steel as a Lightweight Automotive Material," Volvo Cars Body Components, Olofström, Sweden, 2003

10.3 R. Anderson, C. Magnusson, and E. Scheiden, Using Stainless Steel for Energy Absorbing Components in Automobiles, *Innovations in Processing and Manufacturing of Sheet Materials,* M.Y. Demeri, Ed., TMS, 2001, p 97–109

10.4 M.Y. Demeri, High Strength and Stainless Steels for Automotive Applications, *Innovations in Processing and Manufacturing of Sheet Materials,* M.Y. Demeri, Ed., Feb 11–15, 2001 (New Orleans, LA), TMS, Warrendale, PA, 2001, p 83–95

CHAPTER 11

Applications of Advanced High-Strength Steels

ADVANCED HIGH-STRENGTH STEELS (AHSS) are best used for strong structural applications where light weighting enhances the performance of a product. In addition to the automotive industry, which is an important market for AHSS, other industries and potential applications include ship building, aircraft, appliances, and civil and military structures. In this book, the focus is on the automotive applications of AHSS. The automotive industry is benefiting from the unique properties of AHSS because the materials allow automakers to meet tough new emissions standards, safety regulations, fuel efficiency mandates, and customer affordability needs. Automakers responded positively to the influx of new design methodologies, innovative manufacturing processes, and advanced materials by adopting new manufacturing methods and by increasing AHSS content in their vehicles.

11.1 Automotive Applications

Applications of AHSS in the automotive industry revolve around four major areas:

- Improving crash performance (front, rear, side) via increased strength levels
- Reducing vehicle weight via gage reduction
- Improving fuel economy via vehicle weight reduction
- Reducing greenhouse emissions via fuel consumption reduction

Other important applications criteria include improvement in stiffness, fatigue life, corrosion resistance, formability, weldability, and affordability. The currently used AHSS have reduced a vehicle's body structure mass by at least 25%. The third-generation AHSS, under development, are expected to achieve more than 35% in structural mass reduction. This will significantly improve fuel efficiency and reduce greenhouse gas emissions.

It is estimated that for a typical vehicle, replacing its conventional steel with AHSS will save approximately 2.2 tons of greenhouse gases over its life cycle. With millions of cars being produced annually, the total reduction in greenhouse gas emissions will be incredibly high.

Automakers are now using increasing amounts of AHSS in their new vehicles. Figure 11.1 shows that AHSS content per light vehicle increased steadily from 37 kg (81 lb) in 2006 to 88 kg (194 lb) in 2012. A conservative look at the future projects an increase to 159 kg (350 lb) by 2020.

Before discussing the implementation of AHSS in vehicles, a review of the nomenclature of the different vehicle components is helpful in identifying the target applications. The body of a passenger car is divided into two component groups:

- *Body-in-white (BIW) structure* is defined as the assembled skeleton of a vehicle without its moving parts such as the doors, hood, decklid, and fenders.

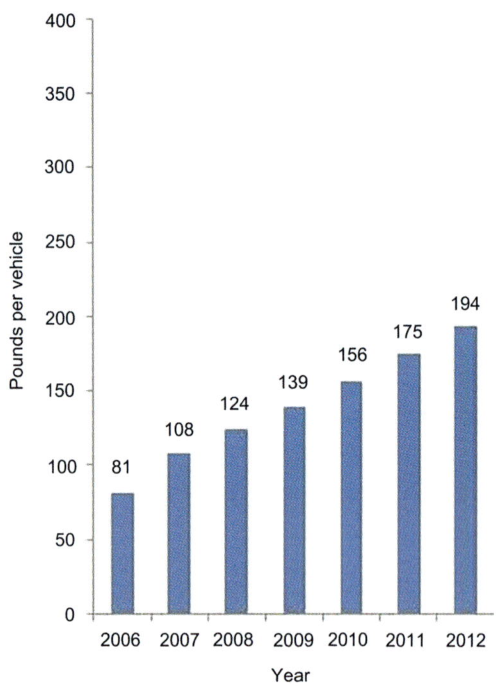

Fig. 11.1 Advanced high-strength steel (AHSS) content in North American light vehicles from 2006–2012. Source: Ref 11.1

- *Closures* are components such as the doors, hood, decklid, and fenders.

Material requirements for each group are different because their functions are different. Body-in-white structural components require materials with high strength and high toughness for energy absorption and anti-intrusion applications. Closures make up the visible style appearance of a vehicle and therefore require good formability, dent resistance, and high-quality finish for paint application. Figure 11.2 shows the strength-ductility chart where component requirements are mapped with corresponding steel grades meeting those requirements. For example, conventional steels are best suited for closures, while AHSS are best suited for BIW structural applications.

Advanced high-strength steels are mainly used for structural members requiring good crash energy absorption and anti-intrusion properties needed for occupant safety. Therefore, BIW components and the steel grades and types used to manufacture them are reviewed here. Figure 11.3 shows the main locations where AHSS are used in a typical vehicle and lists the AHSS grades and types used for each application.

Depending on the car manufacturer, the same component may be designed differently and may be made from the same grade of steel but of

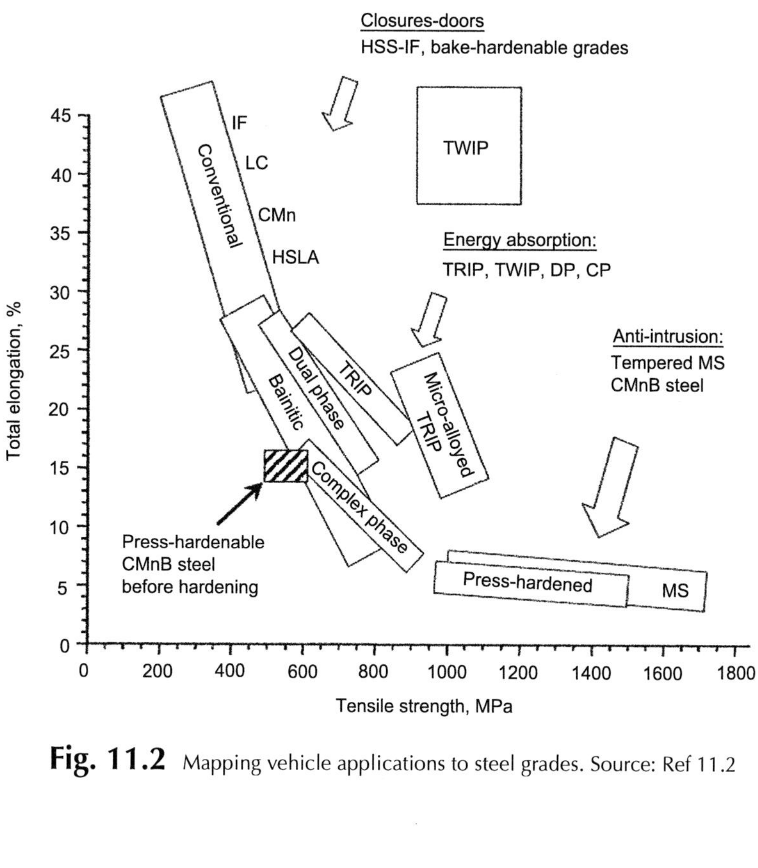

Fig. 11.2 Mapping vehicle applications to steel grades. Source: Ref 11.2

different types (i.e., DP 600 or DP 800) or made from two completely different grades (i.e., DP 600 or TRIP 700). Figure 11.4 shows a photo of a BIW of a typical passenger car.

The names and locations of the structural members of the BIW of a Porsche Cayenne are shown in Fig. 11.5. Also listed are the AHSS grades used to build the vehicle structure.

Automakers are actively pursuing the increase of AHSS content in their vehicles and adopting innovative designs and manufacturing processes to meet fuel consumption and safety regulations. Automakers advertise their achievements by focusing on improved vehicle performance and the minimal adverse effect using advanced steel technologies has on the environment.

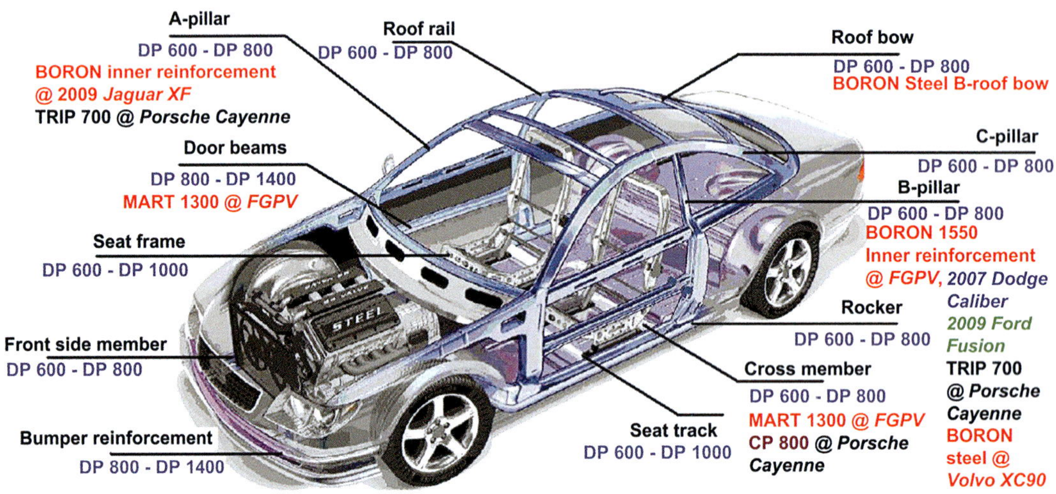

Fig. 11.3 Structural components in a typical car. Source: Ref 11.3

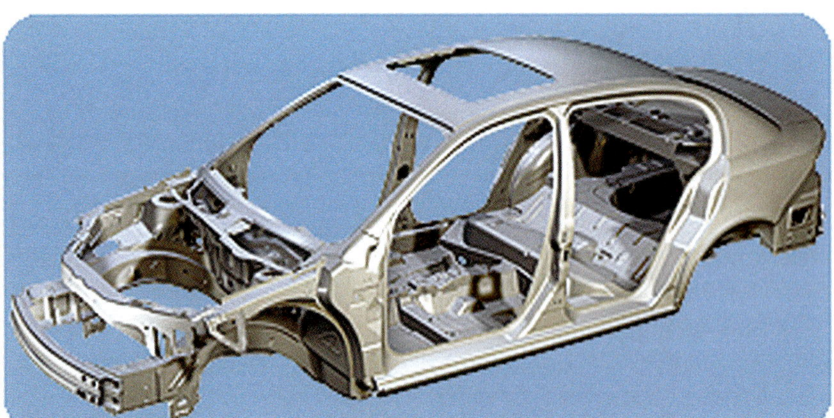

Fig. 11.4 Body-in-white (BIW) of a vehicle

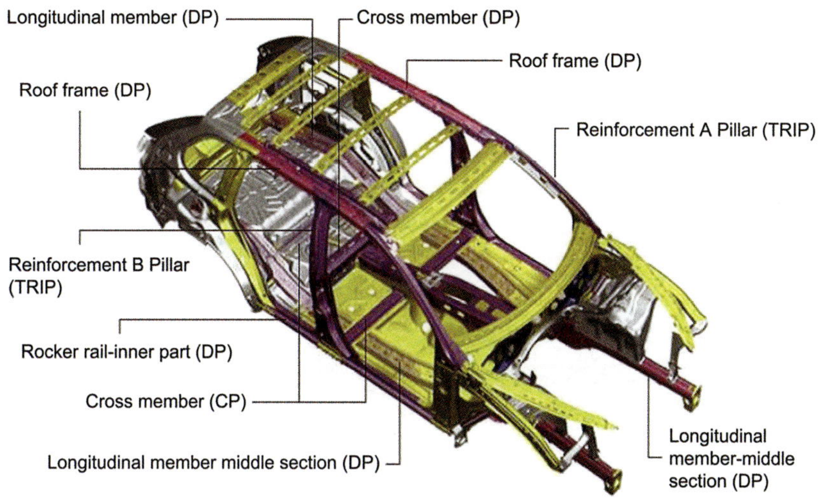

Fig. 11.5 Body-in-white (BIW) of Porsche Cayenne. Source: Ref 11.4

Representative cases for AHSS utilization in five new vehicles are presented, to demonstrate the extent light weighting has established itself as an indispensable strategy for automakers. One of the vehicles, the FutureSteelVehicle (FSV), is a concept vehicle built by the global steel industry to help automakers optimize vehicle body structures through AHSS design concepts, with the purpose of reducing gas emissions for the entire life cycle of the vehicle. Utilization of AHSS is demonstrated by their increased use in building BIW structures and other vehicle components. Selection of the cases is intended to show that all domestic automakers are serious and committed to the full utilization of AHSS in their vehicles. The five representative cases were selected from the 2011 to 2013 model years of domestic automakers:

- 2012 Chrysler Dodge Dart (Ref 11.5)
- 2013 GM Cadillac ATS (Ref 11.6)
- 2013 Ford Fusion (Ref 11.7)
- 2013 GM Chevrolet Sonic (Ref 11.8)
- 2011 FutureSteelVehicle (Ref 11.9)

2012 Chrysler Dodge Dart (Ref 11.5). A color-coded illustration in Fig. 11.6(a) and (b) shows that the BIW structure of the 2012 Dodge Dart contains 30% HSS and 30% AHSS.

2013 GM Cadillac ATS (Ref 11.6). A color-coded illustration in Fig. 11.7 shows the type of materials used to build the BIW structure of the 2013 Cadillac ATS. Advanced high-strength steels are shown in yellow and orange colors.

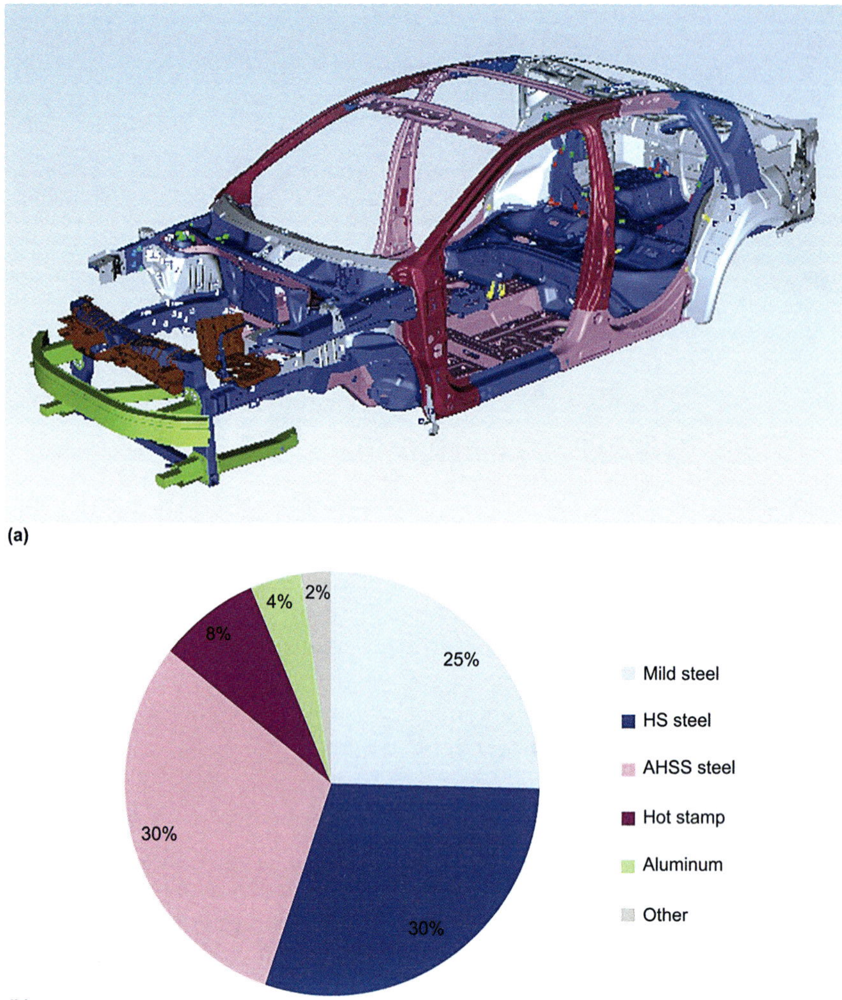

Fig. 11.6 (a) 2012 Dodge Dart BIW color-coded components and (b) material content. Source: Ref 11.5

Material breakdown by mass of the BIW structure, shown in Fig. 11.8, indicates that AHSS account for 34.4% while high-strength, low-alloy (HSLA) steel accounts for 22.6%.

2013 Ford Fusion (Ref 11.7). A color-coded illustration in Fig. 11.9 shows the type of materials used to build the BIW structure of the 2013 Ford Fusion. Advanced high-strength steels account for 36.2% while HSLA steels account for 33.3%.

A color-coded material breakdown for the 2013 Ford Fusion closures is shown in Fig. 11.10. Because of the class (A) surface, formability, and dent resistance requirements in closures, mild steel, bake-hardenable (BH),

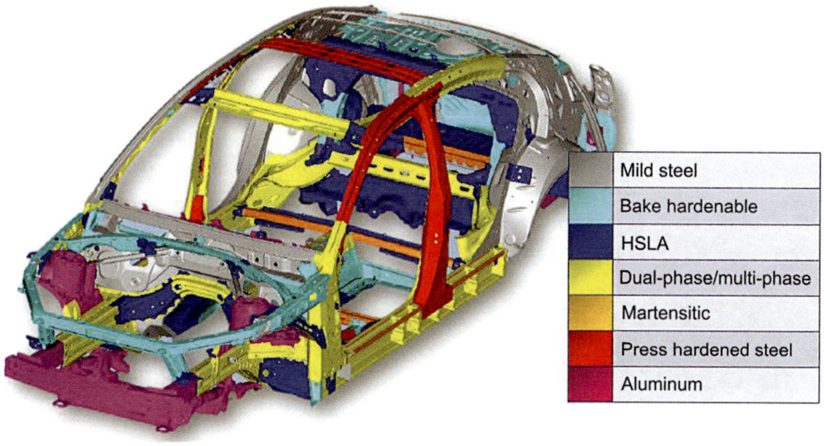

Fig. 11.7 Color-coded material breakdown for BIW of the 2013 Cadillac ATS. Source: Ref 11.6

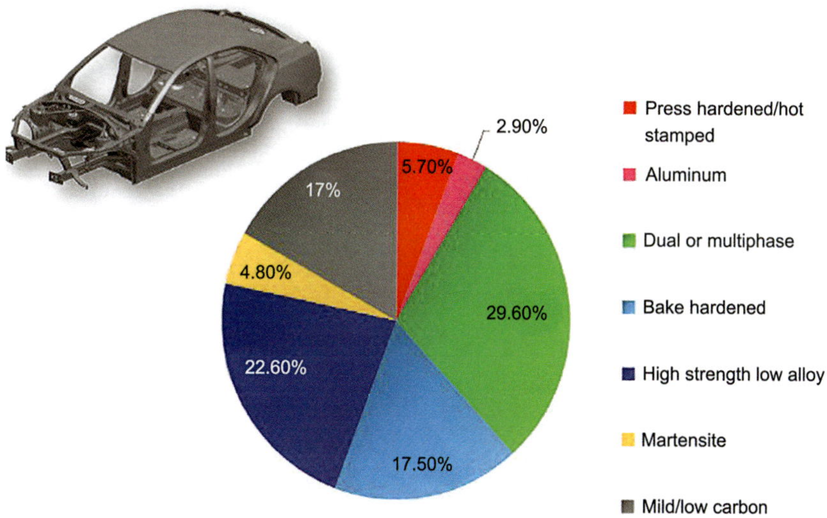

Fig. 11.8 Color-coded material breakdown by mass for BIW structure of the 2013 Cadillac ATS. Source: Ref 11.6

and HSLA steels dominate the content. Advanced high-strength steel accounts for only 8.1%.

The 2013 Ford Fusion subframe front structure, which provides a lower load path into the structure in case of frontal impact load or crash, is shown in Fig. 11.11. The A-pillar/roof rail and the B-pillar were made of DP 1000 steel for the tubes and DP 800 for the brackets. The rocker reinforcement was made of martensitic steel.

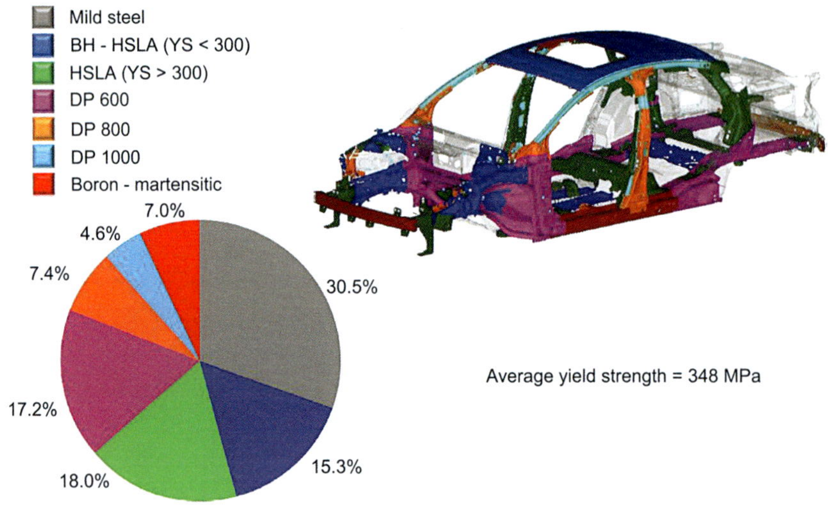

Fig. 11.9 Color-coded material breakdown for BIW of the 2013 Ford Fusion. Source: Ref 11.7

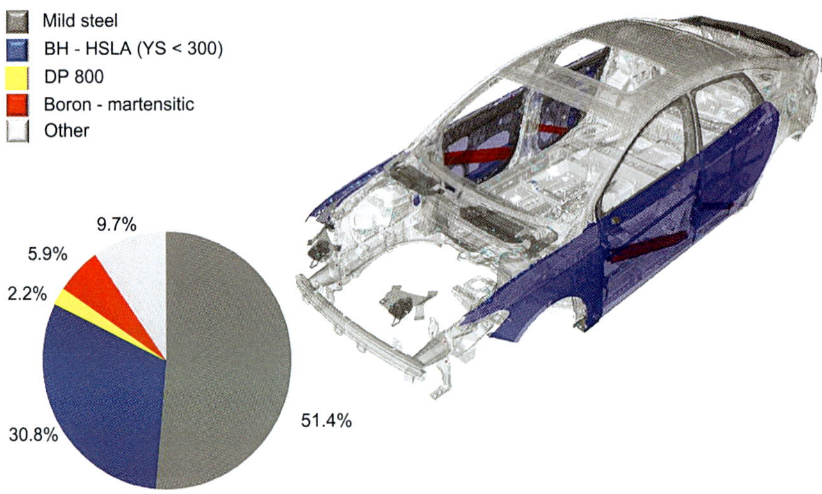

Fig. 11.10 Color-coded material breakdown for closures in the 2013 Ford Fusion. Source: Ref 11.7

The 2013 Ford Fusion roof/side impact structure, which provides a load path for side impact, is shown in Fig. 11.12. The hydroformed A-pillar/roof rail and the dual tube B-pillar were both made of DP 1000.

2013 GM Chevrolet Sonic (Ref 11.8). A color-coded illustration in Fig. 11.13 shows the type of materials used to build the BIW structure of the 2013 Chevrolet Sonic. Advanced high-strength steels are shown in yellow and orange colors.

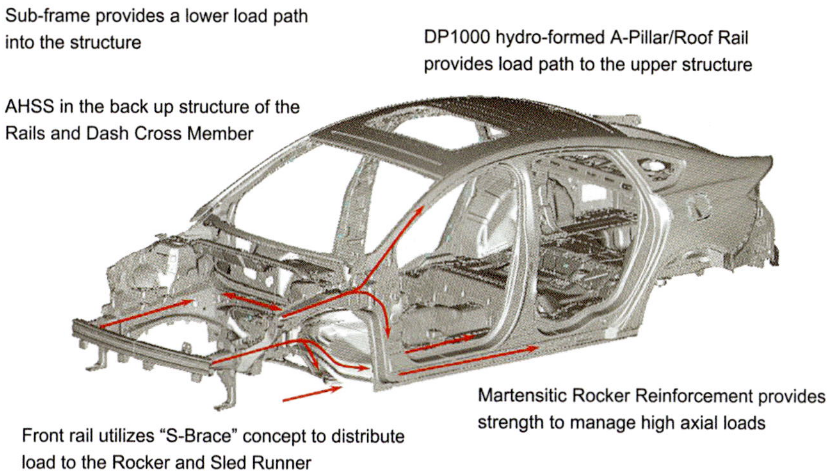

Fig. 11.11 Load paths and AHSS used for the 2013 Ford Fusion subframe front structure. Source: Ref 11.7

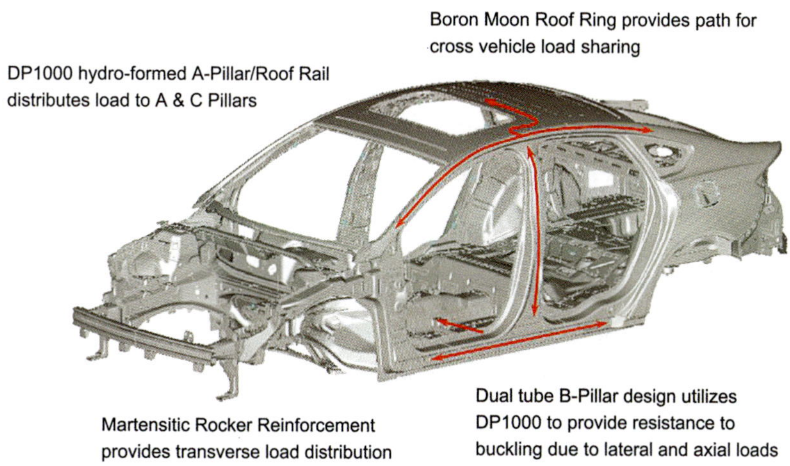

Fig. 11.12 Load paths and AHSS used for the 2013 Ford Fusion subframe roof/side structure. Source: Ref 11.7

Material breakdown by mass of the BIW structure, shown in Fig. 11.14, indicates that AHSS account for only 10% while HSLA steel accounts for 42%. However, the total HSS content comprises 66% of the total body mass of the BIW structure.

2011 FutureSteelVehicle (FSV) (Ref 11.9). A color-coded illustration in Fig. 11.15 shows the type of materials used to build the body structure of the 2011 FSV. Advanced high-strength steel accounts for 53.7% of the body structure mass. This percentage is higher than any new commercially produced vehicle. The high AHSS content demonstrates that such

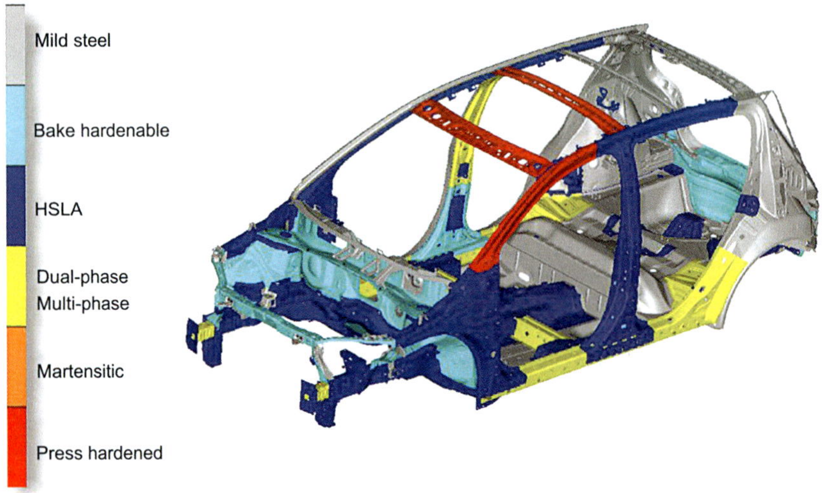

Fig. 11.13 Body-in-white of 2013 Chevrolet Sonic color coded for materials used. Source: Ref 11.8

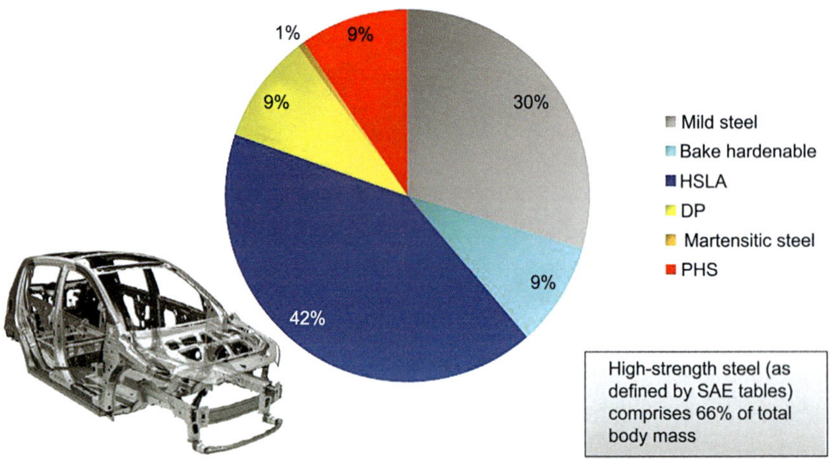

Fig. 11.14 Color-coded material breakdown for BIW structure of the 2013 Chevrolet Sonic. Source: Ref 11.8

materials are key to achieving reduced gas emission targets for the entire life cycle of the vehicle.

International automakers have similar strategies and goals. Mazda reported that its 2008 Mazda2 weighed 100 kg (220 lb) less than its predecessor due to using more than 40% HSS. Mercedes also reported that its 2008 C-Class contained 70% HSS. Examples like these are numerous among automakers today and suggest that vehicles are becoming lighter, safer, more fuel efficient, and friendlier to the environment as more HSS, especially AHSS, are incorporated.

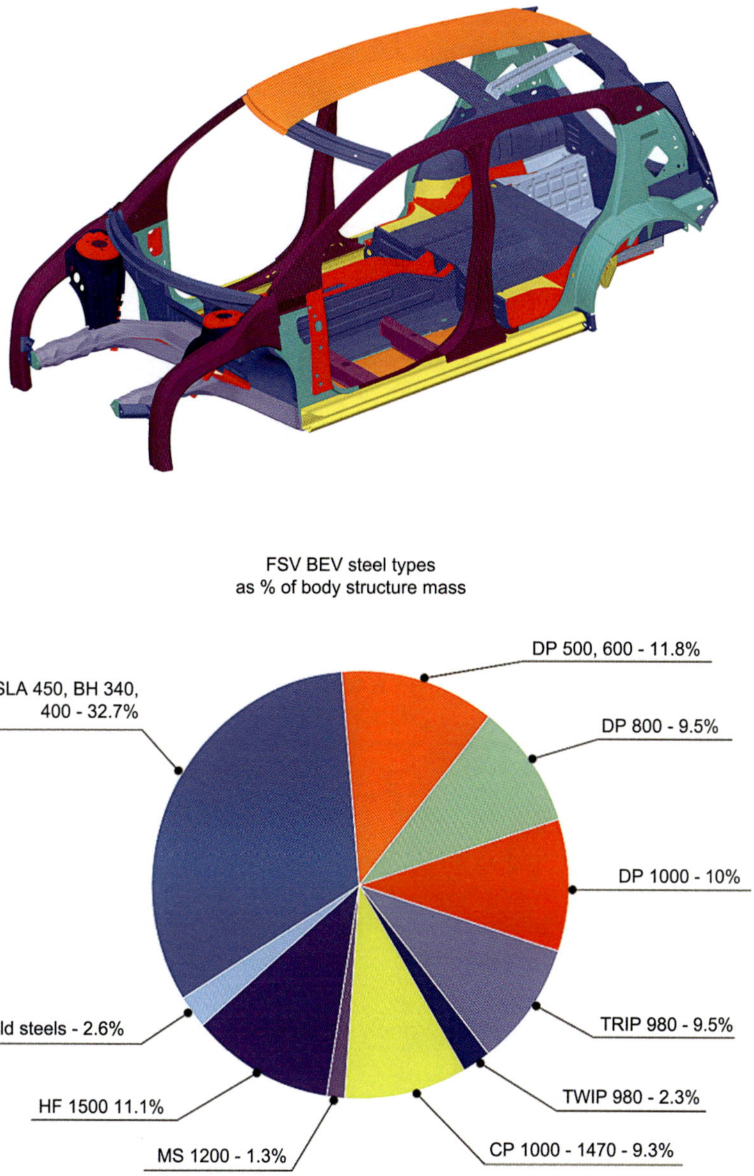

Fig. 11.15 Color-coded material breakdown for the body structure of the Future Steel Vehicle (FSV), battery electric vehicle (BEV). Source: Ref 11.9

11.2 Nonautomotive Applications

Steel has the widest range of applications of any material because of its diversity and versatility of types, properties, shapes, and forms. Applications of steels are found in automotive, construction, aerospace, power generation, ship building, railway, oil extraction, and manufacturing. Figure 11.16 shows how steel touches all aspects of daily life. In 2011, 1518 million tons of steel were produced. The construction industry uses 51% of the produced steel, followed by the automotive industry at 16%.

Advanced high-strength steels, however, have only been used in the automotive industry and have not yet penetrated other industries or markets. But, the benefits of weight reduction can be extended to ship building, construction, military, aircraft applications, and the construction industry.

Examples of where AHSS can be used to improve performance and reduce cost are:

- *Off-shore platforms:* structures and plates for decking
- *Bridges:* structure, railing, and deck plates
- *Military:* lighter tanks, armored vehicles, submarines, aircraft carriers
- *Ship building:* bigger ships and supertankers
- *Aerospace:* high strength and reduced weight are essential

The American Iron and Steel Institute, Ford Motor Company, and the U.S. Army are collaborating on a project to design the next-generation tactical vehicle. The Improved Materials and Powertrain Architecture for 21st Century Trucks (IMPACT) project makes optimal use of AHSS to develop lighter tactical vehicles while maintaining performance, cost, and safety. For all of the examples above, and for many other applications, replacing low-strength steel with higher strength leads to saving costs by using less steel.

Steel packaging protects contents from water, oxygen, and light. It offers maximum convenience, and is 100% recyclable. Steel is an excellent packaging material due to its strength, formability, and durability. It enables high-speed filling operations, resulting in excellent cost efficiencies. Through innovation, the weight of steel packaging has been reduced, and today's solutions include easy-open ends on a multitude of packaging shapes and sizes.

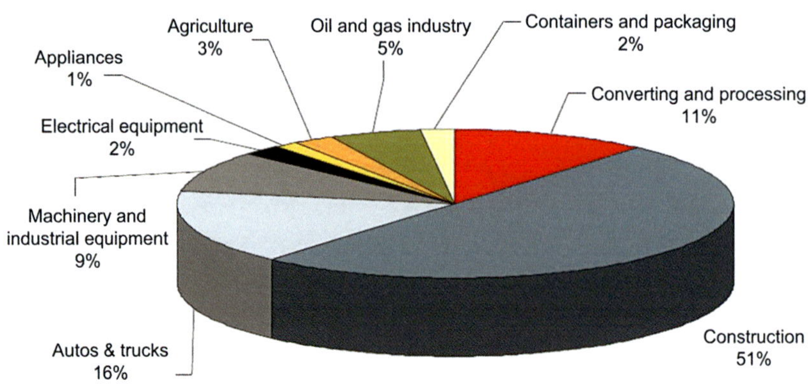

Fig. 11.16 Applications of steel. Source: Ref 11.10

Packaging steels are currently made from low-carbon cold rolled steel strip with gages (thicknesses) ranging from 0.13 to 0.50 mm (0.01 to 0.02 in.). They can be produced with different surface finishes and are available with a variety of metallic and organic coatings. Replacing low-carbon packaging steels with higher strength steels allow for downgaging and result in weight and material savings.

11.3 Utilization and Trend

Steels have been and will continue to dominate the metallic materials available for vehicle manufacturing because of their excellent properties and low cost. For the finished vehicle curb weight, Ducker Worldwide predicts that in the future, the mix of materials in North American light vehicles will shift from high-density and moderate-strength materials to lower-density and higher-strength materials. This shift will occur mainly by replacing mild, BH, and HSLA steels with AHSS, polymers, aluminum, and magnesium alloys. Figure 11.17 shows historical and predicted material share of finished vehicle curb weight from 2008 to 2025. The figure shows that by 2025, flat rolled AHSS and aluminum alloys will increase their share in finished vehicle curb weight while mild, BH, and HSLA steels will decrease that share by a proportional amount. A more detailed chart of the material share in the North American light vehicle curb weight is shown in Fig. 11.18. The percentage of flat rolled AHSS increased from 3% in 2008 to 5% in 2012 and is predicted to reach 7% in 2015. On the other hand, the percentage of mild steel de-

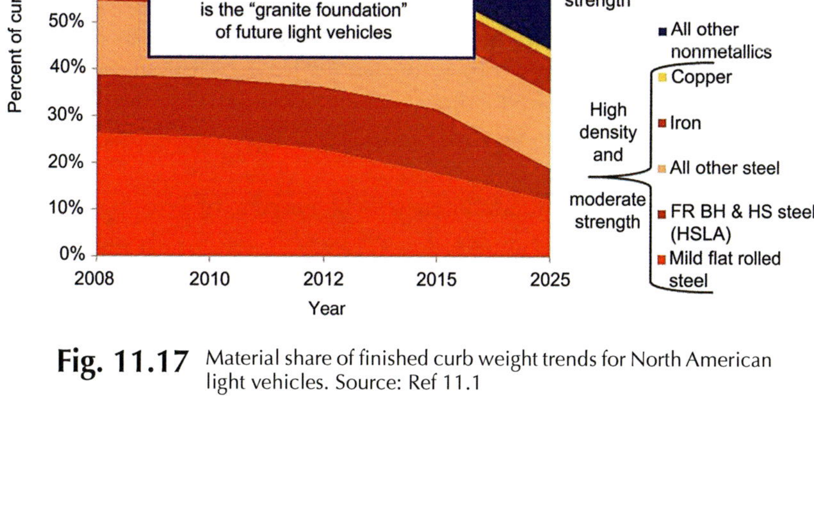

Fig. 11.17 Material share of finished curb weight trends for North American light vehicles. Source: Ref 11.1

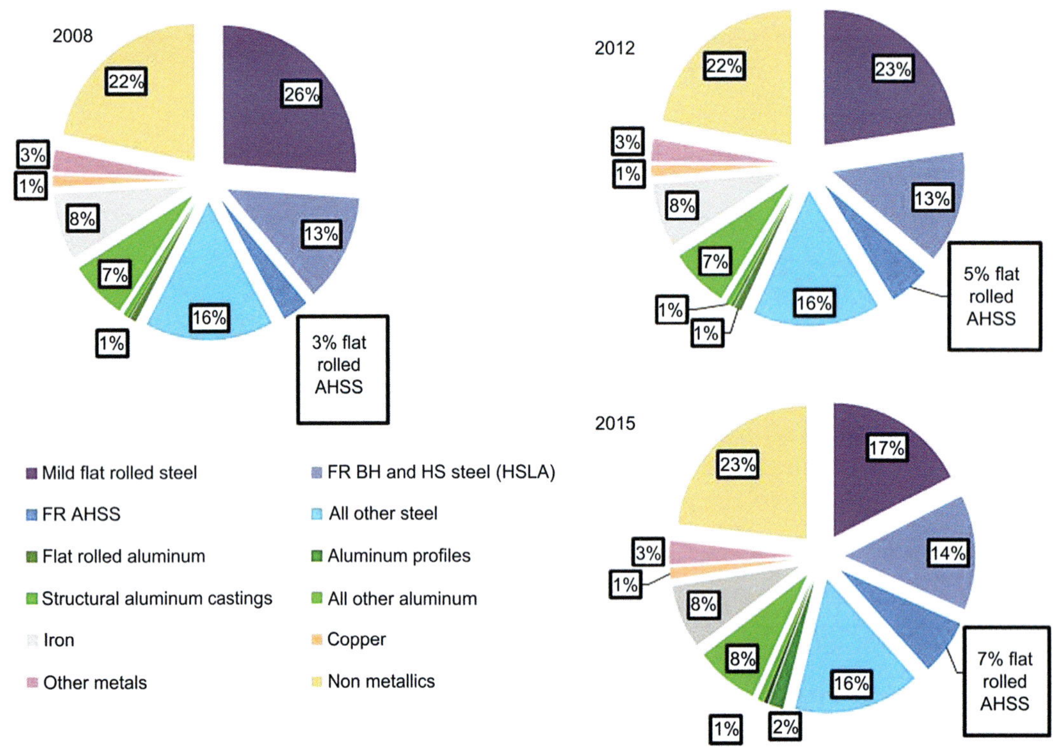

Fig. 11.18 Material share percent of finished curb weight for North American light vehicles. Source: Ref 11.1

creased from 26% in 2008 to 23% in 2012 and is predicted to fall to 17% in 2015.

Figure 11.19 shows the increasing trend of using flat rolled AHSS in vehicles. The amount of flat rolled AHSS used in 2008 was 52 kg (115 lb) per vehicle. In 2010, this amount increased by an additional 34 kg (75 lb), and in 2015 it is predicted to increase by an additional 65 kg (144 lb) per vehicle.

The trend toward increasing AHSS in vehicle bodies and closures applies equally to all global car manufacturers. Figure 11.20 shows historical and predicted metallic material mix for North American light vehicle body, bumper, and closure. The figure shows that by 2025, AHSS and aluminum alloys will increase their share in vehicle body, bumper, and closure while mild, BH, and HSLA steels will decrease that share by a proportional amount.

Figure 11.21 shows an estimated weight percent of North American light vehicle total body, bumper, and closure in 2025. Advanced high-strength steels are predicted to increase their weight percent from 17% in 2012 to 41% in 2025.

Ducker Worldwide provided a report on steel usage trends and the impact of Corporate Average Fuel Economy (CAFE) regulations on the fu-

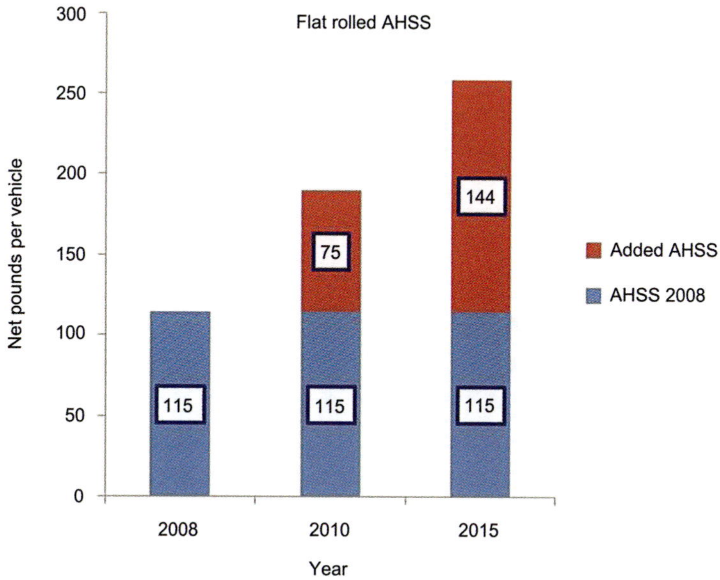

Fig. 11.19 Increasing trend of using flat rolled AHSS in vehicles. Source: Ref 11.1

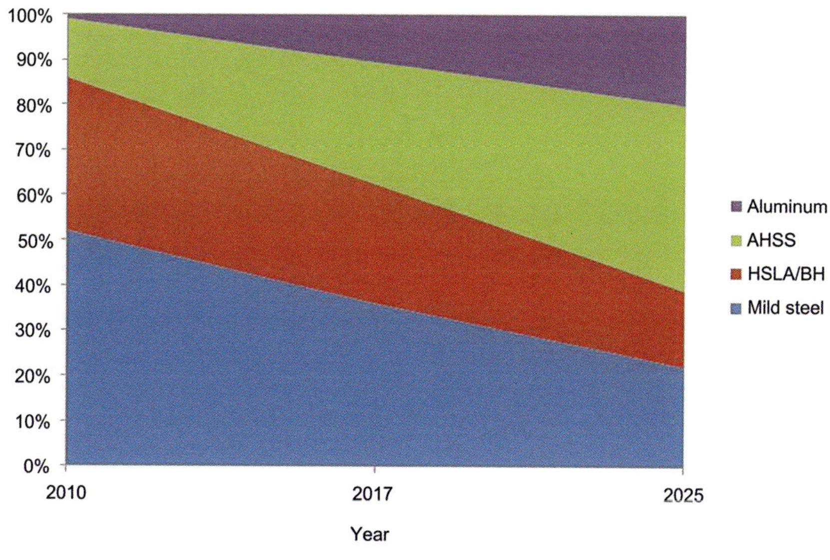

Fig. 11.20 Metallic material trends of North American light vehicle body, bumper, and closure. Source: Ref 11.1

ture of AHSS content in vehicles. In 2006, AHSS content in North American light vehicles was 37 kg (81 lb); this number increased to 71 kg (156 lb) in 2010 and is projected to increase to 157 kg (346 lb) by 2020. Figure 11.22 shows historical and conservatively projected AHSS content in vehicles for the period 2006 to 2020. New innovations in manufacturing

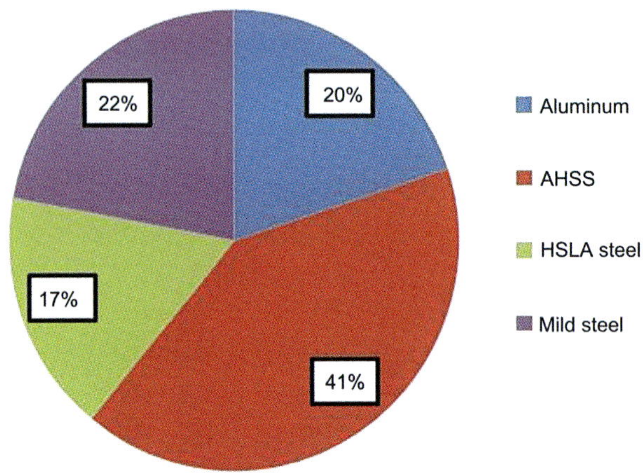

Fig. 11.21 Metallic material weight percent share of vehicle body, bumper, and closure predicted for 2025. Source: Ref 11.1

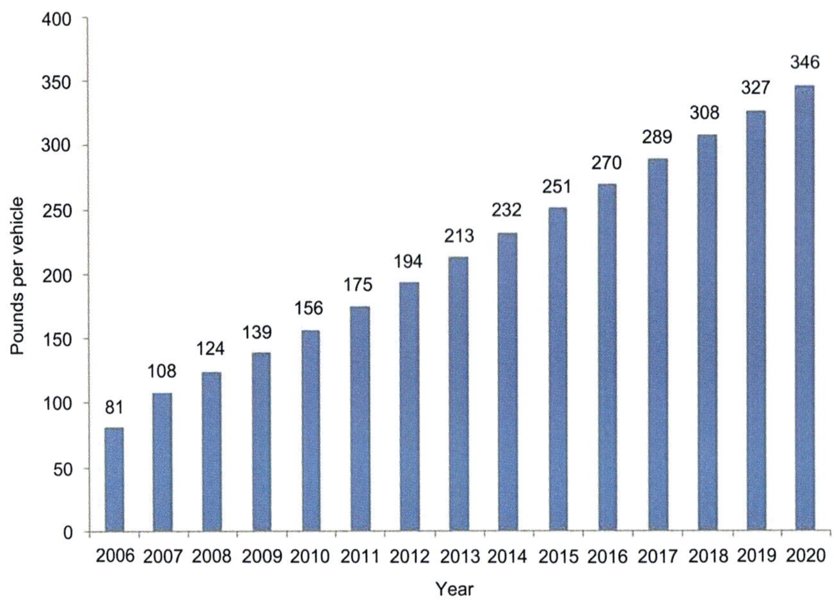

Fig. 11.22 Historical and projected AHSS content in North American light vehicles (trend line only). Source: Ref 11.1

processes are expected to contribute positively to the projected growth of AHSS (Ref 11.1).

The FutureSteelVehicle (FSV) program developed a concept vehicle with a proposed AHSS body structure for the 2011 to 2020 model years. The FSV body structure achieved 35% mass savings, mainly by using 20 new AHSS grades; some of them are not yet in commercial production. The FSV paved the way and set the stage for more use and further development of AHSS. Automakers are convinced that meeting their goals of improved vehicle performance at no cost penalty can only be achieved by increasing the AHSS content in their vehicles. In most cases, the increase in AHSS content comes at the expense of HSLA steels that they replace.

REFERENCES

11.1 R. Schultz, Ducker Worldwide, personal communication, Nov 2012

11.2 E. De Moor, "Assessment of Quenching and Partitioning as a Fundamentally New Way of Producing Advanced High Strength Martensitic Steel Grades with Improved Ductility," Ph.D. dissertation, Gent University, 2009

11.3 Uddeholm Tooling and SSAB Swedish Steel, "Tooling Solutions for Advanced High Strength Steels," presented at Uddeholm Swedish Rally, 2005

11.4 M. Mehrkens, "Advanced High Strength Steel Technology in the Porsche Cayenne," Porsche Engineering presentation at Great Designs in Steel Seminar, 2004 (Livonia, MI), AUTOSTEEL

11.5 J.P. McGuire, "Advanced Stamping for High Strength Steel," Chrysler Group presentation at Great Designs in Steel Seminar, May 16, 2012 (Livonia, MI), AUTOSTEEL

11.6 W.J. Parsons, "Lightweighting the 2013 Cadillac ATS Body Structure," General Motors presentation at Great Designs in Steel Seminar, May 16, 2012 (Livonia, MI), AUTOSTEEL

11.7 S. Morgans, "AHSS Technologies in the 2013 Ford Fusion," Ford Motor Company presentation at Great Designs in Steel Seminar, May 16, 2012 (Livonia, MI), AUTOSTEEL

11.8 K. Mclean, "Chevrolet Sonic," General Motors presentation at Great Designs in Steel Seminar, May 16, 2012 (Livonia, MI), AUTOSTEEL

11.9 WorldAutoSteel, "FutureSteelVehicle (FSV)," Final Engineering Report, Steel Market Development Institute, May 2011, www.worldautosteel.org

11.10 E. Basson, "The Steel Industry in a Sustainable Society," SBB Green Steel Strategies Conference, April 19, 2012 (Berlin), Platts

CHAPTER **12**

Consequences of Using Advanced High-Strength Steels

THE WIDESPREAD USE of advanced high-strength steels (AHSS) in the automotive industry is hindered by many difficulties including limited formability, reduced weldability, increased springback, elevated press tonnage, and accelerated die wear. Even with the expected increase of AHSS in future vehicles and other applications, concerns about the consequences for using AHSS remain. Modifications to the current forming system such as higher-tonnage presses, appropriate die materials, effective tool coatings, efficient lubricants, and higher binder forces should be considered. The implications of using thinner gages on the elastic stiffness and corrosion resistance need to be deliberated or discussed. Because the modulus of elasticity is the same for all steels, downgaging reduces component stiffness. Also, because corrosion resistance of AHSS (except for austenitic stainless steel) is similar to that of mild steel, thinner gages rust faster under the same corrosion conditions.

Problems arising from using AHSS can be circumvented by innovative solutions. This chapter describes the nature of the problems and discusses potential remedies to minimize the adverse effects that may limit the adoption of AHSS in the automotive industry.

12.1 Press Requirements

The energy required to plastically deform a metal depends on its strength and ductility. Uniaxial tensile testing is used to obtain the stress-strain curve of a material. The area under the curve determines the amount of energy required to plastically deform the tensile sample up to a specified strain. The shape of the curve is determined by the work-hardening rate of the material; its tensile strength level increases with increase in that rate. Increased work hardening creates higher loads on press, tools, lubricant, and other load points.

In a forming operation, the flow stress determines the magnitude of the forming load while the area under the stress-strain curve determines the energy required for forming. This energy is simply the load multiplied by the distance travelled by the press ram. Figure 12.1 shows the extra stress, more tonnage, and the extra energy needed to form DP 600 steel over the lower-strength HSLA 450 steel.

Although AHSS have higher strength and require increased tonnage to form a part, they are reduced in thickness, which counteracts the load increase. Thickness reduction, however, increases the tendency for the sheet

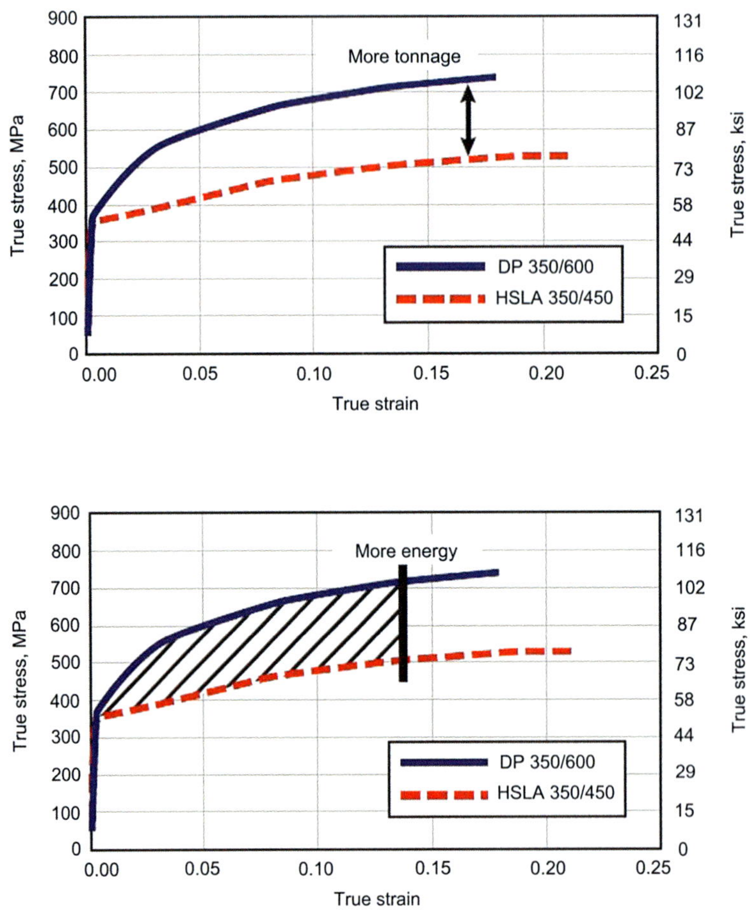

Fig. 12.1 Higher-strength steel increases load and energy required for forming. Source: Ref 12.1

metal to wrinkle and this requires higher binder force to suppress the wrinkles. In the balance, higher loads will be required to form AHSS.

For simple geometries, press load is directly proportional to the tensile strength of the steel. Figure 12.2 shows the measured press loads for the production of a hat-section cross member made of HSLA 350/450 and DP 300/500 steels of the same thickness.

A computer simulation to determine the press energy required to stamp a part was performed. Three different grades of steel—mild; high-strength, low alloy (HSLA); and AHSS—were used in the simulation. Results, showing the press energy required to continue punch travel until the part was completely formed, are presented in Fig. 12.3. The press energy re-

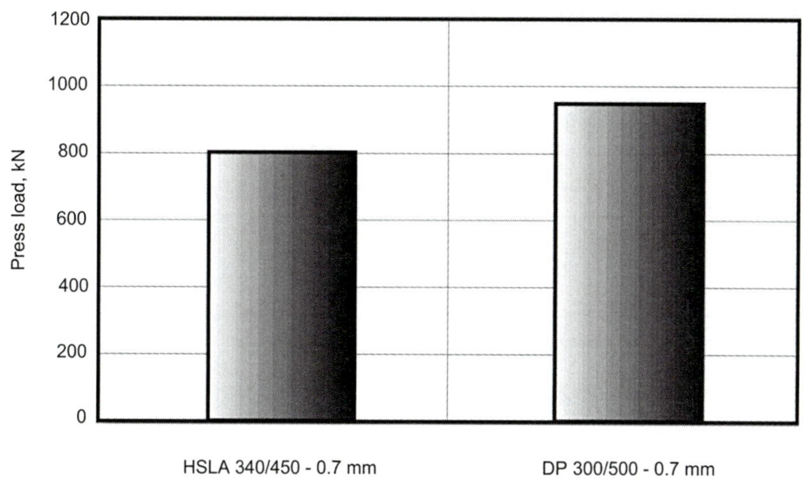

Fig. 12.2 Measured press load for a hat-section cross member. Source: Ref 12.2

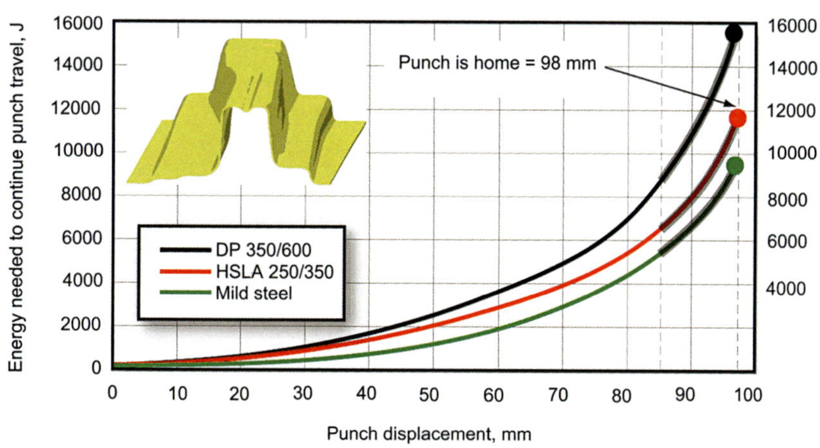

Fig. 12.3 Energy needed to form a part from three different grades of steel. Source: Ref 12.2

quired to form the part was found to increase with increasing the tensile strength of the steel. Forming the part using DP 600 steel required approximately 15,500 J, while forming the same part using mild steel required only 9,500 J.

Estimating press force and press energy required to make a part is important for stamping engineers because press capacity selection and availability must be determined at an early stage to carry out a successful stamping operation. Incorrect estimation of press tonnage capacity may result in damage of the press and production delays.

12.2 Springback and Residual Stress

Springback is the change in the shape of the part from that of the die after forming and unloading. It is, therefore, the deviation of the final product from the desired geometry. Springback occurs in drawing operations because they contain stretching and drawing components. The drawing component is responsible for springback. During bending the outer surface of the bent sheet will be in tension and the inner surface will be in compression. As bending progresses, the neutral axis shifts from the center of the cross section of the sheet metal toward the inner surface (compression side). When the load is removed, springback occurs due to elastic recovery. Springback is practically negligible in stretch forming because stretching produces tensile deformation in both the outer and the inner surfaces of the sheet thickness. The key to removing springback is to change the state of stress on the outside and inside surfaces of the sheet metal from tension/compression to tension/tension. This is usually done by superimposing a small stretch at the end of the forming cycle.

Advanced high-strength steels have large amounts of springback compared to mild and HSLA steels. For example, aluminum-killed drawing quality (AKDQ) steels exhibit springback of 2 to 3%; HSLA exhibits springback of 8 to 12%; and AHSS exhibits springback of 16 to 24%. Springback allowance must, therefore, be increased as material strength increases; a three-degree allowance is needed for mild steels, while a six-degree or more allowance is necessary for HSS and AHSS.

Springback is of great concern to sheet metal forming tool designers, because it can cause serious problems in the assembly of parts and can lead to expensive modifications of the forming tools. Springback and side-wall curl after unloading from stamping dies are very important technical barriers to the widespread use of high-strength metals, especially AHSS. Figure 12.4 shows the profile of a formed U-channel before and after springback.

Springback is explained by the phenomenon of elastic recovery that occurs after unloading a plastically deformed tensile sample of steel (Fig. 12.5). The figure shows a tensile sample that is plastically deformed to a stress-strain level defined by point *A*. The total strain under load at that

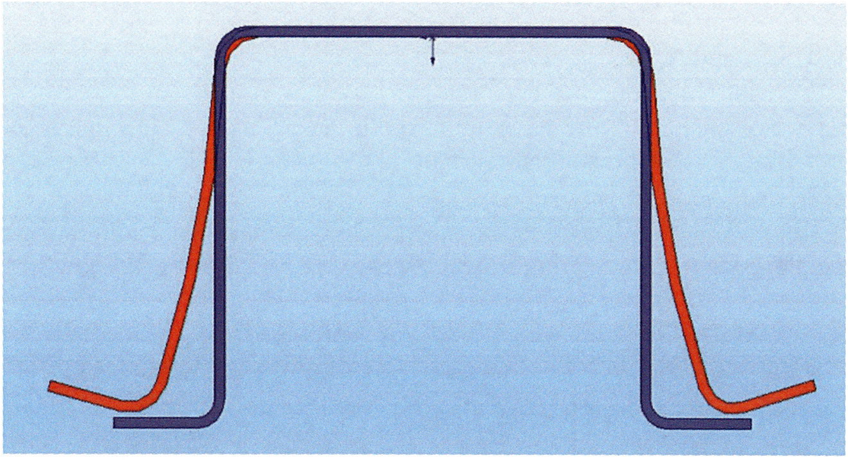

Fig. 12.4 U-channel profiles before (blue) and after (red) springback

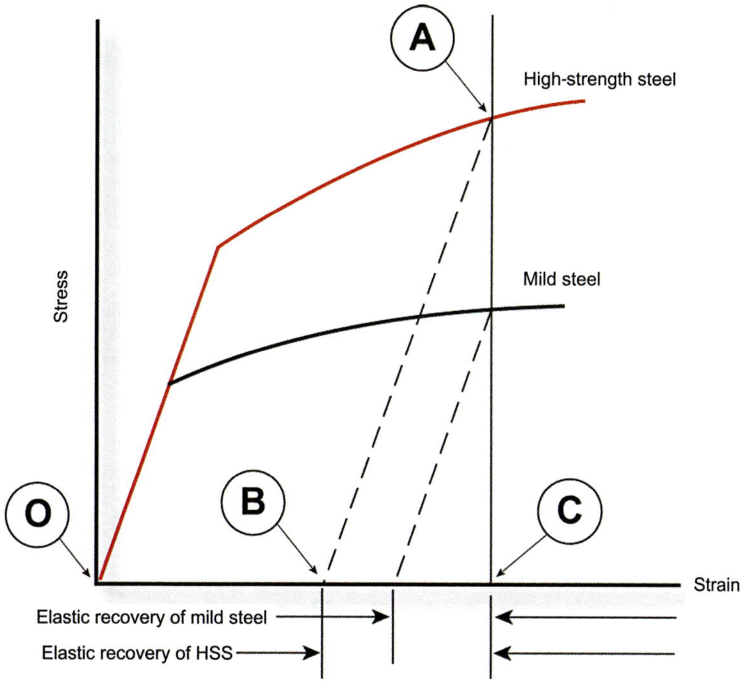

Fig. 12.5 Elastic recovery (springback) in steel. Source: Ref 12.2

point is represented by *C*. If the load is removed, the sample recovers the elastic portion of the strain and springs back to point *B*. The amount of elastic recovery or springback increases with increase in the flow strength of steel.

The effect of yield strength on springback for three steels is illustrated in Fig. 12.6. Springback increases with increase in yield strength of the steel. Martensitic AHSS exhibits the highest level of springback compared to HSLA and AKDQ steels.

U-channel forming has been routinely used to evaluate springback behavior of metals. Figure 12.7 shows tooling used in the U-channel draw test. The test has three variables: tool radii (punch and die), tool gap, and draw bead penetration.

Figure 12.8 shows the two parameters, angular change and curl, that are employed to measure springback in U-channels. The figure also shows the

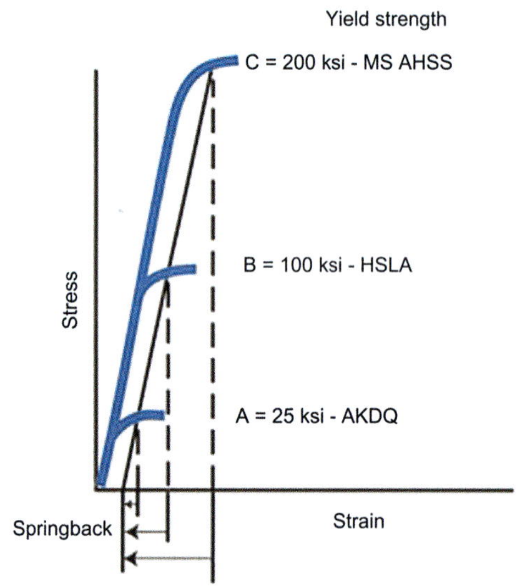

Fig. 12.6 Effect of yield strength on springback in steel. Source: Ref 12.2

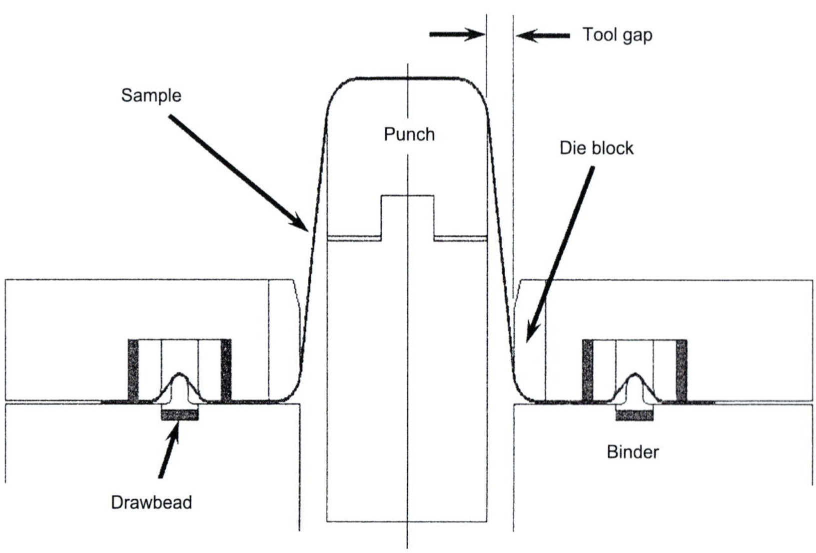

Fig. 12.7 Tooling used for the springback U-channel draw test. Source: Ref 12.3

springback profiles for HSLA 350/450 and DP 350/600 steels. Although the two steels have the same yield strength of 350 MPa (50 ksi) and are formed under the same conditions, dual-phase (DP) has much greater springback than HSLA because U-channels made from DP have greater yield strength after forming. The increase in strength of DP after forming results from its greater work-hardening rate compared to HSLA steel.

Due to the high strength of AHSS, compared to mild steel and conventional HSS, a larger amount of springback develops in the formed high-strength U-channel. Figure 12.9 displays springback profiles of interstitial-free (IF), HSLA 350, and DP 600 steels.

Studies of springback in U-channel forming of HSLA 350, DP 600, and TRIP 700 steels showed that springback depends on the strength level of the stamped part. Steels with a high *n*-value, such as DP 600 and TRIP 700, showed increased strength after forming that resulted in higher springback values, compared to HSLA 350 steel with lower *n*-value.

Other studies showed that reduction of sheet thickness or increase in die gap increases springback. It was found that the most influential factors

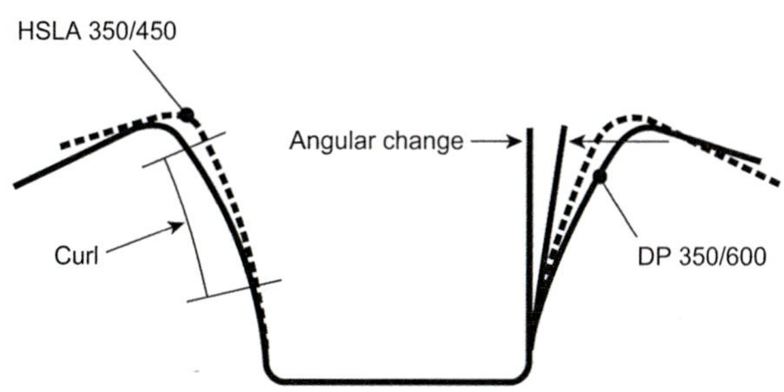

Fig. 12.8 Angular change and curl. Springback parameters in U-channels. Source: Ref 12.2

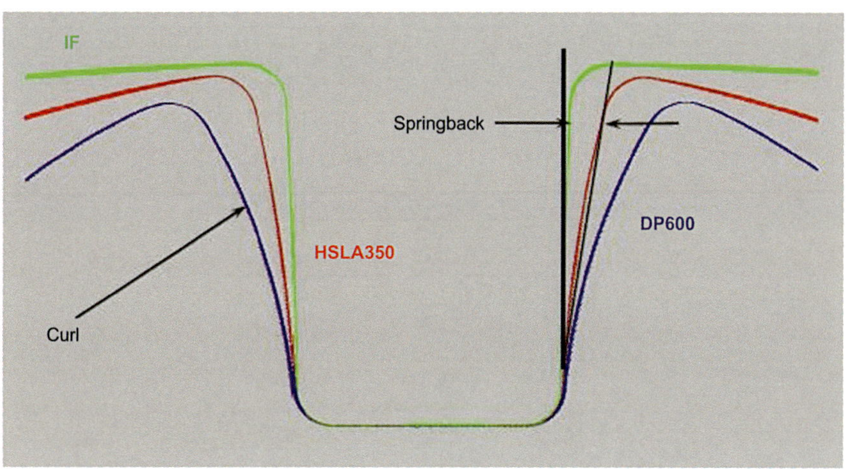

Fig. 12.9 Angular change and sidewall curl for three steels. Source: Ref 12.3

controlling springback and curl were the tooling radius and draw bead penetration. Figure 12.10 shows that draw bead penetration and punch radius have more effect on springback than tool gap.

Higher-than-normal binder pressure and press tonnage is necessary with HSS and AHSS to maintain process control and to minimize buckles on the binder. Double-action press or hydraulic press cushion may be required to achieve the necessary binder forces required to control sheet metal flow into the die cavity. Air cushions or nitrogen cylinders may not provide the required force for setting the draw beads or maintaining binder closure. Controlling metal flow of AHSS can increase the binder tonnage requirements by 20% compared to conventional HSLA steel grades.

The effect of draw bead penetration, tool gap, and die radius on curl radius of HSLA 450 and DP 600 steels is shown in Fig. 12.11. Draw bead penetration and die radius have more effect on curl radius than tool gap.

In general, springback decreases with increase in die radius because of diminishing bending forces in larger radii. Reducing bending forces reduces work hardening and lowers springback angle and sidewall curl (Ref 12.4). Punch radii must be fairly sharp, approximate sheet thickness ~1*t* for HSS and greater than sheet thickness >1*t* for AHSS. Punch radii, however, should be kept as small as practical to reduce springback in the sidewalls.

Springback can be reduced through the proper design of the forming tool and die to generate the desired shape. It is also reduced by controlling the binder forces that act on the drawn sheet. Springback may necessitate using variable binder force control with the ability to achieve at least 25%

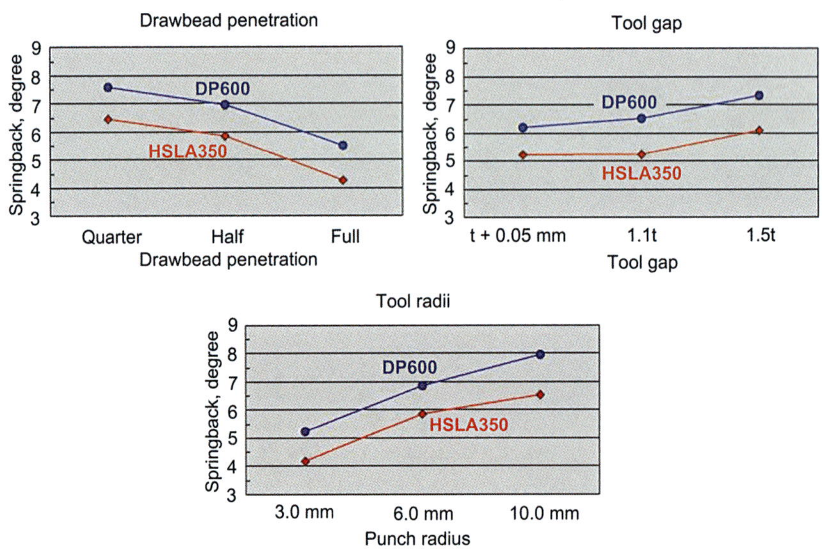

Fig. 12.10 Effect of draw bead penetration, tool gap, and punch radius on springback of steel. Source: Ref 12.3

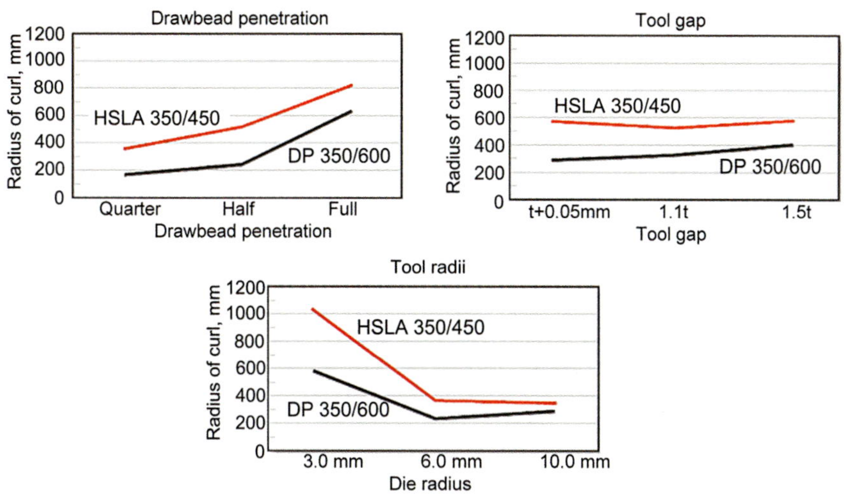

Fig. 12.11 Effect of draw bead penetration, tool gap, and die radius on curl radius of steel. Source: Ref 12.2

higher total binder force. In body panels, springback can be reduced by compensation through the following countermeasures:

- Overforming, so that on unloading, the part springs back to the intended shape
- Bottoming or shape set to induce a minimum of 2% stretch in the part near the bottom of the press stroke
- Stretching by using higher binder forces

Traditionally, press shops and process engineers have developed guidelines, based on experience, for necessary tooling and process modifications needed to control springback. Some of the guidelines for reducing springback in stamped components are:

- Avoiding right or acute angles
- Using larger open-wall angles to allow overbending
- Avoiding large transition radii between two walls
- Reducing die radii to less than $2t$
- Using stiffeners, ribs, and step flanges where possible

An ASTM test standard titled "Standard Test Method for Evaluating Springback of Sheet Metal Using the Demeri Split Ring Test" was issued, under the designation E2492-07, to evaluate the springback behavior of metals in a test that simulates a stretch-draw forming process. The test method can also be used to calibrate computer simulation codes by selecting appropriate control parameters to achieve satisfactory correlation between simulation and test results.

The test measures springback behavior of sheet metal by comparing the diameter of a ring extracted from the wall of a flat-bottom cup and the diameter of the same ring after splitting to release residual stresses and produce springback. Examples of the diameter of the rings before and after splitting are shown in Fig. 12.12 for three sheet metals: drawing quality steel (DQSK), bake-hardenable steel (BH 210), and aluminum alloy (AA 6022).

The magnitude of springback is a convolution of elastic modulus, flow stress, thickness, and amount and type of cold work introduced by the forming process. Because cup drawing contains features of many forming operations such as stretching, drawing, and bending, the amount of springback measured by the test is indicative of the behavior of metal in many stamping operations.

Many computer simulation programs have been developed to predict springback in complex stampings made from HSS. Accurate springback prediction allows springback compensation to be included in the initial design stage thus reducing tool development time. Springback simulation predictions were not completely successful because of the complexities encountered in modeling material behavior, friction properties, and draw bead restraining forces.

Residual stresses in metals are caused by a number of processes such as inhomogeneous deformation, phase changes, and temperature gradients. For example, rapid cooling introduces residual stresses. This discussion focuses on residual stresses caused by plastic deformation. Springback is a manifestation of residual stresses in metals. Residual stresses are in-

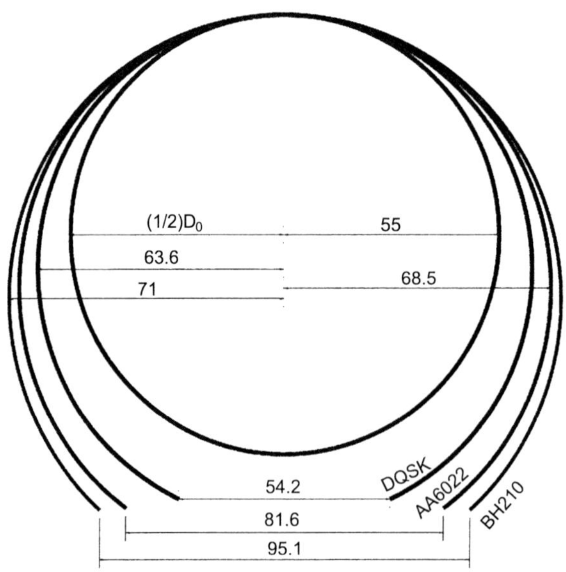

Fig. 12.12 Split rings measuring springback for three metals. Source: Ref 12.5

duced in parts during their manufacture. Such stresses do not disappear when parts relax and all external constraints are removed. Some stamped parts produce significant residual stresses that create large springback when allowed to relax. Springback in a stamped part results from essentially stretch-bend forming. The stretch portion varies with the amount of the clamp force applied by the binder.

An effective way for reducing springback in stampings, especially U-channels and rails, is to use a die process where some form of "post-stretch" or "shape-set" is applied (Ref 12.6). An effective shape-set process involves the activation of a set of lock-beads, in the binder area, just before the end of the punch stroke. This condition produces pure stretch of approximately 2% in the formed channel and causes an appreciable reduction of springback. Figure 12.13 shows a sketch of the binder force profile in a shape-set process.

Tensile residual stresses may lead to stress cracking and, in some alloys, to stress-corrosion cracking that may ultimately lead to premature product failure. The residual stress potential of metals can be evaluated by using the ASTM standard test as discussed earlier in this section.

12.3 Binders and Draw Beads

Binders, sometimes called blankholders, are essential parts of metal forming systems. They control the amount of metal that flows into the die cavity by providing a restraining force on the periphery of the metal blank prior to forming. Figure 12.14 shows a schematic of a simple sheet metal forming system. The basic components are a punch, a die, and a set of binders that may, or may not, include draw beads. The punch draws the blank to form the shape, while the binder controls the flow of metal into the die cavity.

"Draw beads" control metal flow into the die cavity by a combination of friction and bending forces. "Lock beads" completely prevent metal

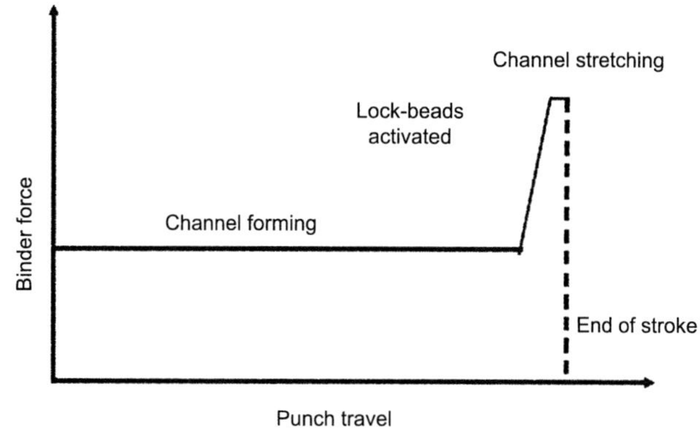

Fig. 12.13 Binder force profile in a shape-set process to reduce springback in U-channels

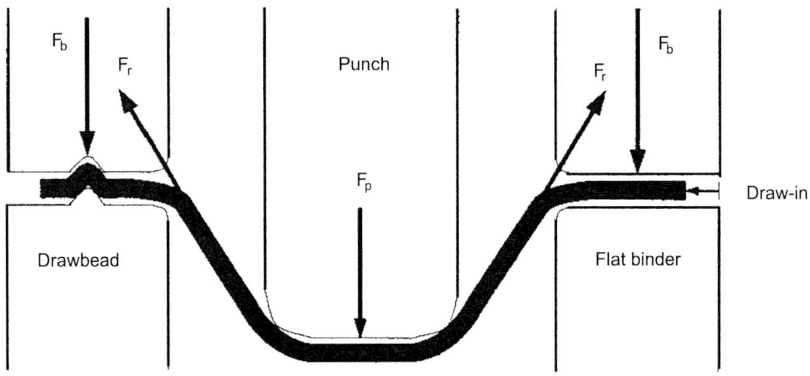

Fig. 12.14 Sheet metal forming system showing the location of the binder. F_b, binder force; F_p, punch force; F_r, restraining force. Source: Ref 12.7

flow into the die cavity by designing rectangular rather than round beads to completely lock the metal. This produces a stress condition in the sheet metal comparable to stretch forming.

The binder, with or without draw beads, is a critical part of the stamping system. While flat binders rely entirely on friction to generate binder restraining forces at the edge of the sheet metal blank, draw beads rely on bending and unbending of the sheet metal and, to a lesser extent, on friction to regulate the amount of the blank material that flows into the die cavity. The bending force is inversely proportional with the radius of curvature; whereas the friction force is directly proportional with the coefficient of friction and the force on the binder. Flat binders are more sensitive to variations in friction conditions, die misalignment, and blank thickness variations.

Draw bead action occurs in two stages (Ref 12.8):

1. *Binder closes:* The draw bead protrusion pushes the metal into a matching groove on the other binder face to form a bead. The depth of movement of the protrusion determines the amount of bead penetration.
2. *Sheet drawn:* Sheet is pulled over the bead and into the die cavity.

The restraining force of a draw bead increases with increase in bead penetration because the sheet metal is subjected to more bending. The bead penetration is determined by the magnitude of the normal force on the binder.

In traditional stamping, the binder force remains constant throughout the forming cycle. The magnitude of the force is determined based on experience or trial and error. If the restraining force is too low, too much material flows into the die causing wrinkling in the part. If, however, the restraining force is too high, not enough material flows into the die causing splitting in the part. The binder determines the amount of restraining force on the sheet metal and this also affects its springback. Figure 12.15 shows the type of forming problems caused by inadequate binder forces.

Draw beads provide more control of metal flow into the die cavity because they can produce higher restraining forces than flat binders. Draw beads for HSS and AHSS should not extend around corners of the draw die because this will result in locking out metal flow and causes splitting in the corners of the part. Draw beads should "run out" at the tangent of the corner radii to minimize metal compression in those corners, as shown in Fig. 12.16 (Ref 12.4).

12.4 Tool Material and Die Wear

Tool or die wear is caused by friction resulting from contact and relative motion between die surface and sheet metal. It reveals itself as damage to the surface of the die causing a progressive loss of material. Tool

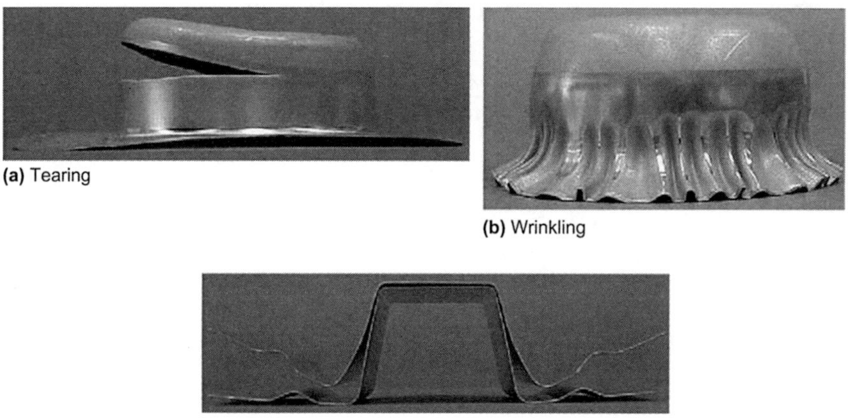

(a) Tearing

(b) Wrinkling

(c) Springback

Fig. 12.15 Forming problems caused by inadequate binder control. Source: Ref 12.7

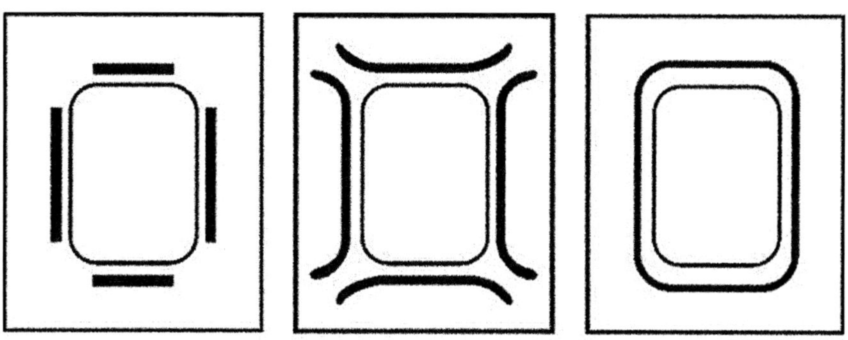

(a) (b) (c)

Fig. 12.16 Draw bead types: (a) conventional draw bead, (b) runout draw bead for high-strength steel, and (c) lock draw bead for stretch forming. Source: Ref 12.4

wear is affected by a wide range of parameters including strength of material, contact pressure, surface finish of contacting material, sliding velocity, type of motion, temperature, die coating, and lubrication. Increasing the yield strength and hardness of the material results in increased die wear, which leads to surface quality and dimensional accuracy problems for the stamped part. In addition, die wear requires the frequent replacement of the dies and this leads to downtime and production losses.

Forming AHSS requires high forming loads and high contact pressures thereby causing elevated frictional forces that may lead to significant die wear. Local contact loads can be very high for part features with small radii. Additionally, AHSS require elevated binder pressures for stamping to maintain process control and to minimize buckles in the binder area. The higher compressive forces on the binder will require better die material, surface treatments, and lubricants. Tools must have good wear resistance for long life operation and production reliability and also must have sufficient toughness to avoid fracturing.

To prevent excessive wear in forming AHSS, new die materials and better die coatings have been developed. New die materials include specially developed wrought and cast tool steels and powder metallurgy tool steels. These materials have been developed to extend the hardness available while maintaining toughness.

Hard material coatings and nitriding have been used to improve the tribological properties of die surfaces. The effect of surface treatment on tool wear is demonstrated by drawing a hat section over different surface treatments of a die surface (Fig. 12.17). Results indicate that hardened

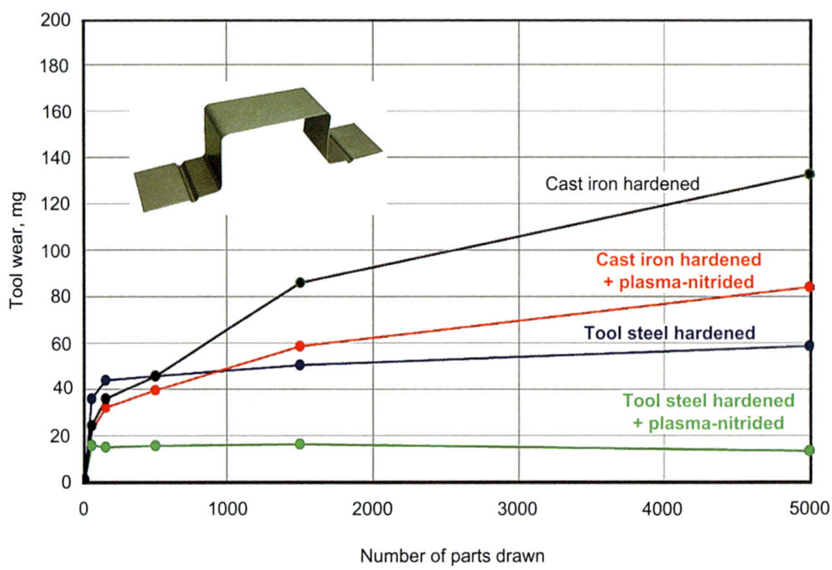

Fig. 12.17 Effect of die surface treatment on tool wear. Source: Ref 12.2

tool steel with plasma-nitrided coating has the lowest tool wear for any number of parts drawn.

Another example demonstrating the effect of tool material and coating on tool life, as measured by the number of good stamped parts made from DP 980, is shown in Fig. 12.18. It is clear that both good tool material and high-performance die surface coating are required for increased tool life.

Coatings are used to reduce friction between sheet metal and die and lower the shear forces that lead to wear. They are applied onto the tool substrate in micron-level thicknesses. Coating processes include thermal diffusion (TD), chemical vapor deposition (CVD), physical vapor deposition (PVD), plasma-assisted chemical vapor deposition (PACVD), and nitriding plus physical vapor deposition. Dual treatment combinations of nitriding and PVD coatings are widely used to provide a harder substrate to the hard but brittle PVD coatings.

Coating processes can either be performed at high temperature, as in TD and CVD, or relatively low temperature, as in PVD (200 to 550 °C, or 390 to 1020 °F, under vacuum). A new CVD coating was developed to reduce abrasive wear during stamping of high-strength materials. It employs selective combinations of individual coating layers of titanium carbide (TiC), titanium carbonitride (TiCN), and titanium nitride (TiN). Current forming dies coated with thermal diffusion of vanadium carbide or coatings (AlTiCr+MoS2, TiN) performed relatively better than conventional tool material, but their wear resistance and durability were not satisfactory. Multilayer structures such as TiAlCN and AlTiCrN+CN are found to work better (Ref 12.4, 12.9).

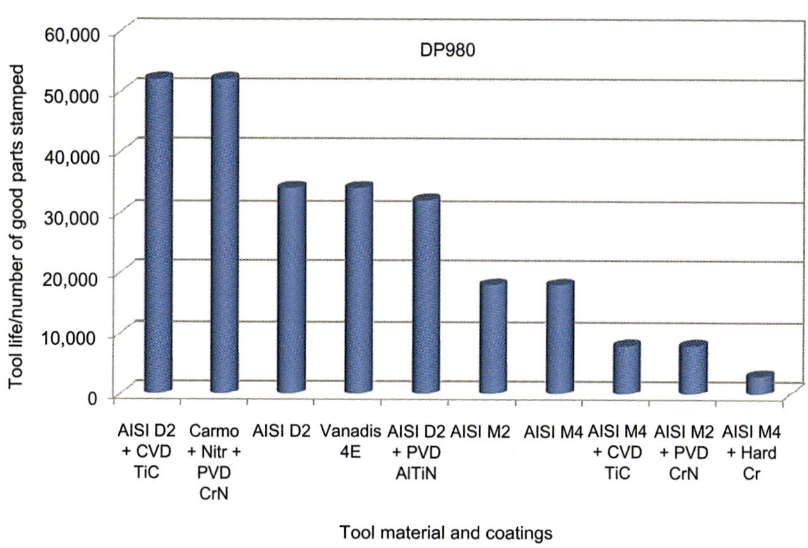

Fig. 12.18 Effect of tool material and die surface coating on tool life. Source: Ref 12.1

Chemical vapor deposition coating costs double that of PVD because the high-temperature processing of CVD includes pre- and post-treatments. The high cost of high-performance tool steels and coatings for AHSS are usually offset by reduced maintenance and increased lifetime and performance.

12.5 Hot Forming

Formability and springback are common problems in conventional stamping of parts made from HSS. Advanced high-strength steels have lower formability and higher springback compared to conventional steels and this has been a major obstacle for the application of those materials to complex automotive parts. For example, some martensitic steels have a total elongation of only 5%, and this has limited stretchability that restricts its application to simple stampings. An advanced forming technique, known as hot forming, has been used to overcome such problems. Figure 12.19 shows a schematic of the hot forming process.

In hot forming processes, a part is either heated to a high temperature above 900 to 950 °C (1650 to 1740 °F) prior to forming, or it is partially formed and then heated prior to final forming to the required shape. The formed part is then quenched in the stamping tool at about 50–150 °C/s (122–302 °F/s). Quenching the formed part causes a significant increase in its strength and dimensional accuracy. The increase in strength is due to phase transformation, on quenching, of austenite to harder phases of bainite or martensite. The increase in formability at high temperature is due to the softness and ductility of austenite. The increase in dimensional accuracy results from the absence of residual stresses due to forming at elevated temperature. Also, plastic deformation at high temperature occurs under much lower yield stress; springback will be minimal because elastic recovery (springback) decreases with the decrease in the yield strength of the steel. When springback decreases, dimensional accuracy of the part should increase.

Some of the drawbacks of hot forming are:

- Extra expenses related to heating the sheet blank
- Special tooling for in-die part quenching

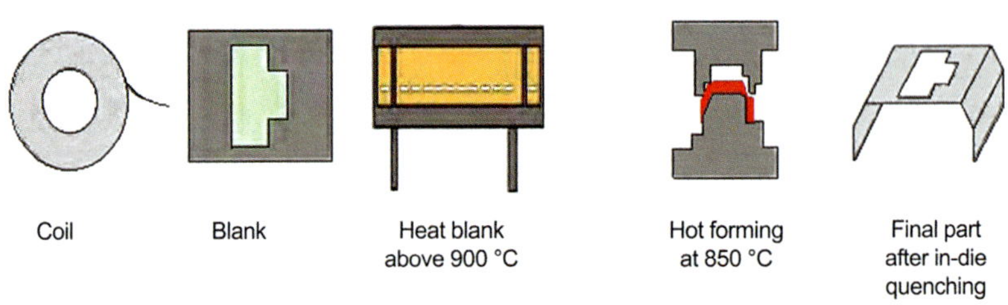

Fig. 12.19 Hot forming process. Source: Adapted from Ref 12.2

- High cycle time for production
- Thermal fatigue in the forming dies due to heating and cooling cycles applied to the dies
- Special lubricants must be used
- Metals oxidize at high temperature and removal of the scale leaves a rough surface.

For these reasons, hot forming is used in limited applications.

12.6 Downgaging Limits

The automotive industry has adopted lightweighting as a greenhouse reduction strategy. Lightweighting is achieved by downgaging vehicle components and using high-strength materials. Advanced high-strength steel designs take advantage of their higher strength-to-weight ratios, compared to conventional steels, by reducing the thickness of components to lower their weight while maintaining or improving their performance. The substitution of AHSS for lower-strength grades and the use of thinner gages to reduce weight have the following implications on automotive components and structures:

- Reduction in the elastic stiffness and the possibility of elastic buckling of the component
- Increase in corrosion problems that may reduce component life and durability
- Decrease in higher-strength ductility that may lead to reduction in component formability

The performance index for selecting a sheet metal panel meeting the constraints on bending strength and thinness is given by Ref 12.10:

$$t >= [6M/W\sigma_y]^{1/2}$$

where, for a panel of width W subjected to a moment, M, without failing, thickness, t, is minimized by selecting steel with the largest value of the yield strength, σ_y.

Stiffness is defined as the ability of a material to resist elastic deformation. It is directly proportional to the elastic modulus, E, and the second moment of inertia, I. Because the elastic modulus is constant for all steels, changing the steel grade for a specific component will not influence its stiffness. Therefore, in steel, stiffness is directly proportional to the second moment of inertia, I, which depends on component geometry (size and shape of the cross section including thickness). For a steel panel, the stiffness, S, the length, L, and the width, W, are usually specified; only the thickness is a free variable that can be changed. The weight of a panel or component can, therefore, be reduced by decreasing its thickness, t, but only so far that the stiffness constraint on the part is still satisfied. This means that there is a threshold above which substitution of lighter mass material cannot exceed due to structural performance issues related to diminishing part stiffness. In order to maintain or improve the stiffness of a part, downgaging must be offset by changes in part design.

One proven method for increasing stiffness of sheet steel is corrugation, which increases the moment of inertia of the sheet section ($I = t^3/12$) about an axis normal to the corrugation. In this way, resistance to bending is increased only in the normal direction to the corrugations. To stiffen the sheet metal in all directions, a hexagonal grid of dimples is usually used. Dimpling has been used in the auto industry in applications requiring downgaging, such as seat frames and side impact bars. Dimples can be applied during the final roll pass through specially designed rolls (Ref 12.11).

Corrosion resistance in components that are made of thinner gages of AHSS is more severe than those made from thicker gaged steels. Assuming that AHSS have about the same corrosion resistance as those of lower-strength steels, except austenitic stainless steels, corrosion is a bigger problem in thin gages because rust will go through them faster and they will suffer a greater percent loss of thickness. Therefore, the thinner gages employed to reduce weight necessitate the use of corrosion protection to prevent structural degradation and loss of durability.

Another concern, especially in the first generation of AHSS, is the loss in ductility with every gain in tensile strength. This trend compromises formability and limits the range of potential weight saving applications for those steels. Because the strength and ductility of the second generation of AHSS are both high, the formability issue is not relevant for those steels.

In 2010, the average North American light vehicle weighed approximately 1910 kg (4200 lb), of which 1130 kg (2500 lb) contain different types of steels. This amounts to a light vehicle steel content of about 60%. Vehicle weight is distributed almost equally between body structure, chassis, and power train. The biggest potential for weight saving is in the body structure and closure panels. Advanced high-strength steels are mostly used in structural applications. In 2010, AHSS content in body structure was approximately 11% of its mass. This percentage is estimated to increase to 35% by 2020. Replacing mild steel with lightweight materials or downgaging very strong steels, within tolerable limits on stiffness, will reduce vehicle weight and improve fuel economy.

12.7 Welding

Welding is a group of joining processes where two or more parts are combined at their contacting surfaces by the application of heat and/or pressure. Welding processes are divided into two main categories:

- *Fusion welding*, such as arc welding, resistance spot welding, oxyfuel gas welding, electron beam welding, and laser beam welding
- *Solid-state welding*, such as friction welding, diffusion welding, ultrasonic welding, roll welding, forge welding, and explosion welding

In fusion welding, joining is accomplished by melting the location of the two parts to be joined, in some cases adding filler metal to the joint. Because melting of the base metals occurs in fusion welding, a heat-affected zone develops around the weld. In solid-state welding, heat and/or pressure are used to join the two parts, but no melting of base metals occurs and no filler metal is added.

The welding methods used by the automotive industry include resistance spot welding, laser welding, and arc welding. Resistance spot welding is the main joining process used in sheet metal fabrication and automotive body assembly. For example, the body-in-white (BIW) automotive structures have 3000 to 5000 spot welds, while the whole vehicle requires approximately 6000 spot welds. Resistance spot welding is a fusion welding process that uses a combination of heat and pressure to accomplish coalescence of the joined metals. Heat is generated by electrical resistance to current flow at the junction to be welded (Fig. 12.20). Spot welding can be used to weld different sheet metals with thicknesses ranging from 0.5 to 10 mm (0.02 to 0.4 in.) and it can be performed at a rate of 60 welds per minute.

Many organizations and manufacturers issue welding guidelines for resistance spot welding. Such guidelines detail important requirements for process conditions. A typical resistance spot welding profile is shown in Fig. 12.21. The schematic includes four time segments: (a) applying pressure on the sheet metal by the electrodes, (b) passing current to heat and

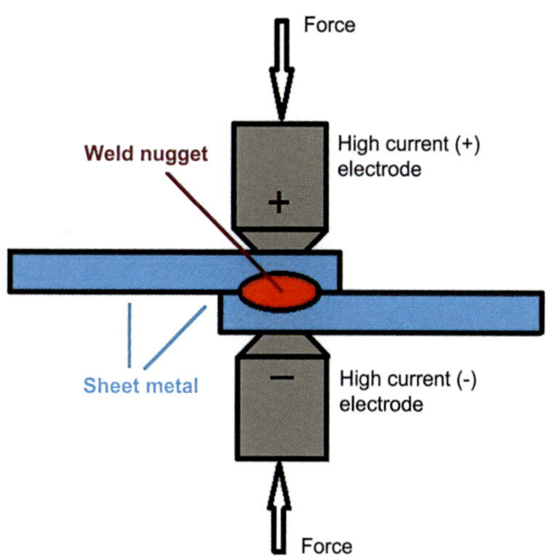

Fig. 12.20 Principal components in resistance spot welding

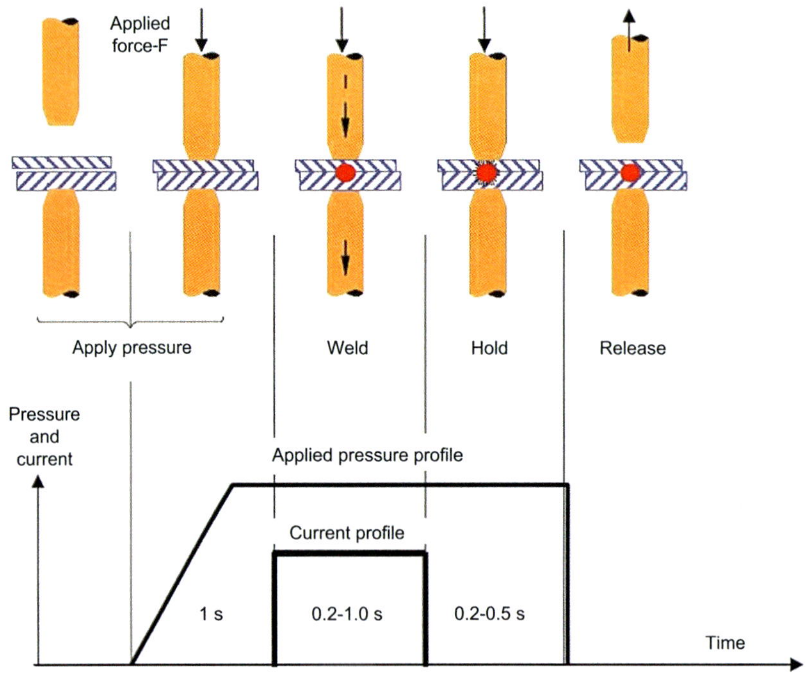

Fig. 12.21 Process conditions for resistance spot welding. Source: Adapted from Ref 12.12

melt the metal, (c) holding to re-solidify the molten metal, and (d) removing the electrodes at the end of the welding cycle. Other important specifications such as electrode type and geometry as well as required flange width are also specified.

Mild and HSLA steels have been used in body structures for many years and are well known to have minimal structural performance degradation due to exposure to elevated temperature during arc welding. Advanced high-strength steels are typically produced by nontraditional thermal cycles and contain microstructural constituents whose mechanical properties can be affected by exposure to elevated temperatures encountered during component welding. This temperature sensitivity could alter the mechanical behavior of AHSS and could seriously affect the structural performance of AHSS components.

The ultimate tensile strength (UTS) of the as-received DP 600 steel dropped from 642 to 515 MPa (93 to 75 ksi) after exposure to a thermal treatment temperature cycle of 650 °C/90 s (1200 °F/90 s). The reason for the drop in strength is attributed to tempering of the martensite phase. The MS 1300 steel exhibited more significant strength loss, also due to tempering, of the martensite phase during heat treatment. The mechanical properties of the DP 780, TRIP 600, and TRIP 780 materials were substantially degraded by exposure to 650 °C for any duration. The transformation-induced plasticity (TRIP) materials were degraded to a slightly lower extent than the dual-phase (DP) materials, probably due to the presence of bainite in the microstructure, which is less susceptible to strength changes during tempering than martensite. Experimental results showed that the mechanical properties of structural components made of DP, martensitic (MS), and TRIP steels of any type, degraded by exposure to 650 °C thermal cycling for any duration. It is, therefore, recommended that such steels should not be processed (welded, straightened, or repaired) by the application of heat (Ref 12.13).

Provided the welding parameters are adjusted, DP, complex-phase (CP), and TRIP steels can be welded using conventional resistance spot welding processes on commercial spot welding machines. Laser welding of the three steels showed no particular difficulties or problems (Ref 12.14). Electrogalvanized twinning-induced plasticity (TWIP) steels were found to be sensitive to liquid zinc embrittlement during welding. Liquid zinc results from melted coating due to the high temperature of the welding process (Ref 12.12).

Most stainless steels can be welded by most methods employed by the industry. It is important to maintain the mechanical properties, minimize distortion, and preserve the corrosion resistance in the weld area. Austenitic stainless steels are the most weldable of the stainless steels. The major problem in welding austenitic stainless steels arises from sensitization in the heat-affected zone. Sensitization occurs when austenitic stainless steels are heated or cooled through the temperature range of 430 to 870 °C (800 to 1600 °F). The chromium along the grain boundaries combines with carbon to form precipitates of chromium carbides. The formation of chromium carbide depletes chromium and lowers corrosion resistance in areas along the grain boundaries. This may lead to loss of toughness and intergranular corrosion of the steel. The sensitization effect can be minimized by reducing the carbon content of the steel. Post-heating after welding is needed to redissolve the precipitates and to relieve the stress caused by the welding operation. The coefficient of expansion of austenitic stainless steels is higher than that of low-carbon steels; therefore, their thermal expansion is greater and precautions are needed to avoid bead cracking and minimize part distortion.

REFERENCES

12.1 T. Altan and E. Billur, "Challenges in Forming Advanced High Strength Steels," paper presented at the International Conference on Developments in Sheet Metal Forming, May 2–4, 2010 (Stuttgart), University of Stuttgart

12.2 "Advanced High Strength Steels Applications Guidelines," Version 4.1, WorldAutoSteel, www.worldautosteel.org, June 2009

12.3 S. Sadogopan, "Formability Characterization of Advanced High

Strength Steels," Ispat Inland Inc. presentation at Great Designs in Steel Seminar, Feb 18, 2004 (Livonia, MI), AUTOSTEEL

12.4 M.Y. Demeri, Forming of Advanced High Strength Steels, *Metal Working: Sheet Forming*, Vol 14B, *ASM Handbook*, ASM International, 2006, p 530–538

12.5 "Standard Test Method for Evaluating Springback of Sheet Metal Using the Demeri Split Ring Test," E 2492-07, *Annual Book of ASTM Standards*, ASTM, 2007

12.6 P. Ulnitz, Springback in High Strength Steel Stamping, *MetalForming Magazine*, April 2009, p 36–37

12.7 C.-W. Hsu, A.G. Ulsoy, and M.Y. Demeri, Development of Process Control in Sheet Metal Forming, *J. Mater. Process. Tech.*, Vol 127 (No. 3), 2002, p 361–368

12.8 M.Y. Demeri, Drawbeads in Sheet Metal Forming, *J. Mater. Eng. Perform.*, ASM International, Vol 2(No. 6) Dec 1993, p 863–866

12.9 C. Escher and T. Henke, Tool Materials for Processing of High Strength Steel Sheets, *New Development in Sheet Metal Forming*, K. Siegert, Ed., IFU, Stuttgart, Germany, 2004, p 103–122

12.10 M. Ashby, H. Shercliff, and D. Cebon, *Materials: Engineering, Science, Processing and Design*, 2nd ed., Elsevier, 2010

12.11 M.F. Ashby, *Materials Selection in Mechanical Design*, 3rd ed., Elsevier, 2005

12.12 C. Beal, "Mechanical Behavior of New Automotive High Manganese TWIP Steel in the Presence of Liquid Zinc," Ph.D. thesis, Lyon National Institute of Applied Sciences, Lyon, France, 2011

12.13 "Advanced High Strength Steel Repairability: Phases 1 and 2," Final Reports, American Iron and Steel Institute, www.autosteel.org

12.14 ArcelorMittal Web site, www.arcelormittal.com

CHAPTER 13

Global Projects on Advanced High-Strength Steels

THE MOST ACTIVE PARTICIPANT in the development and deployment of AHSS technology is the global steel industry represented by WorldAutoSteel (WAS). This global organization develops and implements initiatives offering steel solutions, based on innovative designs and advanced new steels, to address challenges facing automakers to increase fuel efficiency, reduce emissions, increase safety, and maintain affordability of vehicles. The automotive industry responds positively to the influx of new design methodologies, innovative manufacturing processes, and advanced materials by adopting new technologies and increasing AHSS content in their vehicles. Interest in AHSS has increased dramatically in academia and research institutions as demonstrated by the large number of proposals and research projects aimed at improving or developing new generations of AHSS.

13.1 Steel Industry Projects

Steelmakers are at the forefront of developing advanced steels and manufacturing technologies for current and future vehicles. They are heavily involved in the development of AHSS and other manufacturing technologies through partnerships, consortia, and cooperative programs with universities, industrial labs, and national research institutions. A number of forums and collaboration efforts between the Steel Market Development Institute (SMDI), the Auto/Steel Partnership (ASP), and WAS focus on making lightweight steel-intensive vehicles with AHSS. The Steel Market Development Institute is a business unit of the American

Iron and Steel Institute (AISI), and WAS is the automotive group of the World Steel Association. WorldAutoSteel is comprised of 17 major global steel producers from around the world. These collaborations result in the increased use of AHSS grades internationally, resulting in safer, and more fuel efficient and sustainable vehicles. The steel industry continues to develop concepts for steel solutions to meet challenges related to ever-increasing crashworthiness regulations and fuel economy standards. Several subsystems and full-vehicle concept designs demonstrate that including AHSS results in approximately 35% weight savings.

Five major studies were conducted by the global steel industry under the umbrella of WAS, to offer solutions to challenges facing the automakers, namely, increasing fuel efficiency, reducing carbon emissions, improving crash safety, and maintaining affordability. The names and chronological order of these ultra-light steel family research programs are:

- *1998:* The Ultra-Light Steel Auto Body (ULSAB)
- *2000:* The Ultra-Light Steel Auto Closures (ULSAC)
- *2000:* The Ultra-Light Steel Auto Suspension (ULSAS)
- *2002:* The Ultra-Light Steel Auto Body—Advanced Vehicle Concepts (ULSAB-AVC)
- *2008 to 2011:* The FutureSteelVehicle (FSV)

The cost for the first four programs was estimated at $60 million. The steel industry investment in the FSV project was $80 million. WorldAutoSteel invested $6 million on the FSV program.

The 2002 ULSAB-AVC program shifted vehicle material content from a majority of mild and high-strength, low-alloy (HSLA) steel to 80% AHSS, to achieve 25% weight reduction. Since then, automakers have implemented AHSS in structural parts in a wide range of vehicles. As a result, the content of AHSS in an average vehicle jumped from 37 kg (82 lb) in 2006 to 79 kg (174 lb) in 2011. Projections show that the use of AHSS in North American light vehicles will double between 2011 and 2020. The ULSAB family of research studies demonstrates that optimized steel body structure design using AHSS produces structures 25% lighter than those built using conventional steel, with no cost penalty for manufacturers. Figure 13.1 shows a cost comparison of advanced versus conventional steel design results obtained in the ULSAB programs. The total cost included material, forming, and assembly costs.

In 2008, WAS launched a three-year engineering project known as the FutureSteelVehicle program (FSV). The FSV program was aimed at helping automakers optimize body structures through advanced steel design concepts suitable for both conventional and electrified powertrains, with the purpose of reducing gas emissions over the entire life cycle of vehicle. The program was built on the many innovative designs, advanced technologies, and results produced in the previous ULSAB studies. It also included the incorporation of many new grades of AHSS that are expected to be commercially available in the next five to ten years. Many advanced manufacturing processes were used in the FSV program (Fig. 13.2). There was heavy use of tailor welded blanks (TWB) to achieve the FSV weight target. Hot stamping, roll forming, and traditional stamping were also included in the process selection. An example of hot formed TWB used to manufacture the FSV shotgun subsystem is shown in Fig. 13.3; this system is used for frontal and offset crash application. The shotgun is comprised of a three-piece AHSS of varying thicknesses and is manufactured using hot stamping and quenching.

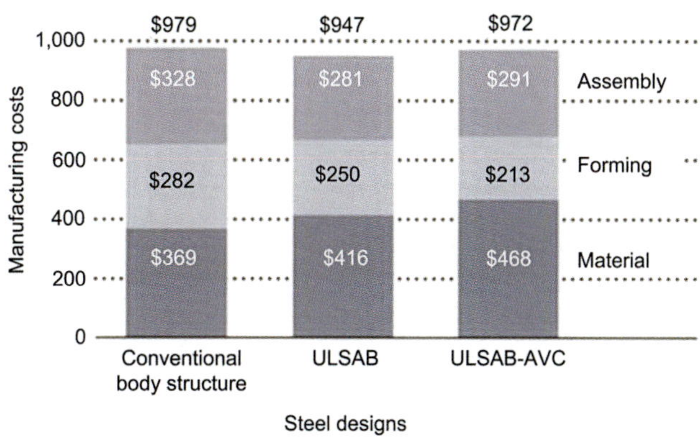

Fig. 13.1 Cost comparisons between conventional and advanced steel designs. ULSAB, Ultra-Light Steel Auto Body; ULSAB-AVC, Ultra-Light Steel Auto Body—Advanced Vehicle Concepts. Source: Ref 13.1

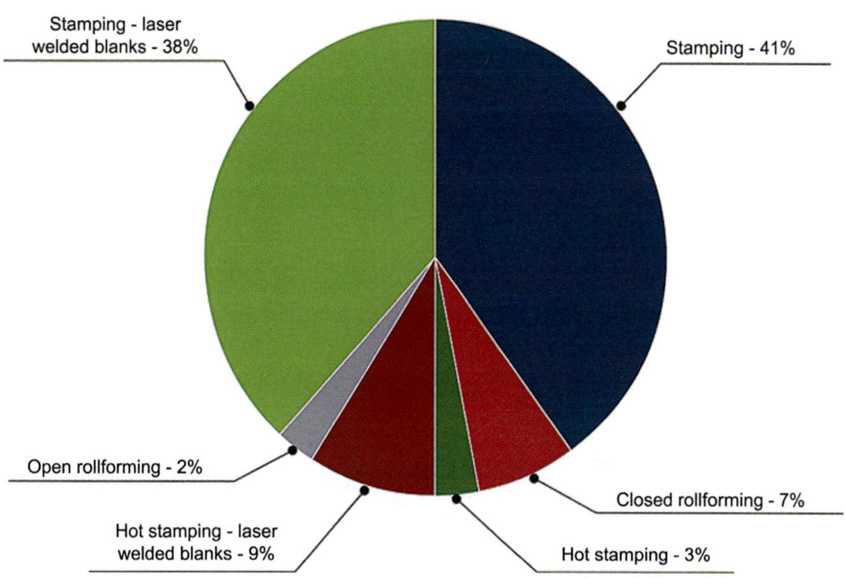

Fig. 13.2 Manufacturing processes used in the FutureSteelVehicle (FSV) program, shown as percent of body structure mass. Source: Ref 13.2

The stated objectives of the FSV program were:

- Achieve the 2015 to 2020 performance criteria.
- Achieve a 35% mass reduction.
- Perform cost analysis.
- Assess life cycle impact.

Results of the program presented car companies with ideas of what mass savings, costs, and carbon footprint can be accomplished with these steel grades and processes. Figure 13.4 shows a photo of the FSV concept car.

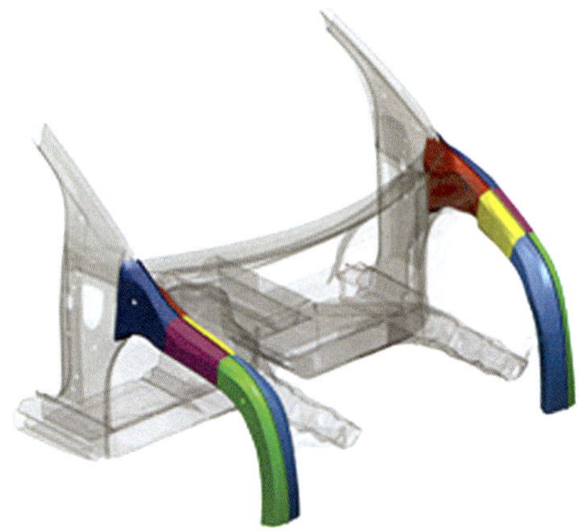

Fig. 13.3 Hot formed tailor welded blanks (TWB) used to manufacture the FSV shotgun subsystem. Source: Ref 13.2

Fig. 13.4 The FutureSteelVehicle (FSV). Source: Ref 13.2

In May 2011, the global steel industry released results of the FSV program. The results (Ref 13.2) were:

- The concept body structure met all targets for crashworthiness and achieved a mass savings of 35%, compared to 25% achieved in the 2002 ULSAB-AVC program. The mass reduction was achieved through the use of AHSS grades and an optimized design.
- The body structure used more than 50% AHSS including dual-phase (DP), transformation-induced plasticity (TRIP), twinning-induced plasticity (TWIP), complex-phase (CP), and martensitic (MS) grades to improve structural performance and reduce mass.
- Mass reduction was achieved at no additional cost over reference vehicle steel structures.
- Total life cycle emissions were reduced by nearly 70%.

Table 13.1 captures the main features and accomplishments of the various light vehicle programs. The FSV program delivered significant weight savings of 35% by incorporating HSS and AHSS in 97% of the body mass in tandem with advanced manufacturing technologies and innovative designs. The manufacturing cost of the FSV body structure was accomplished without any cost penalty relative to the 1995 reference vehicle.

Figure 13.5 lists the steel grades and types used in the ULSAB-AVC program and the extended steel grades and types used in the FSV program. Many of the steel grades used in the FSV program are in the GPa range.

A detailed and final engineering report on the FSV was published by WAS and SMDI (Ref 13.2).

13.2 Government/Industry/Academia Collaboration

United States Council for Automotive Research/U.S. Automotive Materials Partnership (USCAR/USAMP). An important organization for automotive research is the United States Council for Automotive Research (USCAR). It was founded in 1992 as a collaborative technology organization of Chrysler Group, General Motors Company, and Ford Motor Company. Its goal is to strengthen the technology base of the U.S.

Table 13.1 Light vehicle programs and achieved results

Program	Vehicle mass reduction	Average tensile strength of steels used, MPa (ksi)	Body manufacturing cost (225,000 units/year), U.S. dollars	Body mass, kg (lb)
Reference vehicle, 1995	...	271 (39)	997	270 (595)
ULSAB, 1998	–25%	413 (60)	947	203 (448)
ULSAB-AVC, 2002	–25%	758 (110)	979	218 (481)
FSV, 2011	–35%	789 (114)	981	188 (414)

ULSAB, Ultra-Light Steel Auto Body; ULSAB-AVC, Ultra-Light Steel Auto Body—Advanced Vehicle Concepts; FSV, Future SteelVehicle. Source: Compiled from Ref 13.2

Mild 140/270	DP 350/600	TRIP 600/980
BH 210/340	TRIP 350/600	TWIP 500/980
BH 260/370	SF 570/640	HSLA 700/780
BH 280/400	HSLA 550/650	DP 700/1000
IF 260/410	TRIP 400/700	CP 800/1000
IF 300/420	SF 600/780	MS 950/1200
DP300/500	CP 500/800	CP 1000/1200
FB 330/450	DP 500/800	DP 1150/1270
HSLA 350/450	TRIP 450/800	MS 1150/1400
HSLA 420/500	CP 600/900	CP 1050/1470
FB 450/600	CP 750/900	HF 1050/1500
HSLA 490/600		MS 1250/1500

denotes steel included in ULSAB-AVC
denotes steel grades added for FSV

Fig. 13.5 New grades of HSS added for the FSV program. Source: Ref 13.2

auto industry through cooperative research and development. The USCAR focus is to create, support, and direct cooperative research and development to advance automotive technologies and be responsive to sustainability issues. In 1993, USCAR teamed with the U.S. Department of Energy (DOE) and formed the Partnership for New Generation of Vehicles (PNGV) to improve manufacturing competitiveness and develop new technologies for fuel efficient vehicles. This program transitioned to Freedom CAR with the inclusion of five energy and two electric utility companies, to work on hydrogen and fuel cell research. In 2004, the USA-TEAM partnership between the USCAR and the National Institute of Standards and Technology (NIST) was formed to focus on improving the manufacturing competitiveness of the U.S. automotive industry. In 2011, DOE announced the formation of a new partnership, US-DRIVE, whose focus is to drive research and innovation for vehicle efficiency and energy sustainability.

A subsidiary of the USCAR is the U.S. Automotive Materials Partnership (USAMP), whose focus is on developing materials and processes that enable the high-volume production of vehicles with half the current mass, while also making them more affordable, recyclable, and of good quality and durability. The USAMP is responsible for identifying, conducting, and directing noncompetitive research and development in automotive-related materials and processes. Projects on materials include improvements in the production, processing, and performance of lightweight materials: aluminum, magnesium, composites, and steel. Projects on advanced processes include press optimization, flexible binder, warm forming, controllable drawbeads, and welding optimization.

Auto/Steel Partnership/U.S. Automotive Materials Partnership (ASP/USAMP). The USAMP work on steel was conducted mainly through the ASP, which is a consortium of the SMDI's Automotive Applications Council, Chrysler Group, General Motors Company, and Ford Motor Company. The ASP leverages the resources of the automotive and steel industries to help improve vehicle design, safety, fuel economy, and manufacturing. Figure 13.6 shows the ASP linkages with academia, steel suppliers, automakers, national labs, WAS, DOE, and the National Science Foundation (NSF).

The ASP has many projects on AHSS including material evaluation, formability assessment, springback validation, and welding requirements. It reported success in using computer modeling to validate a potential mass savings of 15 to 20% in the future-generation passenger compartment project. The mass savings were achieved by using AHSS for parts typically made of conventional steel and by designs for load paths that ensure crashworthiness of the passenger compartment.

13.3 Academic Research and Development

Interest in AHSS research and development has been growing, especially after the steel and automotive industry developed and implemented DP steels in the 1980s. Most of the current research is on TRIP, TWIP, and third-generation AHSS. The research focuses on the influence of alloy composition on the microstructure evolution during plastic deformation and the resulting mechanical properties. Following are reviews of some of the latest projects on AHSS.

Structure and Mechanical Properties of Fe-Mn Alloys (McMaster University, 2008, Ref 13.4). Advanced high-strength steels are charac-

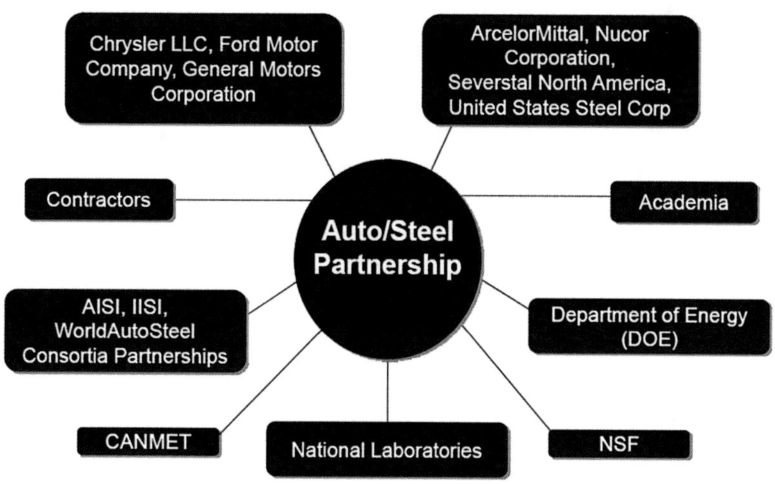

Fig. 13.6 Partnership linkages for Auto/Steel Partnership (ASP) projects. Source: Ref 13.3

terized by a combination of high strength and good ductility. This is achieved in metal alloys possessing high and sustained work-hardening rates during plastic deformation. In TRIP and TWIP steels, these properties are accomplished by the evolution of phase transitions (austenite to martensite transformation and deformation twinning) during plastic deformation. Austenitic high-manganese alloys are known to have such qualities and, therefore, were the subject of this investigation.

Research at McMaster University was focused on the structure and properties of two high-manganese alloys, Fe-24Mn and Fe-30Mn. The high manganese content ensures that austenite is retained fully at room temperature. The chemical composition of the two alloys is listed in Table 13.2.

The high work-hardening rate during plastic deformation of Fe-30Mn is attributed to its low stacking fault energy (SFE). The high strength and ductility result from the complex combination of deformation mechanisms including dislocation glide, martensitic phase transformation (TRIP effect), and mechanical twinning (TWIP effect). Alloy Fe-30Mn possesses an austenite single-phase microstructure after annealing; whereas alloy Fe-24Mn has a mixture of austenite and 50% ε hexagonal close-packed (hcp) martensite. The single-phase alloy Fe-30Mn remains fully austenitic after water quenching to room temperature. It was annealed at 900 °C (1652 °F) for two hours in an inert atmosphere and then quenched in oil instead of water to reduce internal stresses. Alloy Fe-24Mn was subjected to six different heat treatments to evaluate their effect on phase fractions of the steel. Microhardness testing and microstructure evaluation of the six samples indicated that different heat treatments have no effect on the amount of austenite and ε-martensite.

Summary of results for Fe-30Mn alloy (single-phase austenitic microstructure). Scanning electron microscopy (SEM) and x-ray methods were used to monitor the effect of plastic deformation on microstructure and phase transformation in the Fe-30Mn alloy. This was done by measuring the evolution of phase fraction of transformed ε-martensite in deformed tensile samples. Results indicated that rapid phase transformation occurred at true strains of ε = 30% and continued until fracture. Figure 13.7 shows microstructures of annealed and plastically deformed samples at true strain of ε = 37.3%.

Electron backscatter diffraction (EBSD) analysis gives a phase map of the annealed and deformed microstructures (Fig. 13.8). The initial micro-

Table 13.2 Chemical composition of the high-manganese steels

Steel grade	Composition, wt%			
	Fe	Mn	C	S
Fe-24Mn	75.57	24.30	0.0163	0.0052
Fe-30Mn	69.08	30.79	0.0160	0.0060

Source: Ref 13.4

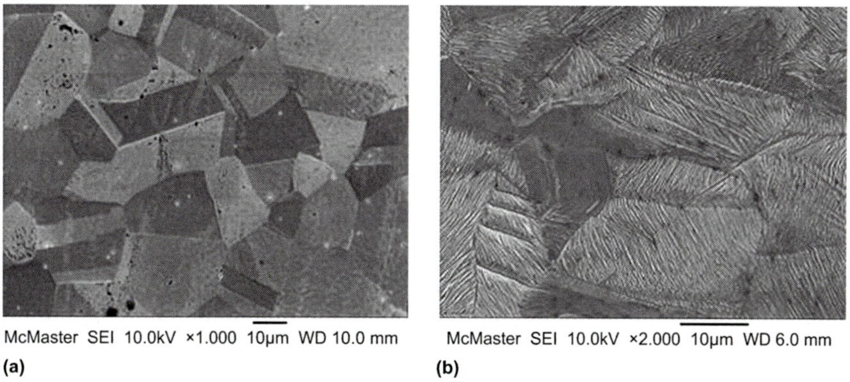

Fig. 13.7 Scanning electron microscope images of (a) annealed and (b) deformed Fe-30Mn. Source: Ref 13.4

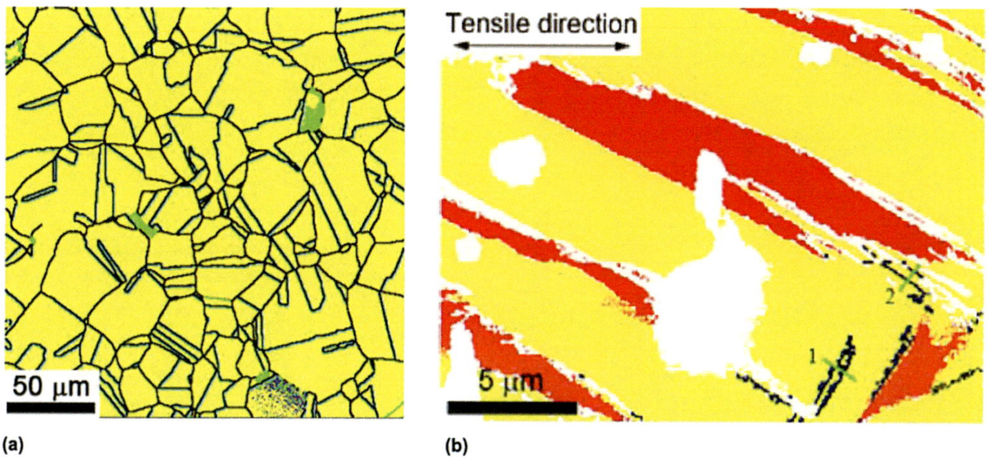

Fig. 13.8 Electron backscatter diffraction mapping of (a) annealed and (b) deformed Fe-30Mn. Source: Ref 13.4

structure of annealed Fe-30Mn steel is fully austenitic with annealing twins (Fig. 13.8a), and the microstructure of deformed samples (ε = 37.3%) contain transformed martensite and mechanical twins (Fig. 13.8b). The yellow color represents austenite, the red represents martensite, and the blue lines indicate mechanical twins.

The microstructure of the deformed sample showed regions of the transformed martensite in multiple grains. The kinetics of martensite transformation were studied by x-ray diffraction on samples deformed uniaxially to different levels of strain. Figure 13.9 shows the variation of the volume fraction of transformed martensite with true strain. Critical phase transformation occurred at true strains of ε = 0.30. The balance of the volume fraction percent is austenite.

Summary of results for Fe-24Mn alloy (complex-phase austenitic and martensitic microstructure). The annealed Fe-24Mn alloy has a complex microstructure consisting of approximately 50% ε-martensite plates of different sizes inside the equiaxed austenite grains. Figure 13.10 shows microstructures of annealed and plastically deformed samples at true strain of 30%. The microstructure shows more pronounced phase transitions, indicating that phase transformation becomes an important deformation mode at that strain level. The initial microstructure showed a high density of stacking faults, indicating a low level of stacking fault energy of 8 mJ/m^2 for Fe-24Mn compared to 15 mJ/m^2 for Fe-30Mn alloy.

Analysis by EBSD gives a phase map of the annealed and deformed microstructures (Fig. 13.11). The initial microstructure of annealed Fe-24Mn steel is composed of austenite and ε-martensite (Fig. 13.11a); the deformed steel (ε= 20%) contains austenite and transformed martensite

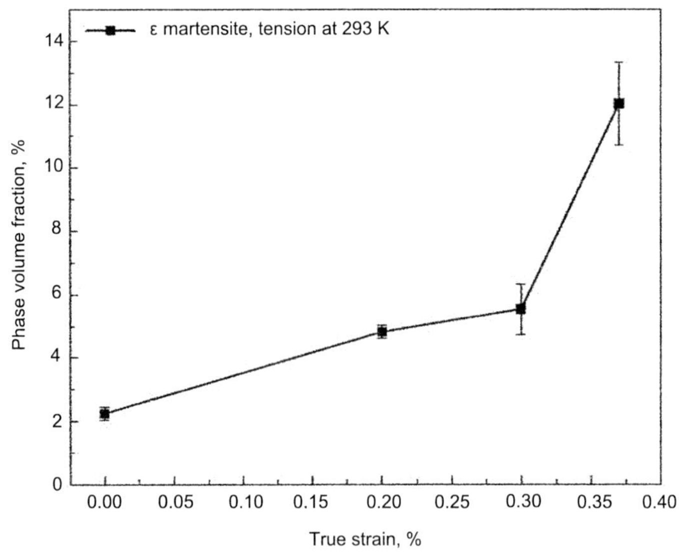

Fig. 13.9 Variation of transformed fraction of martensite with true strain for Fe-30Mn. Source: Ref 13.4

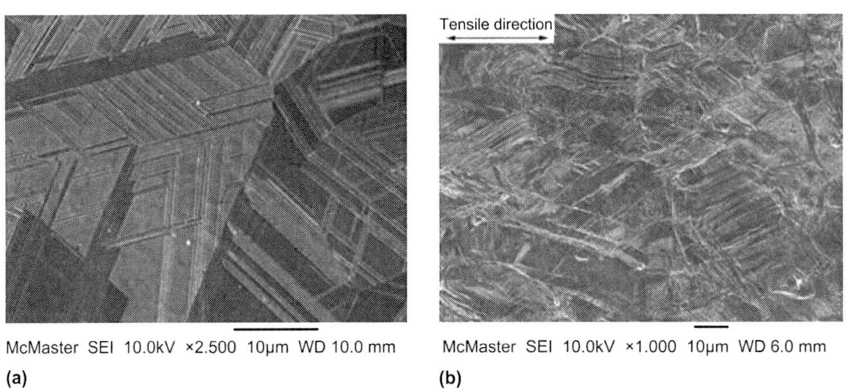

Fig. 13.10 Scanning electron microscope images of microstructure of (a) annealed and (b) deformed Fe-24Mn. Source: Ref 13.4

(Fig. 13.11b). The yellow color represents austenite, the red represents martensite, and the blue lines indicate mechanical twins. The EBSD phase maps indicate that the fine plates in the deformed steel result from deformation-induced transformation of austenite to ε-martensite, whereas the thick plate is a thermally transformed ε-martensite.

In the plastic deformation of the Fe-24Mn alloy, strengthening is accomplished by dislocation accumulation during co-deformation of austenite and ε-martensite. Additional strengthening is produced by the increasing volume fraction of the transformed ε-martensite. This co-deformation process can be modeled by assuming that Fe-24Mn is a composite material consisting of two phases, austenite and ε-martensite, and therefore the flow stress depends on the evolving volume fraction of each component. The evolution of the volume fraction of ε-martensite, as a function of true strain for Fe-24Mn alloy, is shown in Fig. 13.12.

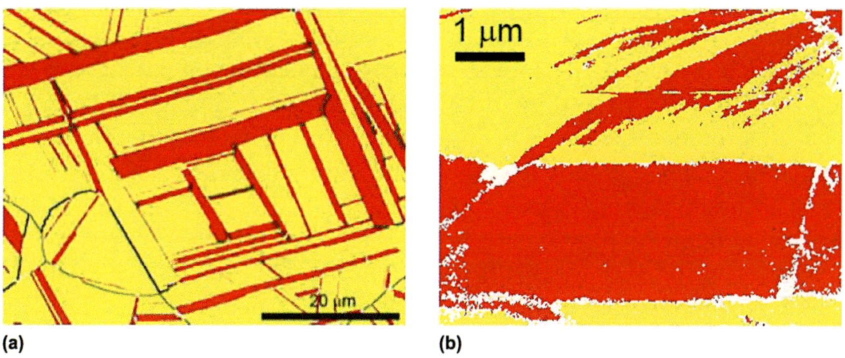

Fig. 13.11 Electron backscatter diffraction mapping of (a) annealed and (b) deformed Fe-24Mn. Source: Ref 13.4

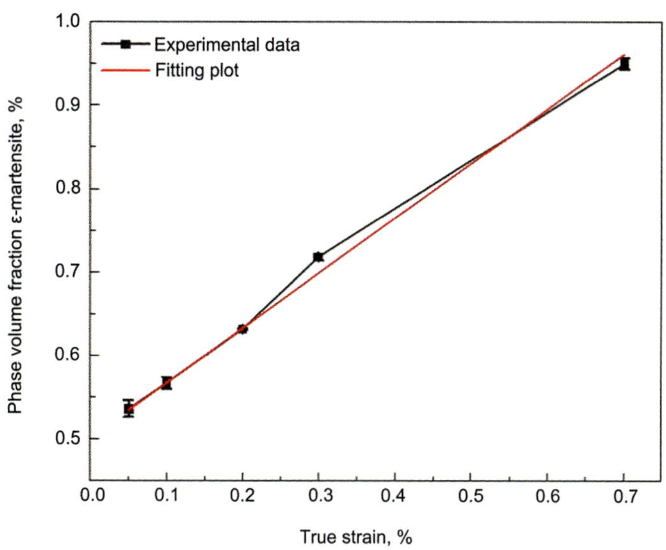

Fig. 13.12 Evolution of phase volume fraction of ε-martensite with strain. Source: Ref 13.4

The composite deformation behavior of Fe-24Mn can be estimated if the stress-strain of the constituents, austenite and ε-martensite, are known. Figure 13.13 shows the stress-strain curves of the modeled ε-martensite constituent and the experimental curves for austenite and the composite Fe-24Mn alloy.

Because Fe-30Mn remains fully austenitic up to 30% tensile deformation, its true stress-strain curve (Fig. 13.13) is a good approximation of the deformation behavior of austenite. The figure also shows that the yield strength of ε-martensite is approximately 450 MPa (65 ksi), whereas that of austenite is approximately 150 MPa (20 ksi).

Microstructure Evolution in TWIP Steel (Tampere University of Technology, Finland, 2009, Ref 13.5). Research was conducted at Tampere University in Finland to study the deformation mechanisms of fully austenitic TWIP steels. Electron backscatter diffraction and transmission electron microscopy (TEM) were used to monitor texture evolution during the uniaxial plastic deformation of a hot rolled Fe-28Mn steel.

In TWIP steels, the presence of high manganese content lowers the stacking fault energy and gives rise to mechanical twinning as the dominant deformation mechanism, in addition to dislocation slip. Deformation twins increase gradually with deformation and act as strong obstacles to dislocation motion. This explains the gradual increase in the work-hardening rate with plastic deformation.

A series of interrupted tensile tests were conducted to study the microstructure at true plastic strains of $\varepsilon = 0.1, 0.2, 0.3$, and at fracture strain of 0.49. Figure 13.14 shows the microstructure at true strain $\varepsilon = 0.1$ and at true strain $\varepsilon = 0.3$. Twins appear as parallel lines within the grains and their intensity is shown to increase with plastic deformation.

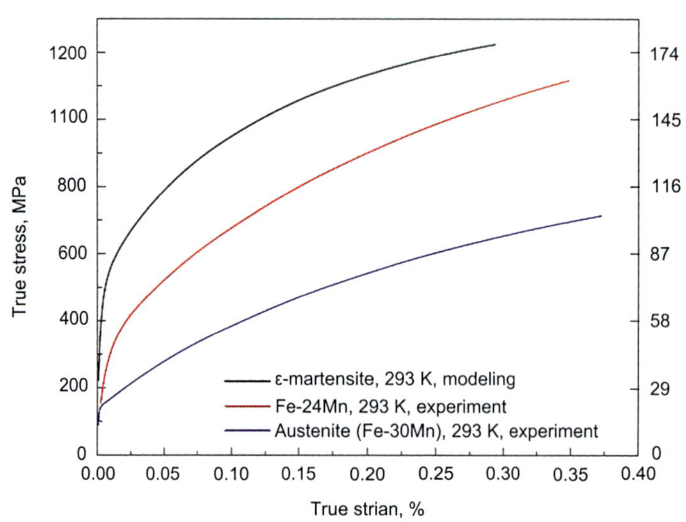

Fig. 13.13 Predicted true stress-strain curve of ε-martensite and experimental curves of Fe-24Mn and Fe-30Mn (austenite). Source: Ref 13.4

Conclusions of the study are:

- Twinning activity depends on texture (grain size and orientation), temperature, and strain rate.
- Lowering deformation temperature to –50 °C (–60 °F) lowers the number of deformation twins because the texture has not yet evolved to a point where enough grains are favorably oriented for twinning.
- The microstructure of the fine-grained TWIP steel showed no evidence of twinning.
- At high deformation rates, a high density of deformation twins is evident, which results in increase in the strength level.

Grain Refinement of DP Steel (Max-Planck-Institute for Metalforming, Germany, 2010, Ref 13.6). The research focuses on using grain refinement as a tool to increase the strength of DP steels without reducing its ductility. Three grain sizes of a ferrite/martensite dual-phase (DP) steel—coarse (12.4 μm), fine (2.4 μm), and ultrafine (1.2 μm)—were prepared and tested under tensile and impact conditions to evaluate the effect of grain refinement on the mechanical properties of the steel. The three materials have the same chemical composition, the same martensite volume fraction, and the same martensite carbon content. Different thermomechanical processing routes were used to obtain the three grain sizes. Table 13.3 lists the microstructure parameters and tensile data for three grain sizes of DP steel.

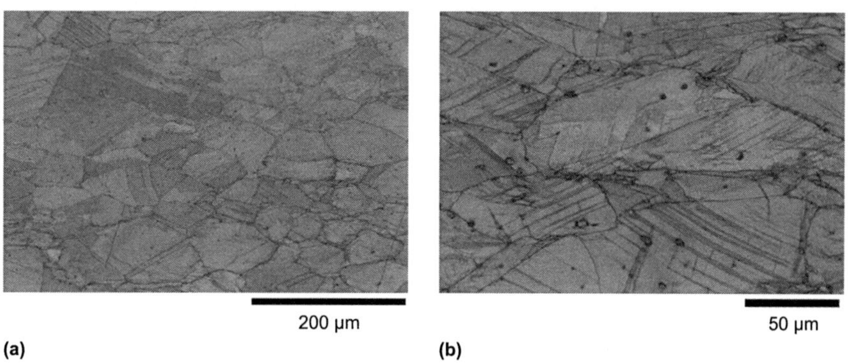

Fig. 13.14 Twin formation after (a) true strain ε = 0.1 and (b) true strain ε = 0.3. Source: Ref 13.5

Table 13.3 Microstructure and mechanical properties of three grain sizes of dual-phase steel

Steel grain	Martensite volume fraction, %	Ferrite grain size, μm	Yield strength, MPa (ksi)	Tensile strength, MPa (ksi)	Total elongation, %	n
Coarse grain (CG)	31.3	12.4	445 (65)	870 (126)	7.7	0.21
Fine grain (FG)	30.1	2.4	483 (70)	964 (140)	8.9	0.18
Ultrafine grain (UFG)	29.8	1.2	525 (76)	1037 (150)	7.3	0.18

Source: Adapted from Ref 13.6

Microstructures of grains used to evaluate the effect of grain refinement on mechanical properties are shown in Fig. 13.15. The microstructure of the coarse grain (CG) material consists of a ferrite matrix with a grain size of 12.4 μm and 31% martensite islands. The microstructures of the fine-grain (FG) and the ultrafine-grain (UFG) material consist of ferrite matrix with grain sizes reduced to 2.4 and 1.2 μm, respectively. The martensite volume fraction remained essentially the same for all three grain sizes. This leads to the conclusion that differences in the mechanical properties listed in Table 13.3 can only be attributed to the different grain sizes and not to martensite volume fraction.

Engineering stress-strain curves of the CG, FG, and UFG DP steels are shown in Fig. 13.16, with ferrite grain sizes in parentheses. As expected, the tensile strength increased with decrease in grain size but changes in total elongation were minimal.

The main conclusion of this research is that grain refinement in DP steel increases strength, improves impact toughness, and promotes ductile fracture.

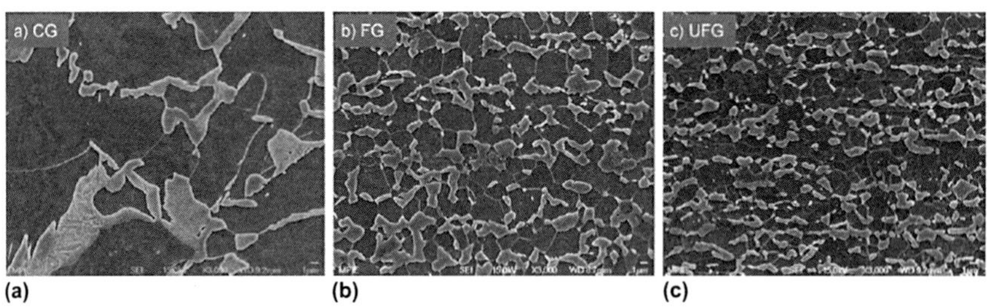

Fig. 13.15 Microstructures of (a) coarse-grained (CG), (b) fine-grained (FG), and (c) ultrafine-grained dual-phase (DP) steel. Magnification: 3000× for all images. Source: Ref 13.6

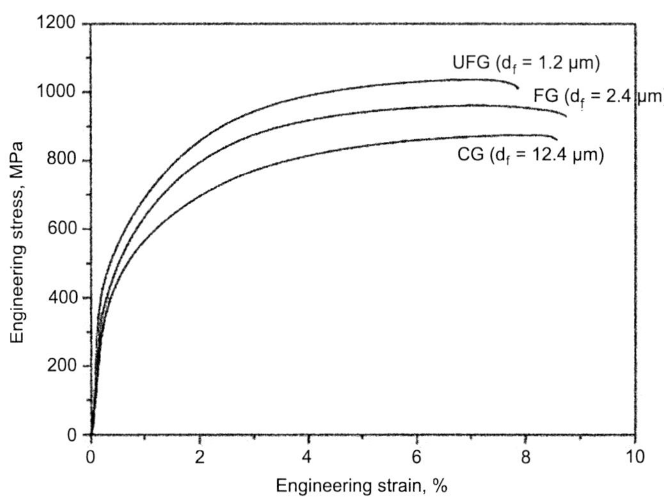

Fig. 13.16 Engineering stress-strain curves of three grain sizes of DP steels. Source: Ref 13.6

REFERENCES

13.1 "Advanced High Strength Steels Applications Guidelines," Version 4.1, WorldAutoSteel, www.worldautosteel.org, June 2009

13.2 WorldAutoSteel, "FutureSteelVehicle (FSV)," Final Engineering Report, Steel Market Development Institute, May 2011, www.worldautosteel.org

13.3 R. Heimbuch, "Advanced High-Strength Steel Research and Development," Auto/Steel Partnership presentation at 2009 U.S. Department of Energy (DOE) Hydrogen Program and Vehicle Technologies Program Annual Merit Review and Peer Evaluation Meeting, May 18–22, 2009 (Arlington, VA)

13.4 X. Liang, "Structure and Mechanical Properties of Fe-Mn Alloys," master's thesis, McMaster University, July 2008

13.5 S. Curtze, V.-T. Kuokkala, M. Hokka, and T. Saarinen, Microstructure and Texture Evolution in High Manganese TWIP Steels, *Proc. DYMAT 2009,* Vol 2, Sept 7–11, 2009 (Brussels, Belgium), EDP Sciences, 2009, p 1007–1013

13.6 M. Calcagnotto, D. Ponge, and D. Raabe, Effect of Grain Refinement to 1 μm on Strength and Toughness of Dual-Phase Steels, *Mat. Sci. Eng. A,* Vol 527 (No. 29–30), Nov 2010, p 7832–7840

CHAPTER 14

Design Guidelines for Advanced High-Strength Steels

THE INCREASED USE OF advanced high-strength steels (AHSS) has created a need to develop a knowledge base of data and experiences aimed at the successful applications of the new materials. Two documents were created to fill that need and to help automakers meet their application goals. The first document, titled "Advanced High-Strength Steels Application Guidelines," Version 4.1 (Ref 14.1), was released in June 2009 by WorldAutoSteel. The second document, titled "Advanced High-Strength Steel Applications: Design and Stamping Process Guidelines" (Ref 14.2), was released in January 2010 by the Auto Steel Partnership (ASP).

The WorldAutoSteel document is a leading information resource for people developing, making, and using AHSS. The document contains 338 pages and hundreds of figures and photos organized in six sections:

- *Section 1:* General Description of AHSS (definitions, metallurgy, high-strength steel)
- *Section 2:* Forming (computerized forming, sheet forming, tube forming, hydroforming)
- *Section 3:* Joining (welding procedures)
- *Section 4:* Glossary
- *Section 5:* References
- *Section 6:* Appendix (Auto Steel Partnership AHSS Case Study Summaries)

The ASP document is a complementing document that focuses on in-depth case studies of selected parts with emphasis on the development and implementation of sheet metal stamping processes for AHSS. The ASP document is a guide for using HSS for body-in-white (BIW) applications. The guide explains the steels characteristics and provides guidelines for die design, construction, and tryout. It also discusses product design and advanced die process concepts for forming HSS. The document contains 348 pages and hundreds of figure, photos, and sketches organized in four sections and an appendix:

- *Section 1:* AHSS Case Studies/Summaries
- *Section 2:* Lessons Learned from AHSS Case Studies
- *Section 3:* Glossary: AHSS Applications Guidelines
- *Section 4:* Die Sketches
- *Appendix:* AHSS General Description and Forming

Both documents serve as the main source for understanding the experience-based guidelines that were developed for forming and welding AHSS. The guidelines are briefly reviewed and highlighted in the next two sections.

14.1 Forming Guidelines

The properties of AHSS are characterized by high strength and somewhat low ductility. Such properties lead to increased tendencies for splitting, buckling, residual stresses, and springback. Forming AHSS require higher tonnages and this leads to higher loads on dies and tools, resulting in die wear and tool failure. In general, AHSS parts are designed with minimum draw depth and reduced stretch to avoid splits. Also, gentle shape transitions are used to avoid wrinkles and large radii are specified to facilitate metal flow.

The following forming guidelines are the result of lessons learned from AHSS case studies (Ref 14.2):

- As steel strength increases, surface transitions must be gradual to minimize splits, bucking, and springback.
- As steel strength increases, part designs must be simplified. Large complex parts should be split into smaller stampings.
- As strength increases, sidewall angles must be designed more open for springback overbend allowance; at least 6° for DP 600, 8° for DP 800, and 10° or more for DP 1000.
- For steels with reduced stretch flangeability, minimize stretch flanges by design. Some AHSS have stretch flangeable grades and types that can be used for stretch flange applications.
- As steel strength increases, part and die radii must also increase. For DP 600, the minimum part radius should be greater than three times (>3t) the metal thickness; whereas the die radius should be greater than five times (>5t) the sheet metal thickness.

- Structural parts such as rails, cross bars, pillars, and so forth, should be designed as open-ended channels. Closed-end draws must be shallow.
- As steel strength increases, higher binder pressures will be required for forming operations.
- As steel strength increases, tools wear rates and flexing increase. This will require upgrades to tool materials (trim and die) and more efficient lubrication.
- Higher press speeds and deeper draws will generate more heat. An appropriate draw lube or die cooling may be needed.
- Form as much of the required shape as possible in the first die. Restrike die should not be used to correct major shape problems; it should be used only to correct wrinkles and springback.
- Perform stretch flanging before going into trimming.
- Use hard weld on all draw beads. High-wear die radii must be hard welded and polished to prevent stripping of steel coatings.
- Trim and pierce angles should be <10° for holes larger than 6 mm (0.24 in.). Piercing angle should be 90° to surface for small diameter punches (<6 mm). Piercing of small-diameter holes, less than 6 mm, should be square to surface only.
- As steel strength increases, binder pressure must be increased to prevent wrinkling and to reduce springback. For example, DP 600 requires 2.5 times greater tonnage than that required for aluminum-killed drawing-quality (AKDQ) steel of the same thickness.
- Maintain die clearance as tight as possible to reduce bending and unbending in the die gap.
- Stretch forming produces a stronger panel with less springback than drawing because of work hardening and superimposed stretching on bending.
- As steel strength increases, potential depth of a panel decreases for drawing and stretch forming. Advanced high-strength steel deep panels require using drawing operations.

14.2 Welding Guidelines

As mentioned in section 12.7, "Welding," AHSS are typically produced by nontraditional thermal cycles and contain microstructural constituents whose mechanical properties can be affected by exposure to elevated temperatures encountered during component welding. This temperature sensitivity could alter the mechanical behavior of AHSS and could seriously affect the structural performance of AHSS components.

The ASP led a project to study the welding performance of AHSS for auto body structural components. Five fusion welding processes (MIG pulse/ac, MIG pulse/dc, MIG/laser-assisted, laser, and laser plasma) were used to study several combinations of HSLA 350Y and AHSS (DP 600, DP 780, DP 800, and DP 980).

Results of the study to determine the welding performance of AHSS for auto body structural components are (Ref 14.3):

- All materials tested were successfully joined using the five fusion welding processes.
- Material strength or thickness had no effect on laser welded joint strength.
- Welding processes using fillers demonstrated better results than those without fillers.
- Laser welded lap joints generally failed in the weld metal, while metal inert gas (MIG) fillet joints generally failed in the heat-affected zone.
- Zinc-coated materials demonstrated high levels of porosity.

Another study was led by ASP to investigate resistance welding performance of seven types of AHSS (DP 600, DP 800, DP 980, RA 830, MS1 300, TRIP 600, and TRIP 800). HSLA 340 was used as a baseline for comparison. Some of the steel types were coated (galvannealed, electrogalvanized, and hot dip galvanized). The main objective of the study was to determine AHSS weld properties produced using conventional welding processes. The project had two phases: (1) welding samples from eight grades and types of steel, and (2) determining mechanical performance of those welds. Static, dynamic, hardness, and fatigue properties of welds were determined. The quality of welds produced and the effect of weld parameters on nugget fracture and fusion zone were evaluated using peel, fatigue, and impact testing. Ultrasonic evaluation of the welds including surface indentation, weld thickness, nugget width, internal void width, and location were determined. Figure 14.1 shows the geometric features of a spot weld. Results of the project produced general information that can be used to determine the performance of resistance welding of AHSS.

Results of the resistance welding performance study are (Ref 14.4):

- Advanced high-strength steels have higher static tensile shear strength, impact tensile peak load, and energy absorption than the HSLA 340 baseline.

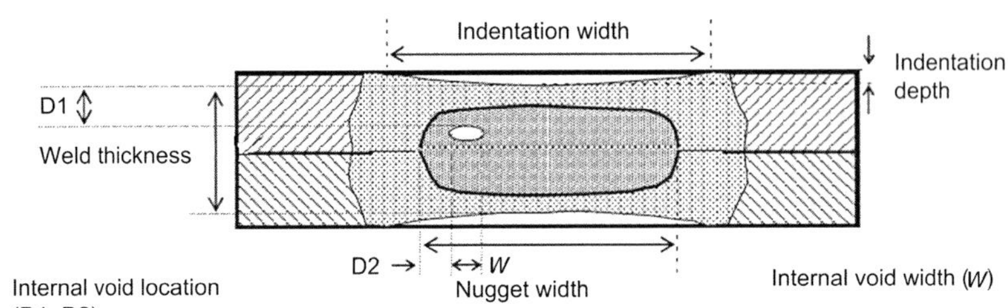

Fig. 14.1 Geometric features of a spot weld. Source: Adapted from Ref 14.4

- Advanced high-strength steels have higher static and dynamic loading gage-for-gage than traditional low-carbon steels with yield strength below approximately 400 MPa (58 ksi).
- Resistance spot welding of AHSS with optimized welding parameters was validated on spot welding of a front structure.

A third study was led by ASP to establish design guidelines for gas metal arc welded (GMAW) chassis structures, with a minimum thickness of 2.0 mm (0.08 in.), employing both conventional and AHSS. The guidelines are based on existing original equipment manufacturer (OEM) arc weld design standards (Ref 14.5).

Gas metal arc welding is a fusion welding process in which coalescence of the metals is achieved by the heat from an electric arc between an electrode and the workpiece. Electric energy from the arc produces temperatures of approximately 5500 °C (9930 °F), hot enough to melt any metal. It uses a consumable bare metal wire as electrode, and shielding is accomplished by flooding the arc with CO_2 gas. A schematic of the GMAW process is shown in Fig. 14.2.

Guidelines for employing the GMAW process for welding chassis structures with a minimum thickness of 2.0 mm (0.08 in.), made from both conventional steel and AHSS, are (Ref 14.5):

- Welding closer than 25 mm (1 in.) to surfaces or edges coated with sealer or adhesives is not recommended.
- The minimum specified weld length should be ≥13 mm (0.5 in.).

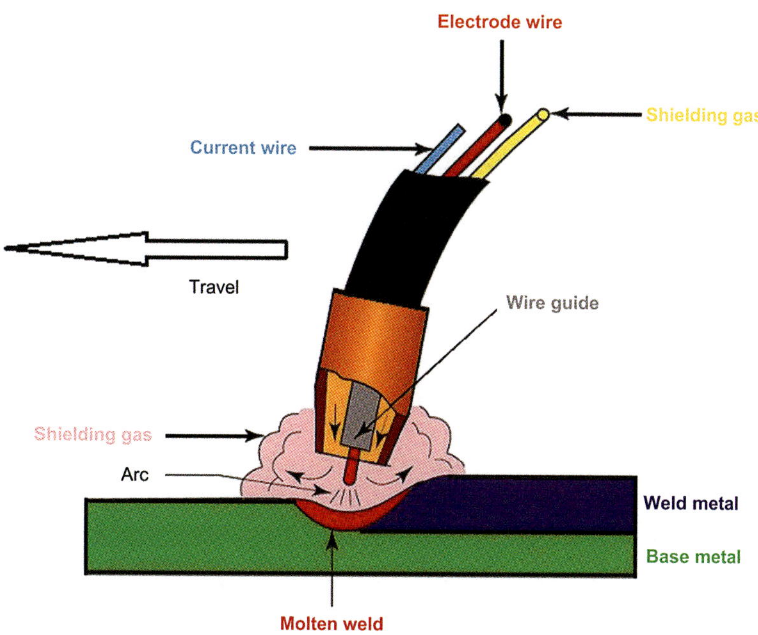

Fig. 14.2 Gas metal arc welding (GMAW) process

- Excessively large weld size is not recommended.
- For a lap joint, it is recommended that the thinner steel be on the top.
- Three-thickness joint design is not recommended.
- Recommended minimum steel thickness is 2 mm (0.08 in.), maximum is 9 mm (0.04 in.), and the maximum combination ratio is 3:1.
- Sulfur content should be ≤0.05 wt%, phosphorus ≤0.06 wt%, and boron ≤0.005 wt%.
- No free-machining steels should be used for chassis structures when GMAW is the welding process.
- AWS A 5.18 solid welding wires ER 70S-6 and ER 70S-3 are recommended for chassis structures. AWS solid wire ER 80S-D2 may also be used if a higher-strength weld metal is needed.
- The recommended filler metal diameter is 1.14 mm (0.045 in.). Other wire diameters, such as 0.89 mm (0.074 in.) and 1.32 mm (0.052 in.), may also be used depending on the specific application.
- Joint types covered by the guidelines are shown in Fig. 14.3.
- Recommended basic welding positions are shown in Fig. 14.4.

More design guidelines for welding chassis structures of all grades of steels (including AHSS) using the GMAW process are detailed in the ASP document (Ref 14.5). The guidelines follow procedures established by the American Welding Society (AWS) and the American National Standards Institute (ANSI) standards and specifications.

The WorldAutoSteel document "Advanced High Strength Steels Applications Guidelines" (Ref 14.1) includes a section on joining. According to the guidelines, AHSS can be successfully welded without major changes in welding operations or modifications to equipment or processing. However, it must be realized that AHSS derive their properties from their mi-

Joint type	Weld type	Combination of joint and weld type
Lap	Fillet	
Lap	Songle/double-sided slot weld	
T-joint	Fillet	
Corner/butt	Flare bevel/flare V-groove weld	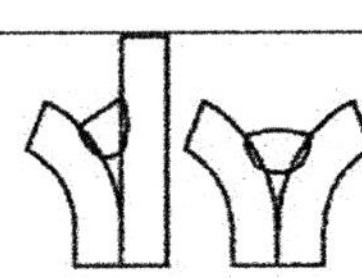

Fig. 14.3 Examples of welds for chassis structures. Source: Ref 14.5

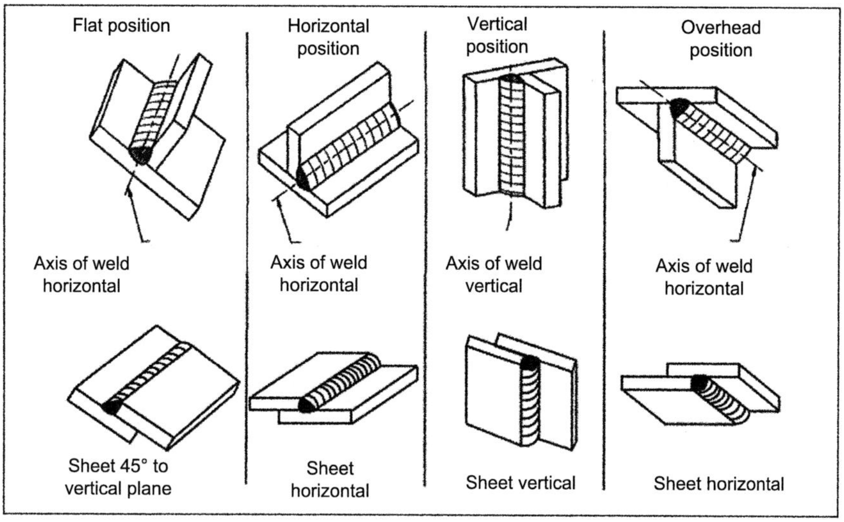

Fig. 14.4 Basic welding positions. Source: Ref 14.5

crostructure, which changes during welding. For example, martensite and/or bainite form due to the fast cooling encountered in welding.

Welding schedules for AHSS may require the following modifications to resistance spot welding, compared to welding of mild steel:

- Electrode force must be increased.
- Weld time must be increased.
- Operating current must be decreased.
- Electrode tip diameter must be increased.
- Type of electrode may need to be changed.
- Minimum weld size must be increased.

For a given button size, spot weld strength was found to increase with increase in the tensile strength of AHSS. Figure 14.5 shows the load-carrying capacity in shear of various grades and types of steels.

Arc welding (GMAW, TIG, and plasma) can be used in a similar way to mild steel including the shield gases. Higher-strength filler wires are recommended for higher-strength AHSS grades. Results of in-field repairs of AHSS indicate that GMAW is acceptable as a repair method; but flame straightening should be avoided because the heating cycle in flame straightening (650 °C, or 1200 °F, for 90 s) softens the repaired part.

14.3 Performance Evaluation (Ref 14.6)

The benefits of using HSS in car body structures and components can be analyzed by using the performance indices that were developed for materials selection. Table 14.1 shows the performance indices that must be considered for selecting the best materials for car body applications.

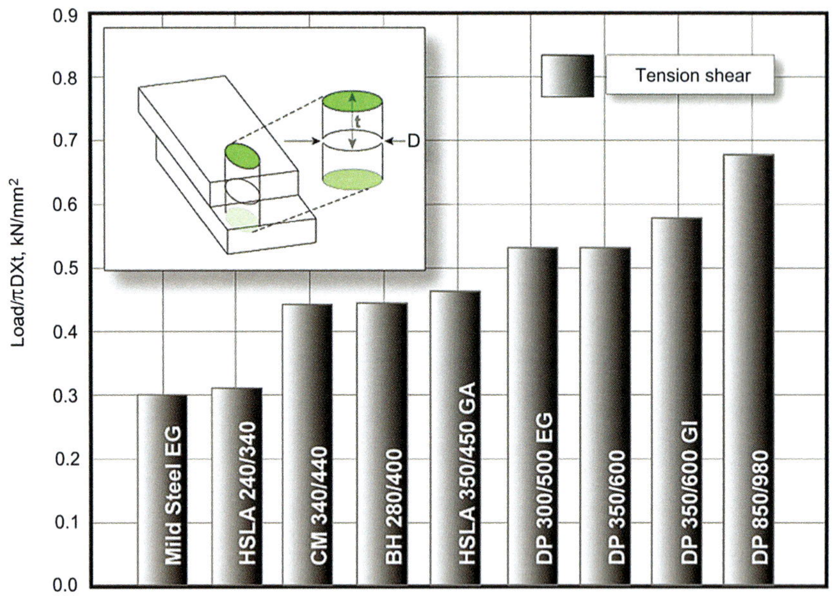

Fig. 14.5 Load-carrying capacity of spot welds of various grades and types of steels. Source: Ref 14.1

Table 14.1 Performance indices for car body applications

Application requirements	Performance index
Stiff and light panel	E/ρ
Dent- and crash-resistant light panel	$(\sigma_y)^{1/2}/\rho$
Tensile/compressive strength of a light panel	σ_y/ρ
Strong and light beam	$(\sigma_y)^{2/3}/\rho$
Strong and minimum-cost beam	$(\sigma_y)^{2/3}/C_v\rho$
Bending stiffness of a light panel	$(E)^{1/3}/\rho$
Buckling stiffness of a light panel	$(E)^{1/2}/\rho$

E, modulus of elasticity; σ_y, yield strength; ρ, density; C_v, relative cost per unit volume

Performance indices are groupings of material properties developed by Ashby (Ref 14.7 to 14.9) to maximize some aspect of the performance of a component. Some of the performance indices are stiffness-specified (specific stiffness) and others are strength-specified (specific strength). Materials with large values of performance indices are the best materials for the intended application.

In the table, E is the modulus of elasticity, σ_y is the yield strength, ρ is the density, and C_v is the relative cost per unit volume. The following example shows how the performance index is applied for selecting the best material for a strong automotive panel with minimum cost.

Consider a sheet metal panel loaded in bending (Fig. 14.6). The panel length is L, its width is w, and its thickness is t. Thickness is a free variable that can change for lightweighting applications. Reducing thickness reduces weight, but the panel must support the bending load, F, without yielding. The panel stiffness is $S = F/\delta$, where F is the total load and δ is the bending deflection.

Following the procedure outlined in Ref 14.10, the performance index, M_p, for a light and strong panel is given by:

$$M_p = (\sigma_y)^{1/2}/C_v\rho$$

where C_V is the relative cost per volume and ρ is the density of the material. The higher the value of the performance index, M_p, the better is the performance of the panel.

The selection chart with the minimum cost guidelines is shown in Fig. 14.7. Assuming a minimum strength of 800 MPa (116 ksi) and a slightly

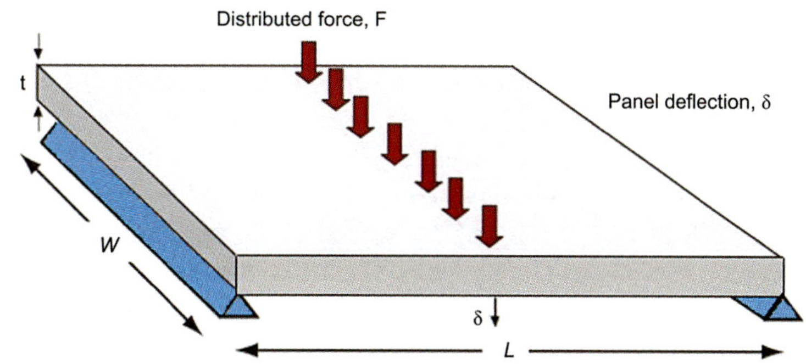

Fig. 14.6 A panel loaded in bending. Source: Adapted from Ref 14.10

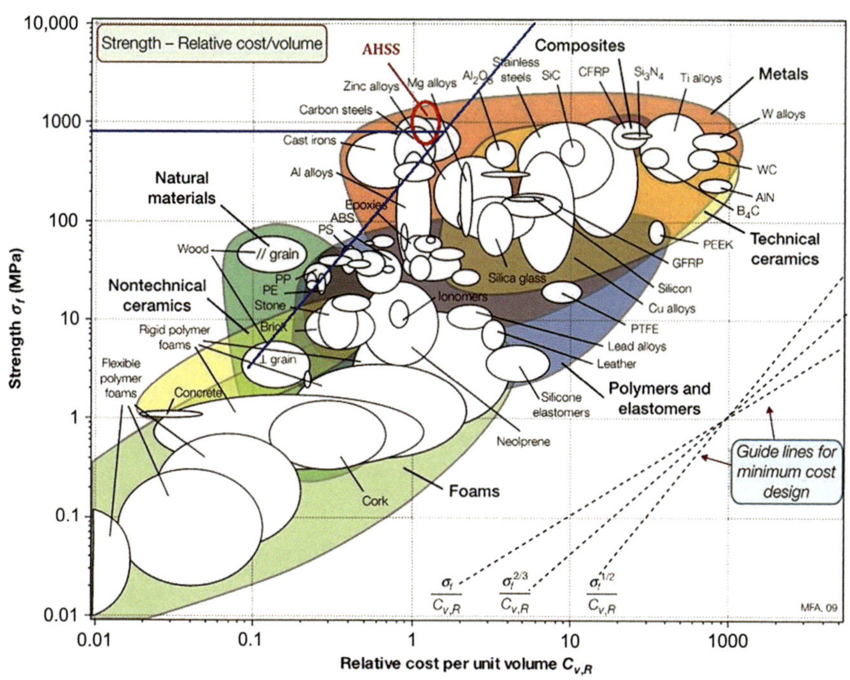

Fig. 14.7 Selection chart for a strong panel with minimum cost loaded in bending. Properties template courtesy of Granta Design. Source: Ref 14.11

higher cost relative to mild steel, AHSS are the best choice for a strong and cost-competitive automotive panel.

REFERENCES

14.1 "Advanced High Strength Steels Applications Guidelines," Version 4.1, WorldAutoSteel, www.worldautosteel.org, June 2009

14.2 "Advanced High Strength Steel Applications: Design and Stamping Process Guidelines—AHSS Case Studies," Auto/Steel Partnership, Jan 2010, www.a-sp.org

14.3 "Advanced High Strength Steel (AHSS) Weld Performance Study for Autobody Structural Components," Auto/Steel Partnership, www.a-sp.org, 2007

14.4 "An Investigation of Resistance Welding Performance of Advanced High Strength Steels," Auto/Steel Partnership, www.a-sp.org, 2007

14.5 "GMAW Weld Design Guidelines for Chassis Structures," Auto/Steel Partnership, www.a-sp.org, November 2007

14.6 M.Y. Demeri, Forming of Advanced High Strength Steels, *Metal Working: Sheet Forming,* Vol 14B, *ASM Handbook,* ASM International, 2006, p 530–538

14.7 M.F. Ashby, Performance Indices, *Materials Selection and Design,* Vol 20, *ASM Handbook,* ASM International, 1997, p 281–290

14.8 M.F. Ashby, Material Property Charts, *Material Selection and Design,* Vol. 20, *ASM Handbook,* ASM international, 1997, pages 266–280

14.9 M.F. Ashby, *Materials Selection in Mechanical Design,* 3rd ed., Elsevier, 2005

14.10 M. Ashby, H. Shercliff, and D. Cebon, *Materials: Engineering, Science, Processing and Design,* 2nd ed., Elsevier, 2010

14.11 Granta Design, www.grantadesign.com

Advanced High-Strength Steels—Science, Technology, and Application
M.Y. Demeri

CHAPTER 15

Innovative Forming Technologies

STAMPING OF AUTOMOTIVE PANELS such as doors, fenders, hoods, roofs, quarter panels, and liftgates is essentially a sheet metal drawing process where sheet blanks are drawn into a die cavity by an advancing punch. Conventional stamping is the most commonly used manufacturing process to produce parts for all types of vehicles. Other manufacturing processes such as roll forming, hot forming, and tube hydroforming are also used.

Incoming sheet materials for such processes come in one of these forms:

- Sheet blank
- Sheet coil
- Laser welded blank
- Laser welded coil
- Flexible (tailor) rolled blank
- Flexible (tailor) rolled coil

Tailor welded blanks (TWB) are sheet metal blanks that are made up of parts of different strengths and thicknesses, joined at the factory mainly by laser welding. Tailor welded blanks and tailor welded coils (TWC) have been used in the automotive industry for about 30 years. Tailor welded coils are usually fed into a progressive die for blanking before moving to the forming stations. The benefits of TWB include weight and cost savings. Over fifty parts have been identified for TWB/TWC applications where optimal steel strengths and thicknesses are placed to improve product performance. In many applications TWB eliminate the need for reinforcements. Examples of automotive applications of TWB include side frames, door inners, B-pillars, cross members, shock towers, floor pans, and rails. The Ultra-Light Steel Auto Body (ULSAB) and the FutureSteel-Vehicle (FSV) programs use TWB in many components of their technology demonstration vehicles. Figure 15.1 shows a sketch of a two-segment TWB for the body side of the FSV vehicle.

Another example for the application of TWB in the FSV program is the unique design of the front rail subsystem for front crash structures (Fig. 15.2). It is designed to achieve minimum weight and best crash performance and is manufactured using TWB with varying gages of transformation-induced plasticity (TRIP) steel.

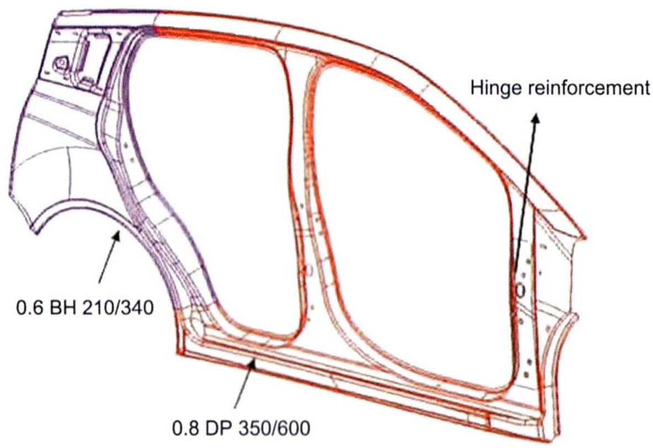

Fig. 15.1 Tailor welded blank for the body side of the FutureSteelVehicle. Source: Ref 15.1

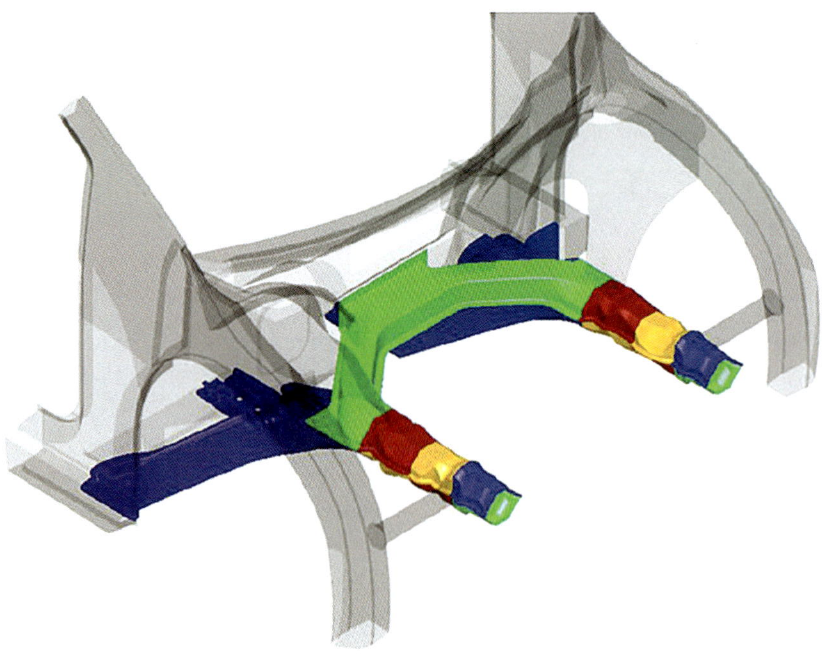

Fig. 15.2 FutureSteelVehicle front rail subsystem. Source: Ref 15.1

A schematic of the tooling used in a conventional stamping process is shown in Fig. 15.3. The basic components of a stamping tool are punch, die, die cushion, and binder. F_b is the binder force and F_p is the punch force. The amount of sheet metal drawn in the die cavity is controlled by binders through the application of restraining forces on the blankholder.

In traditional stamping, sheet metal panels are drawn using rigid binders with constant preset binder forces under open-loop machine control. Binders can be flat or a combination of a fixed drawbead and a constant binder displacement. In flat binders, restraining forces in the drawn sheet are provided by friction between the binder plates and the sheet metal. In fixed drawbead binders, restraining forces result from a combination of plastic deformation and friction. Traditional stamping leaves no flexibility for using difficult-to-form materials and for responding to process variations (lubrication, material, die wear, blank placement) that lead to stamping inconsistencies or even failure.

High-strength steel and AHSS are less formable than mild steel, and parts made from them lack dimensional control because of the significant amount of springback that they produce after forming. The increased use of AHSS to achieve weight reduction in automobiles has led to the development of new tool designs and innovative manufacturing processes. Among these technologies and processes are:

- *Real-time process control:* Adjusting the binder force in-process based on tracking a reference punch force
- *Active drawbeads:* Varying drawbead depth during forming
- *Active binders:* Varying binder force during forming

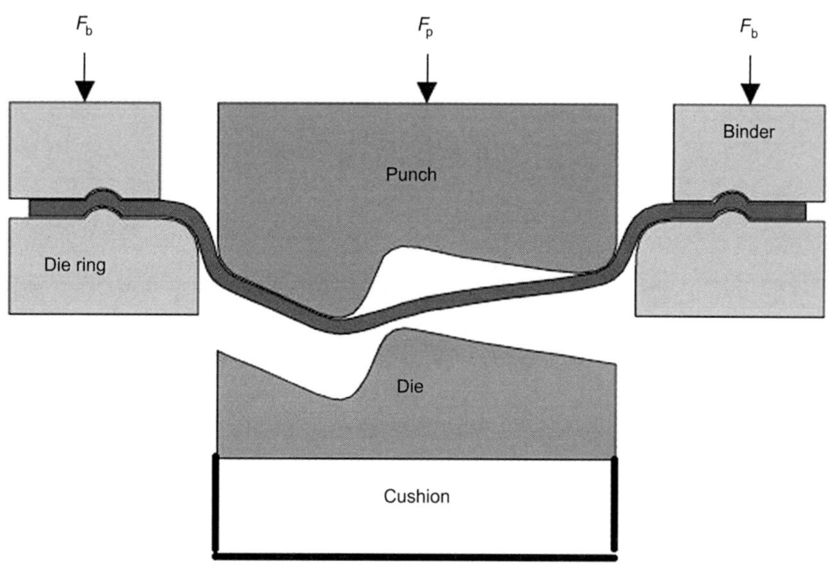

Fig. 15.3 Sheet metal forming system showing the location of the blankholder (binder). Source: Ref 15.2

- *Flexible binders:* Local control of the binder area
- *Flexible rolling:* Varying thickness of flat rolled sheet metal

Because the binder is at the heart of the forming process, three of the five advanced technologies—namely, active drawbeads, active binders, and flexible binders—deal with innovative ways of controlling and manipulating the binder. Real-time process control deals with making sure disturbances to the forming system are neutralized. Flexible rolling provides the advantage of TWB in reducing weight and cost but without the problems associated with welding. These processes and technologies are presented in the next five sections of this chapter.

15.1 Real-Time Process Control (Ref 15.2)

In sheet metal forming processes, the binder force controls material flow into the die cavity, which is critical to producing a good part. The quality of stamped parts is important in avoiding assembly problems and in achieving product performance. Major problems with vehicle quality such as wind noise, water leaks, squeaks, and fit and finish are directly related to the quality and dimensional accuracy of stamped components. The development of production dies and the fine-tuning of the stamping process to produce high-quality parts with low scrap rate is a very expensive and time-consuming process. After production begins, disturbances to the process such as material property variations, changes in friction conditions, and tool wear may cause many parts to be scrapped due to tearing, wrinkling, or poor dimensional accuracy. These problems can be minimized by having a good control of the stamping process.

It has been established that failure by wrinkling or tearing is highly dependent on the magnitude and trajectory of the binder force. Dynamic variation of the binder force during the forming stroke has also been shown to affect formability, strain distribution, and springback. Optimal forming trajectories can be obtained under constant and variable binder force conditions but there is no guarantee that process variables will remain constant during the stamping process. Specifying a binder force trajectory is not easy because the part shape changes during forming. Also, stresses in the part cannot be determined because the coefficient of friction is not a controllable quantity and it varies from location to location. Therefore, the forming process must be controlled and a closed-loop system with an appropriate local control parameter (friction, draw-in) must be used to track a predetermined optimal control parameter trajectory.

Process control can be used to adjust the binder force in real time based on tracking a reference punch force trajectory to improve part quality and consistency. Key issues in process control include process controller and reference punch force trajectory design. Previous research has shown that varying the binder force during forming can improve part quality and consistency.

One strategy for controlling sheet metal forming processes is to use feedback process control. In this strategy, a measurable process variable (e.g., punch force) is controlled by following a predetermined (e.g., punch force-displacement) reference trajectory through manipulation of the binder force. Research was conducted to test and verify this strategy by developing a controller that can be used in forming consistently high-quality cylindrical cups. The developed process model has the binder force as input and the punch force as output (Ref 15.3). The idea behind the controller is to produce cylindrical cups with optimal height, without splitting, regardless of the initial blankholder force and friction condition. Figure 15.4 shows a diagram of the control system used in the experiments.

The feedback control system was applied by modulating the binder pressure to produce a punch force consistent with a desired reference punch force trajectory while rejecting disturbances in the system. Disturbances in the forming system may include variations in the manufacturing parameters such as frictional behavior, blank thickness, material properties, blank placement errors, and die wear. The binder force-generating unit is used as the actuator.

Successful implementation of the feedback control system entails the successful forming of a group of ten cups. Results showed that by using the feedback controller, successful cup drawing increased by almost 50% compared to drawing without the controller. Figure 15.5 shows a photo of a successful and a failed cup.

Although this experiment was performed on mild steel, the concept can be applied to any other metal including AHSS. This research and others that followed proved that real-time adjustment of the binder force can lead to better part quality and improved consistency.

Another example of applying closed-loop control in sheet metal forming was demonstrated in a series of tests to form steel pans with and without closed-loop control. The tests demonstrated that closed-loop control, with wrinkle height sensor, can produce wrinkle-free pans by modulating the binder force in real time. In the first set of experiments, wrinkles in the

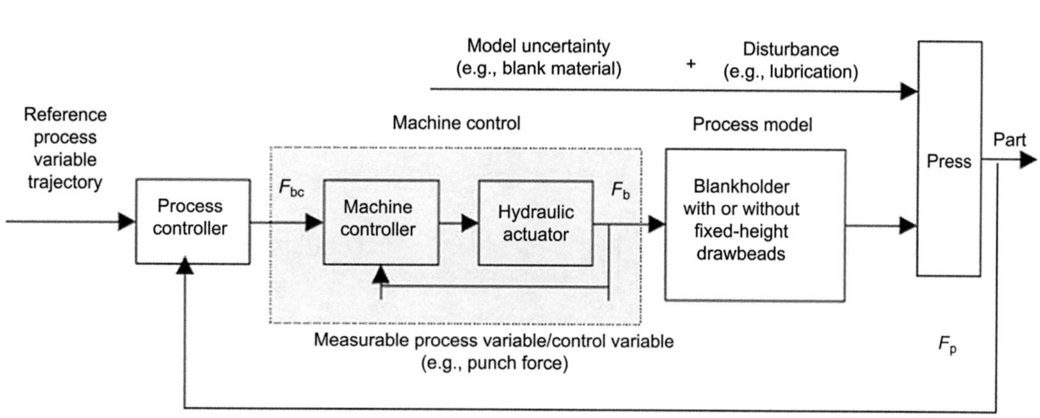

Fig. 15.4 Block diagram of feedback control system. Source: Ref 15.2

steel pan occurred when a small binder force of 6 kN was applied to each of the six activated cylinders under the binder. The wrinkles were over 10 mm (0.4 in.) high and were considered to be severe to the extent that the drawing process was stopped at a depth of 100 mm (4 in.) for fear of damaging the die. In the second set of experiments, the 6 kN binder force was maintained and the closed-loop system was activated. A wrinkle height limit was set at 6 mm (0.2 in.) and the closed loop controller was able to maintain the imposed limit by modulating the binder force on the six cylinders. Figure 15.6 shows steel pans produced with and without closed-loop control. The difference in the quality of the two pans confirms the importance of closed-loop control in sheet metal forming processes.

In conclusion, real-time process control can be used to change a process/control variable, such as the binder force, in such a way that a desired process/control variable can be maintained. Also, it can react to changes in process uncertainties and disturbances such as lubrication, sheet thickness, mechanical properties, blank placement, and die wear.

15.2 Active Drawbeads

Drawbead geometry and position have major influence on the strain distribution in sheet metal forming. Drawbeads generate binder restrain-

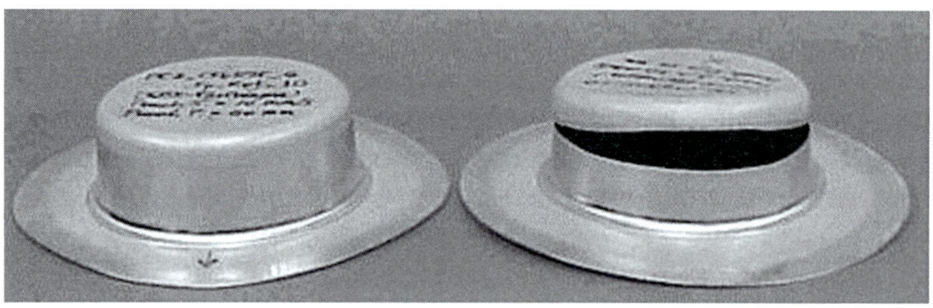

Fig. 15.5 Successful and failed cups. Source: Ref 15.2

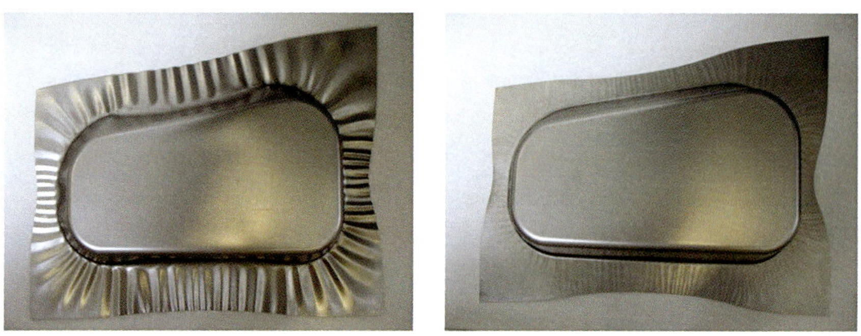

(a) (b)

Fig. 15.6 Steel pans produced (a) without closed-loop control and (b) with closed-loop control. Source: Ref 15.4

ing forces mainly by bending and unbending the sheet metal as it flows into the die cavity. Frictional forces between the sheet and the binder have a relatively smaller contribution in generating the restraining force. The restraining force of a drawbead increases with increase in bead penetration because the sheet metal is subjected to more deformation resulting from bending and unbending. Active drawbeads vary their penetration during the forming stroke; hence varying the restraining force on the sheet metal (Ref 15.5). Figure 15.7 shows a sketch of an active drawbead tooling. The amount of bead penetration, that is, the amount of sheet metal flowing into the die cavity, can be controlled in three situations:

- *Constant binder force:* Bead penetration is constant and determined by a constant value of the binder force (hold-down pressure).
- *Active binder:* Bead penetration varies with the changes in the value of the binder force in real time.
- *Active drawbead:* Bead penetration is controlled by a force transducer that is built into the draw radius of the lower binder.

Active drawbeads add another level of process control in sheet metal forming. They must be used in conjunction with a controller, a sensor, and a force transducer. Extensive work on active drawbeads shows that a controllable restraining force, caused by adjusting the penetration of drawbeads, can improve sheet metal formability (Ref 15.6).

Drawbead restraining forces, at a constant depth, increase with the flow strength of the sheet metal because bending deformation around the bead requires higher stresses for stronger materials. The drawbead restraining forces for AHSS are expected to be higher than those of mild or high-strength, low-alloy (HSLA) steels. All indications show that controlling restraining forces caused by adjusting the penetration of drawbeads can improve formability and dimensional accuracy of stampings.

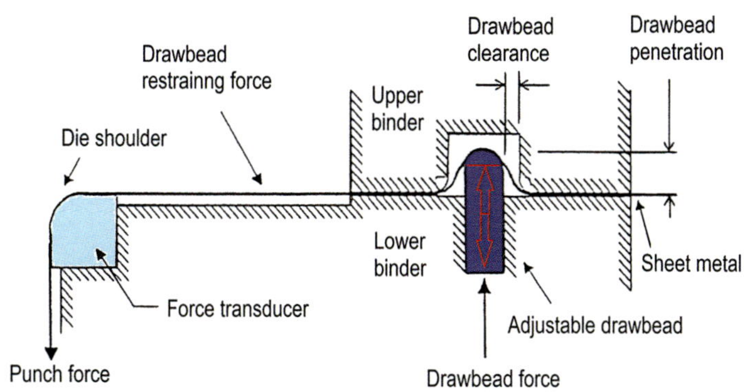

Fig. 15.7 Sketch of an active drawbead tooling.

15.3 Active Binders

Active binders, sometimes called programmable binders or pressure cushions, are designed to provide a variable pressure trajectory in the binder throughout the press stroke. These systems consist of hydraulic cylinders and proportional valves connected to an open-loop programmable controller or a closed-loop control system to manipulate the binder force to meet specific process demands. Active binders differ from air or nitrogen cushion systems in that they can increase or decrease the pressure during the stroke while the others can only increase it.

The binder restraining force has a strong effect on the final part dimension through its effect on springback. Different binder force trajectories (step increase, step decrease, and level change, etc.) produce different strain histories in the blank material, leading to different formability and springback in the final part. Formability and dimensional accuracy can, therefore, be improved by the proper control of the binder force. Representative binder force trajectories are shown in Fig. 15.8.

Experiments were conducted to determine the effect of binder force trajectory on dimensional accuracy of U-channels. In one study, the effects of two binder force trajectories on the springback of DP 590 U-channel forming were investigated (Ref 15.7).

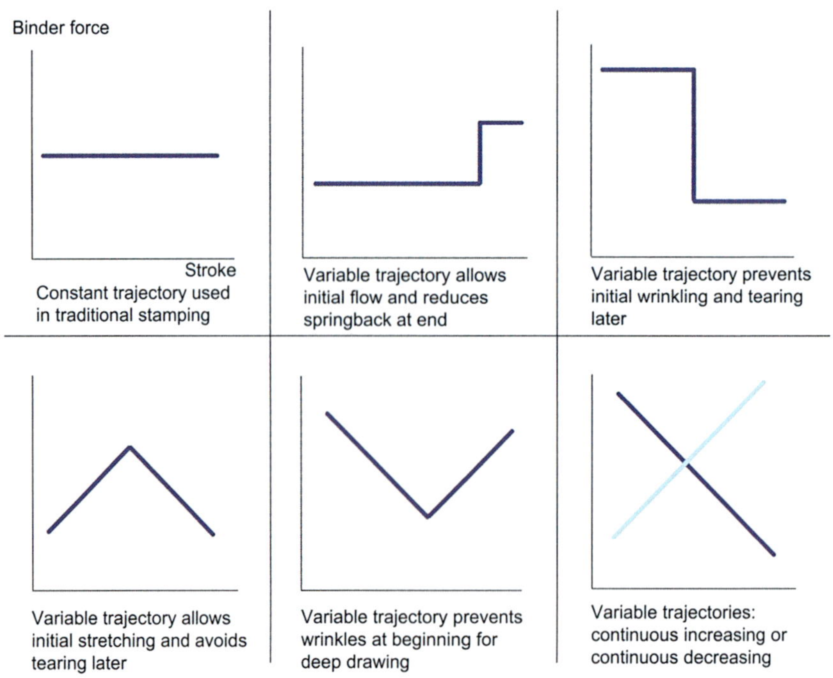

Fig. 15.8 Representative binder force trajectories and their effect on sheet forming

The geometry of the U-channel and its setup are shown in Fig. 15.9. A hydraulic press with a programmable cushion was used. The cushion was activated by four independently programmed cylinders where the tonnage in each cylinder could be adjusted up to 12 times during the stroke. Two binder force trajectories were used in the experiment (Fig. 15.10):

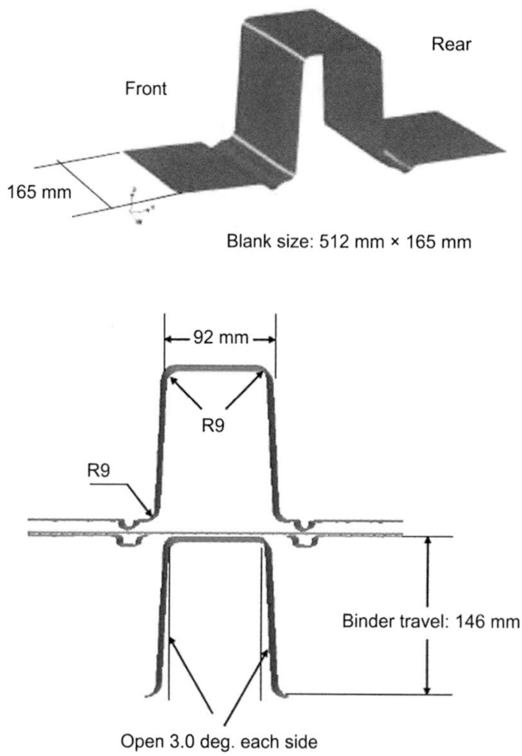

Fig. 15.9 Geometry of the U-channel and its setup. Source: Ref 15.7

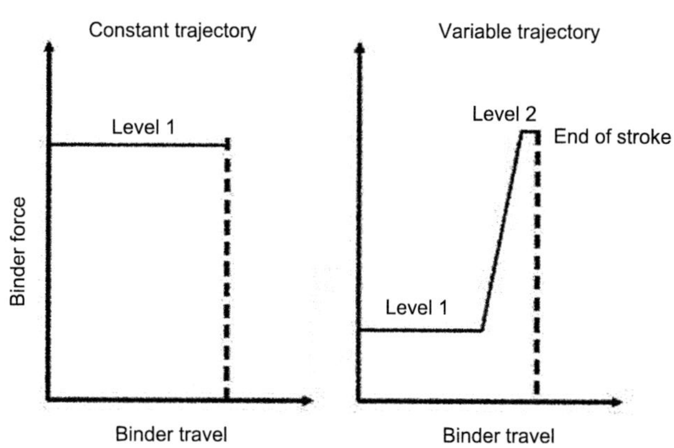

Fig. 15.10 Binder force travel trajectories. Source: Ref 15.7

- *Trajectory 1: Constant binder force (CBF).* The constant binder force (CBF) was applied during the entire stroke. As is known, springback decreases with increasing binder force. The maximum binder force was set to be 3 tons below the force required to cause fracture in the U-channel.
- *Trajectory 2: Variable binder force (VBF).* Low-level binder force, sufficient to prevent separation between the sheet and the binder, was applied during most of the stroke. The binder force was then increased, just before the end of the stroke, to a maximum level of 3 tons below the force required to cause fracture. The low level allowed the material to flow while the high level locked the material.

Results of the experiments are shown in Fig. 15.11 for the constant and the variable trajectories. Profiles of the U-channel indicate that using variable binder trajectory reduced springback compared to the constant trajectory. This result is expected because applying a higher binder force just before the end of the stroke results in stretching the wall of the U-channel, causing the strain to be tensile across the entire sheet thickness thereby reducing the plastic moment that causes springback.

For trajectory 1 (constant binder force), the maximum binder force was 70 tons. Fracture occurred when 3 additional tons were added to the binder force. For trajectory 2 (variable binder force), the minimum low-level force was 25 tons and minimum springback occurred when the binder force was programmed to change from 25 to 145 tons at 38 mm (1.5 in.) off bottom.

Flexible binder technology, where multi-point cushion systems are equipped with independently programmed cylinders, were used in the industry as far back as 1994. Figure 15.12 shows a sidewall panel die with

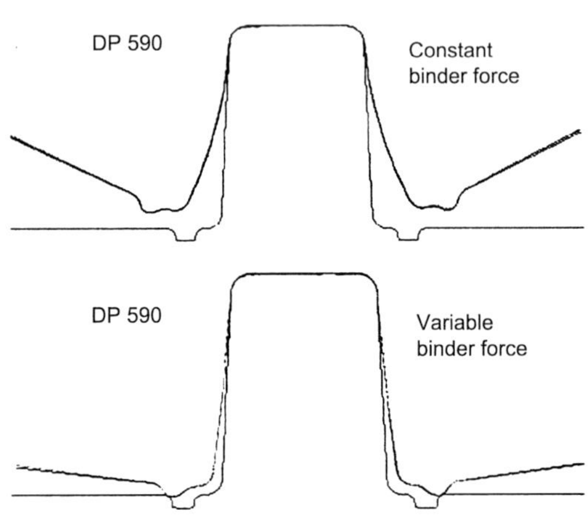

Fig. 15.11 Effect of binder force trajectory on springback of DP 590 steel. Source: Ref 15.7

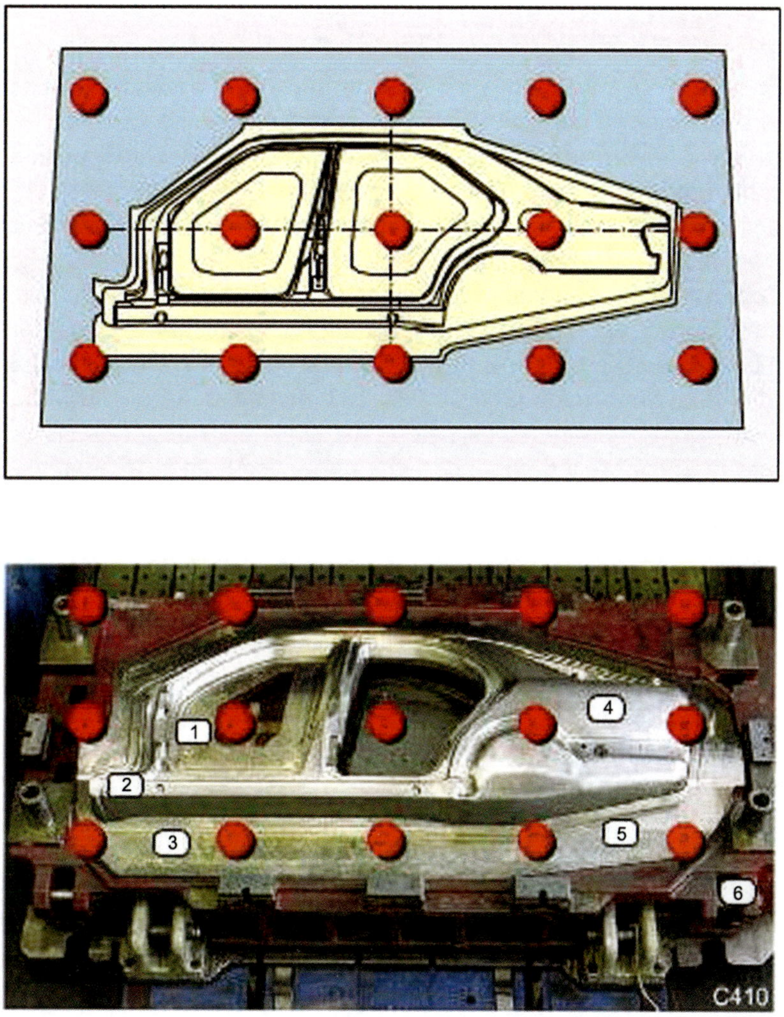

Fig. 15.12 Multipoint active binder. Photo courtesy of Muller-Weingarten

a multipoint cushion system containing 15 individually controlled hydraulic cylinders, installed at the Daimler Benz plant in Sindelfingen.

Although most new presses have the capability to handle multi-point cushion systems, using that capability in stamping has been hampered by the complexity associated with implementation of flexible binder technology.

In conclusion, active binders, with the capability to change binder pressure in real time, can improve formability and dimensional accuracy of all steels, especially AHSS, which are less formable and have higher springback than mild and conventional steels. In the U-channel experiments, the variable binder profile reduced springback by 60% compared to the constant binder trajectory.

15.4 Flexible Binders (Ref 15.8)

The quality of a deep-draw part is determined by the amount of material that flows into the die cavity. An excessive flow causes wrinkles, while insufficient flow results in fracture. The binder plays a key role in regulating the material flow by exerting a predefined binder force. When designed correctly, the binder force profile can prevent wrinkling and delay tearing in the formed part.

Conventionally, the stamping industry uses rigid tooling placed on top of a cushion system composed of a certain number of nitrogen cylinders. The binder force is applied uniformly on the binder surface and held constant during the forming process. Also, the magnitude of binder force for a given part is determined based on past experience and trial and error. Figure 15.13 shows a rigid binder with nitrogen cylinders to align the die and absorb initial forming impact. The system has limited flexibility and requires an elaborate and costly die tryout process to produce quality panels.

Significant effort is spent on fine tuning dies to eliminate wrinkles or splits in formed panels. Fine tuning the die is an iterative process of welding, grinding, polishing, and testing to produce defect-free parts. This process is time consuming and expensive. Also, there is an increasing trend of using strong metals for lightweight applications in the transportation industry. Such metals are less formable than the mild and traditional

Fig. 15.13 Rigid binder with nitrogen cylinders. Courtesy of TDM/Ford Motor Co.

HSLA steels that they are replacing, and this has pushed the stamping industry to consider alternative methods to forming lightweight and high-specific-strength metals. Because sheet metal stamping is controlled by the binder, efforts to modernize the process center on making the binder flexible and controllable. The effort culminated with the development of flexible binders. Figure 15.14 shows a schematic of a flexible binder system with hydraulic cylinders. The system provides individual control of local areas in the binder. It can vary cylinder pressure in space and time, thereby reducing expensive and time-consuming die tryout.

In the previous section on active binders, variable, instead of constant, binder force profiles were used to improve the formability and springback of drawn parts. To implement binder force profiles, a cushion with independently programmed hydraulic cylinders is needed. Tonnages in each cylinder can be adjusted during the stroke. Although different cylinders can have different pressures, the effect of such variation on the binder area is not localized but diffused because of the rigid structure of the binder. So, the effectiveness of active binders in providing full control of the binder area is diminished by the design of their rigid structures.

Innovative binder concepts and tool designs provided a solution to rigid binders and provided more local control of the binder areas. The Institute for Metal Forming Technology in Stuttgart, Germany, developed the idea

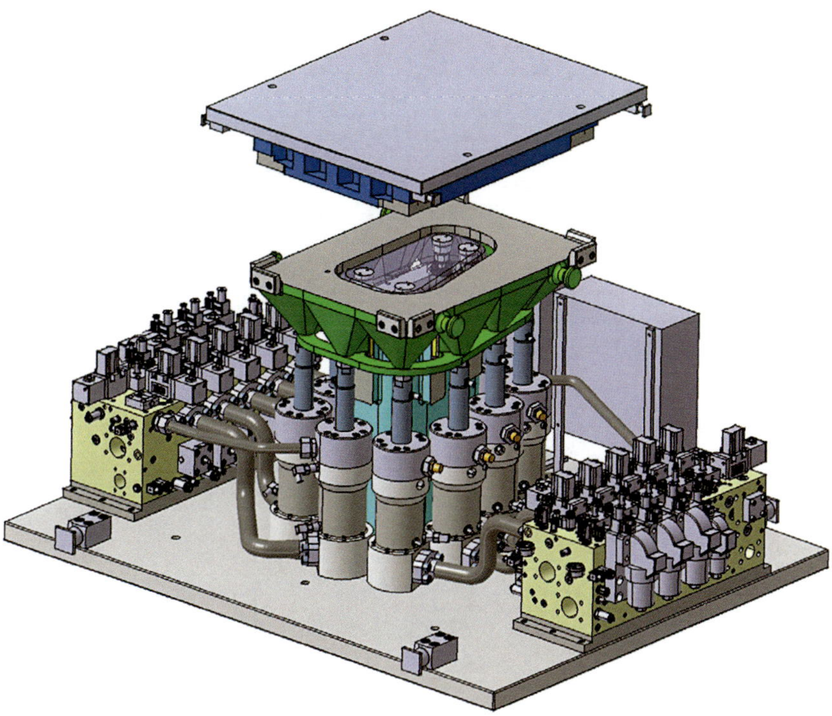

Fig. 15.14 Schematic of a flexible binder system with hydraulic cylinders. Source: Ref 15.9

of segmenting the lower binder to reduce its rigidity and allow for local, instead of global, control of the binder area. The idea of a superflexible binder load control system was implemented first on experimental and later on full-size industrial parts.

The basic components of a flexible binder control unit are shown in Fig. 15.15. The lower binder has cone-shaped segments to provide local binder force control and the upper binder employs honeycomb structures to provide rigidity and lightweight. Building flexible binders requires optimization of the tool designs to achieve flexibility in the lower binder and lightweight rigidity in the upper binder.

The unit is designed to draw an experimental generic pan that captures the main features of industrial stamping. It has ten hydraulic cylinders and is equipped with a closed-loop control system with different types of sensors (wrinkle height, friction force, draw-in, and punch force) to provide effective process control in real time. The closed-loop control system is shown in Fig. 15.16 where friction force and wrinkle height are used as the measurable process/control variables. A user-friendly interface with touch screen capability to select individual cylinder pressures is a key component of the system.

Various experiments were conducted using this unit to test advanced binder control concepts such as active binders with variable force profiles

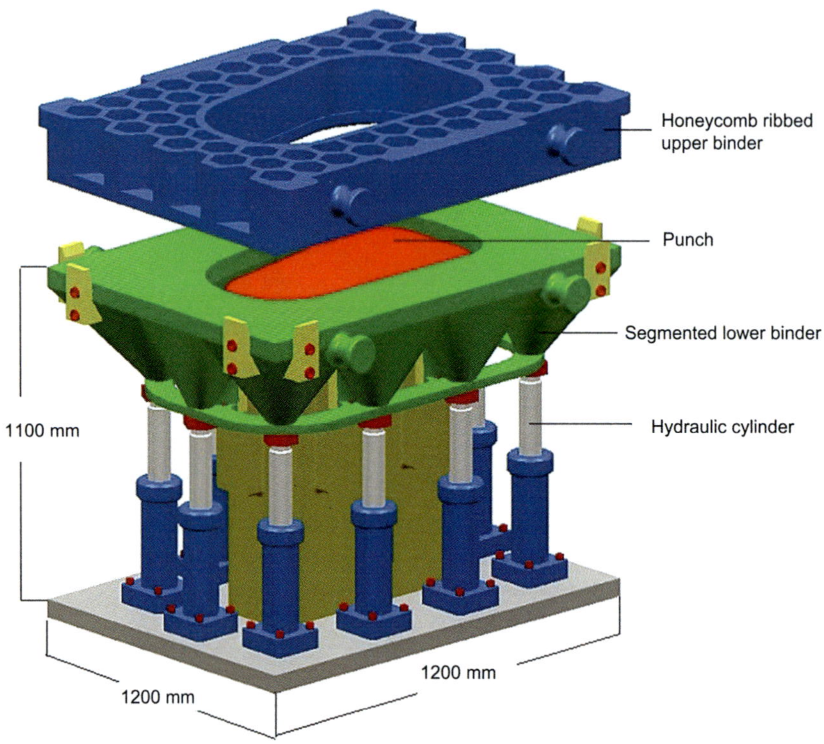

Fig. 15.15 Basic components of a flexible binder control unit. Source: Ref 15.4

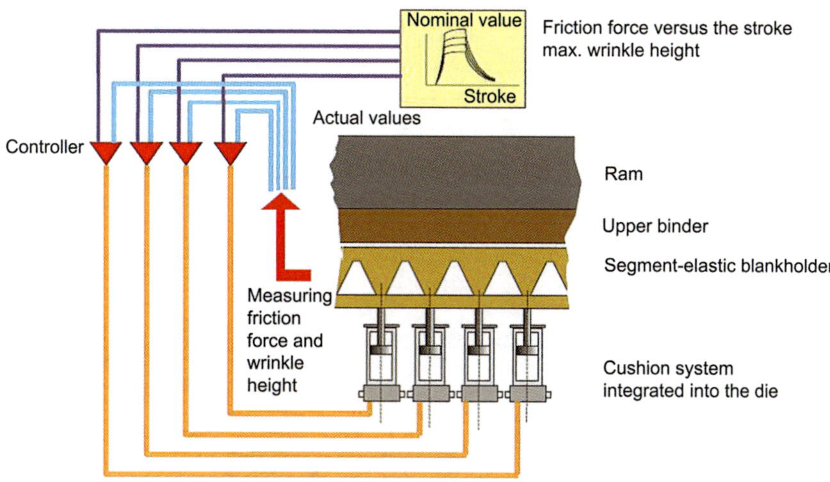

Fig. 15.16 Real-time process control system for flexible binder. Source: Ref 15.9

and closed-loop control. Test results indicate that the flexible binder control system with variable binder force profiles and closed-loop control produced pans with better quality than those produced in rigid tooling.

Segmented flexible binders have been commercially used in Germany to produce kitchen stainless steel double sinks. However, they are implemented in hydraulic, not mechanical presses. Because most of the stamping presses in the automotive industry are mechanical, there was an interest in exploring the fitness of mechanical presses to handle segmented flexible binders with their multi-point cushion systems.

The successes demonstrated by a number of experimental segmented flexible binder systems in improving part quality and dimensional accuracy in hydraulic presses encouraged the United States Counsel for Automotive Research (USCAR) consortium to support a cooperative program to expand this technology to mechanical presses. The program objective was to develop and demonstrate, on an industrial scale, an optimized closed-loop flexible binder control system that can be installed in mechanical presses. The selected part was a liftgate outer and the selected materials were aluminum alloy, bake-hardenable (BH), and dual-phase (DP) steels.

A full-size flexible binder control unit with a hydraulic cushion consisting of 26 independently controlled cylinders with full-size liftgate inner tooling was built and successfully installed in a mechanical press. Figure 15.17 shows the lower segmented flexible binder for the liftgate inner panel.

The actual liftgate die mounted in a mechanical press is shown in Fig. 15.18. The pyramidal structure in the lower binder is clearly seen. The lower die sits on top of a hydraulic cushion with 26 independently controlled cylinders.

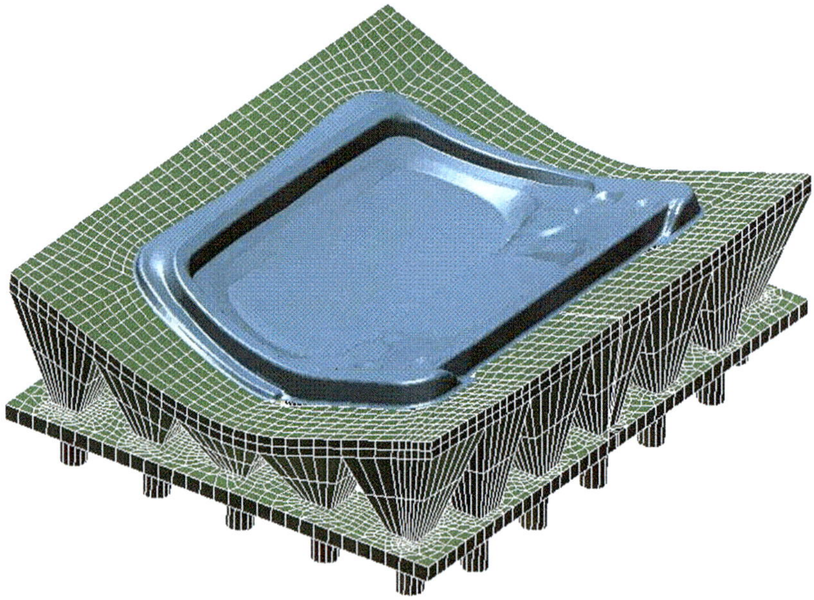

Fig. 15.17 Segmented lower die for the liftgate panel. Source: Ref 15.8

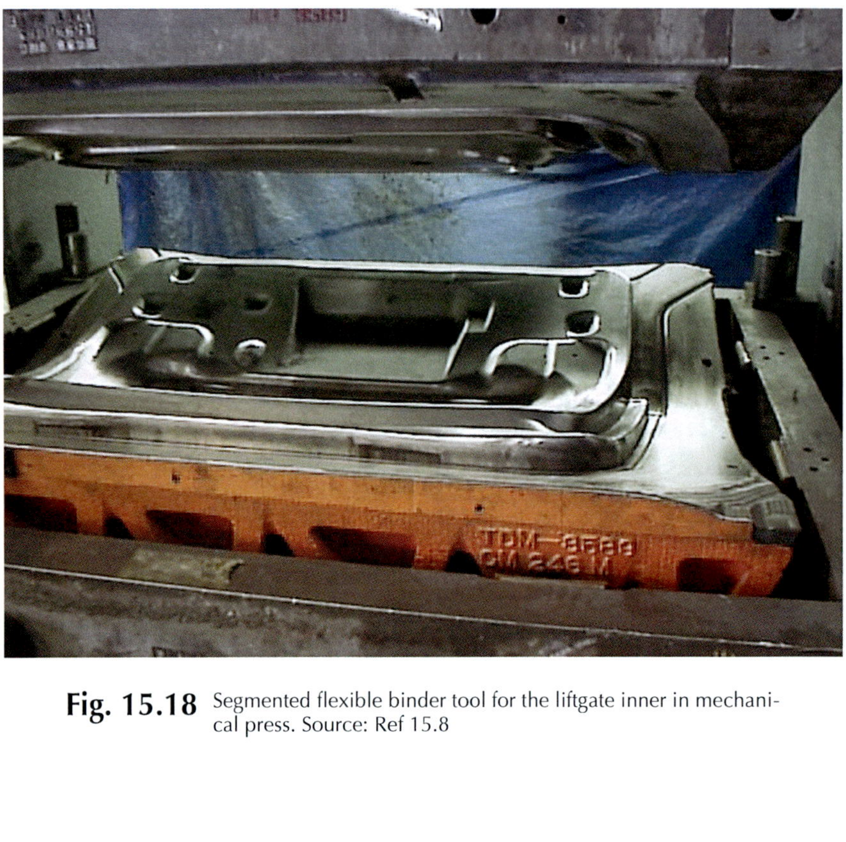

Fig. 15.18 Segmented flexible binder tool for the liftgate inner in mechanical press. Source: Ref 15.8

Constant and varying binder force trajectories generated by both trial and error and finite-element analysis (FEA) simulation were implemented and liftgate panels from BH 210, DP 500, and A6111-T4 were successfully made using the same set of tools. The tryout time using the system was considerably less than that using the conventional method of working the die. For example, the BH part was made in approximately four hours compared to about five days with die work, while the aluminum part was made in about three days. Splits could literally be healed in minutes by adjusting the tonnages of the appropriate cylinders. Prior efforts to make the part with aluminum A6111-T4 using conventional methods were unsuccessful.

Figure 15.19 shows liftgate inners made from 6111-T4 aluminum, BH 210, and DP 500 steel. In all cases, the parts were formed without any splits on the first run. However, there was some wrinkling that was corrected by trial and error adjustment of the cylinder tonnages.

Predicted optimized binder force trajectories, varying in space but constant during the stroke, for A6111-T4, BH 210, and DP 500 are shown in

Fig. 15.19 Liftgate inners made from three different metals using the same flexible tooling. Source: Ref 15.8

Fig. 15.20. The trajectory patterns for the three materials are similar. They only differ in the level of force applied on each pin. This force level follows the same pattern followed by the yield/tensile strength of those materials.

Stamping trials have shown that flexible binder control technology with closed-loop control, process simulation, and optimization can improve part quality and consistency, enhance process robustness, and reduce the

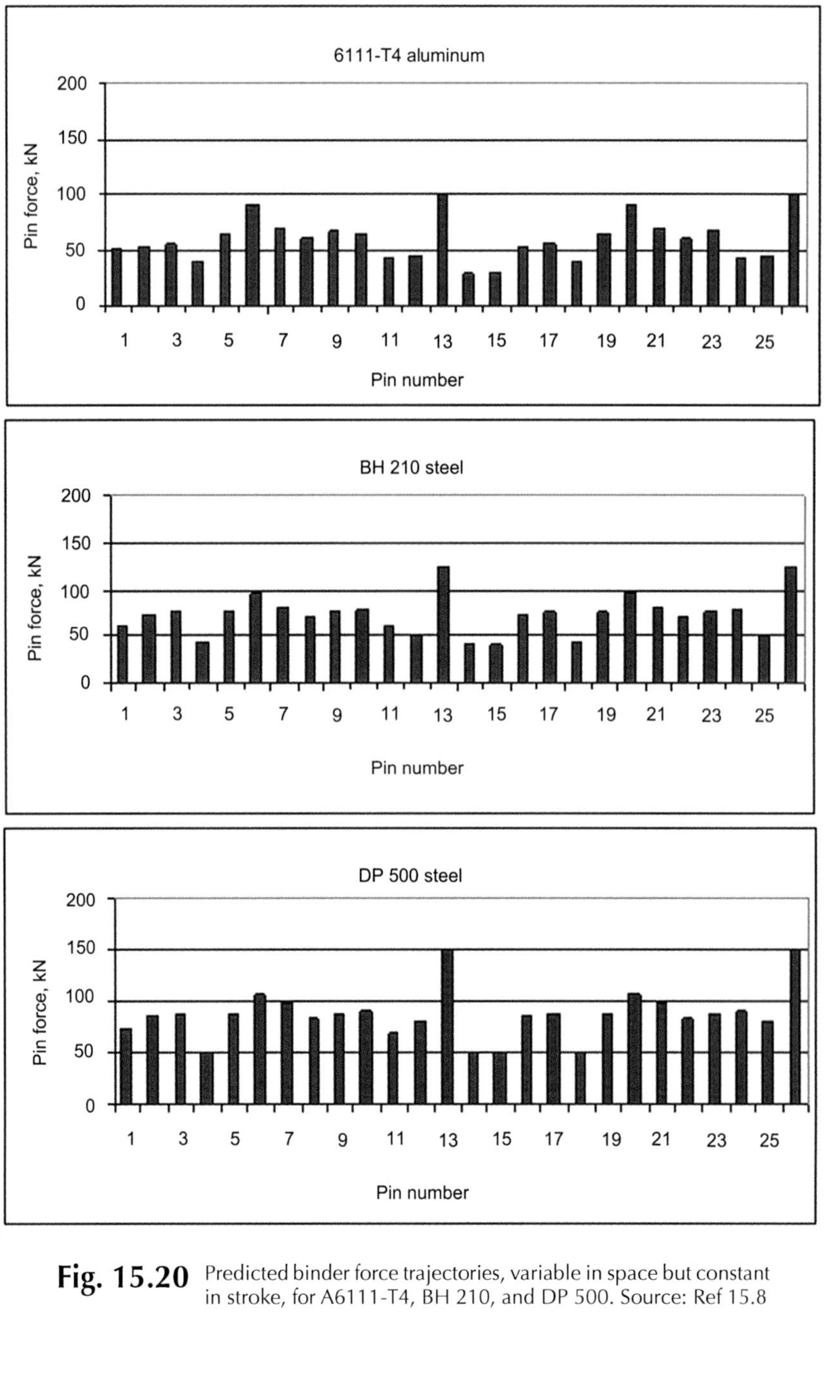

Fig. 15.20 Predicted binder force trajectories, variable in space but constant in stroke, for A6111-T4, BH 210, and DP 500. Source: Ref 15.8

time and cost for die tryout and fine tuning. If implemented, this technology will lead to significant improvement in product development by allowing more aggressive designs and utilization of difficult-to-form materials.

In summary, the following benefits can be realized by using flexible binder control systems to stamp high-strength, low-ductility, and elevated-springback metals such as AHSS:

- Reduce die tryout and development time
- Reduce scrap due to wrinkling and tearing
- Reduce time to tune/spot binder
- Compensate for tool wear by adjusting binder force during production
- Reduce tool rework during production
- Improve part quality and dimensional accuracy
- Change pressure with stroke
- Reduce springback and residual stresses
- Stamp difficult-to-form materials
- Reverse pressure settings using software (unlike grinding and welding)

15.5 Flexible Rolling

Flexible rolling is a cold sheet metal rolling process that competes with the process of producing tailor welded blanks (TWB). It is a relatively new process that uses a rolling mill, with a modified rolling gap control system, to produce sheet metal coils with a varying thickness profile in the rolling direction. This process is similar to flat rolling except that the upper roll in flexible rolling can change the rolls gap by means of a defined movement during rolling to produce variation in thickness along the length of the coil. Figure 15.21 shows a traditional flat rolling process where the rolls gap is kept constant to produce a constant reduction in coil

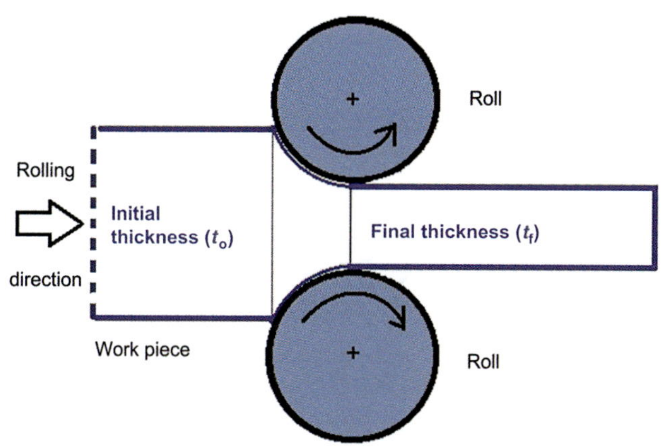

Fig. 15.21 Traditional flat rolling process

thickness. Both traditional and flexible flat rolling are deformation processes where incoming coil thickness is reduced by compressive forces exerted by two opposing rolls.

The principle of producing a flexible (tailor) rolled coil is shown in Fig. 15.22. The flexible rolling system is equipped with a real-time closed-loop control system to ensure that the coil thickness profile is within the tolerances of the process. A close-up of a typical thickness profile is shown in Fig. 15.23. It is important to notice that thickness transitions are smooth so that stress concentrations are avoided.

In flexible rolling, coil sections are compressed differently from location to location. This produces local variation in flow strength and strain hardening along the length of the coil. Figure 15.24 shows the flow strength and thickness profiles in a cross section of a flexible rolled coil. This condition leads to inhomogeneous mechanical properties and requires an annealing process to recrystallize the metal and regain its original mechanical properties (strength and ductility).

The formability of HSLA steel samples taken from (1) regular constant thickness flat rolling blank, (2) two-thickness TWB, and (3) two-thickness flexible rolling and annealed blank were compared using the limiting dome height test. Figure 15.25 shows results for the three samples.

Ductility, as measured by the dome height test, was lower for the tailor welded sample compared to both the regular and the flexible rolled ones.

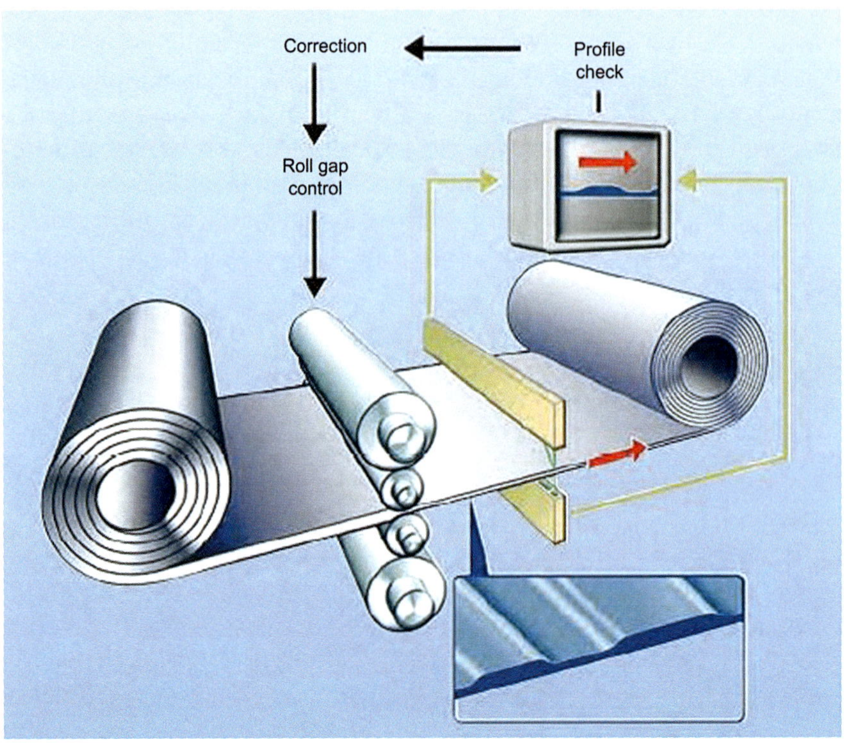

Fig. 15.22 Principle of producing a flexible (tailor) rolled coil. Source: Ref 15.1

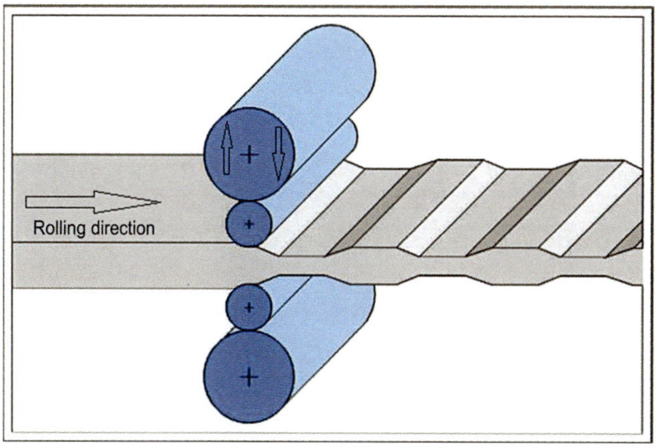

Fig. 15.23 Thickness profile in flexible rolling. Source: Ref 15.1

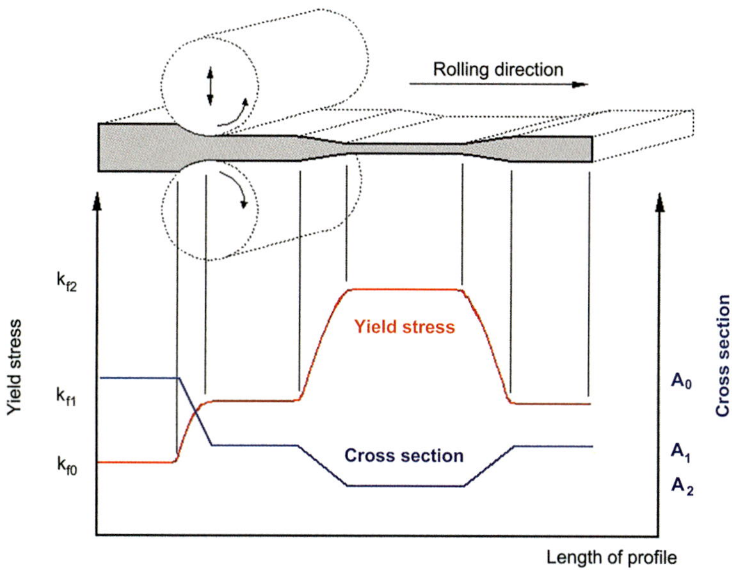

Fig. 15.24 Cross section and flow strength profiles in flexible rolled coil. Source: Ref 15.10

It was also found that the dome height of the flexible rolled and annealed samples increased with increasing length of thickness transition between thick and thin sections. If the thickness transition length is comparatively large, the dome height value approaches that of the constant thickness sample.

Flexible rolling can be used in applications similar to those of TWB. A deep-drawn experimental pan, with two different thicknesses of 0.9 and 1.4 mm (0.04 and 0.06 in.) in three locations, is shown in Fig. 15.26.

Flexible rolling technology has been used by European car manufactures for many vehicle applications ranging from cross members, side

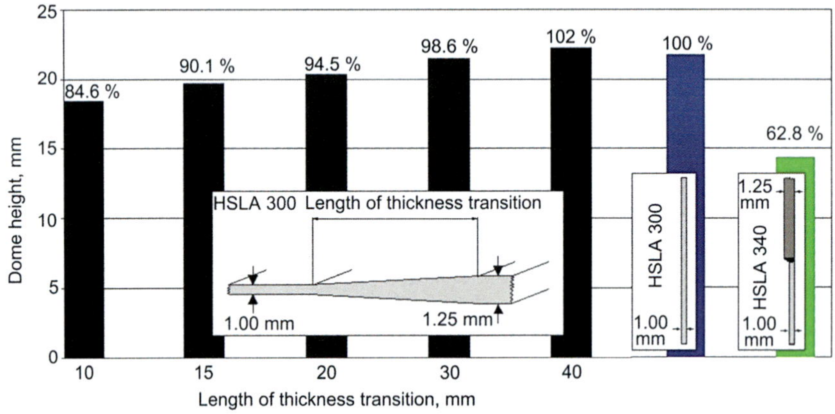

Fig. 15.25 Dome heights for regular, tailor welded, and flexible rolled blanks of high-strength, low-alloy (HSLA) steel. Source: Ref 15.10

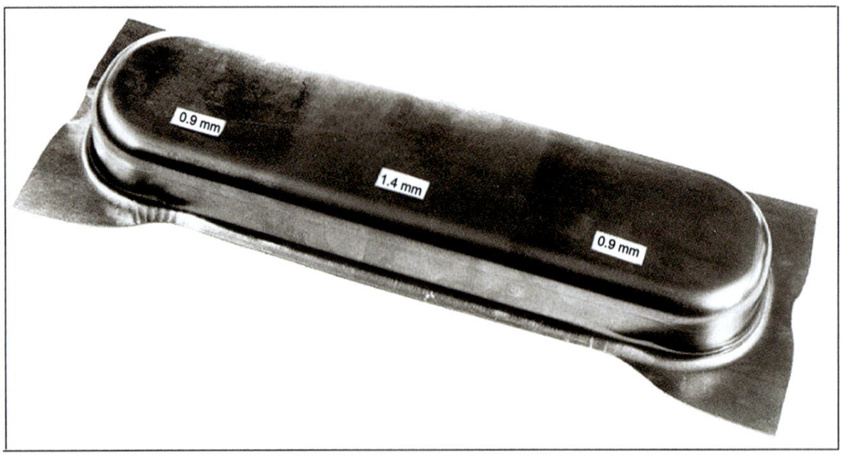

Fig. 15.26 Experimental pan made from flexible rolling blank. Source: Ref 15.10

members, crash boxes, B-pillars, and hydroformed tubes. Figure 15.27 shows photos of the various applications of flexible rolling for automotive applications. In the United States, the excellent crash performance rating of the 2012 Ford Focus is attributed to a newly designed AHSS B-pillar produced from flexible rolling blanks. This technology allows the component to increase strength in areas that are subjected to the highest loads, resulting in improved crash performance of the vehicle.

In conclusion, flexible rolled blanks offer an alternative to TWB for reducing vehicle weight without the problems of stress concentration in sharp thickness transitions, welding seams, and heat-affected zones. Other advantages include simplified manufacturing by offering coil-to-coil production and strength control that varies with thickness variation.

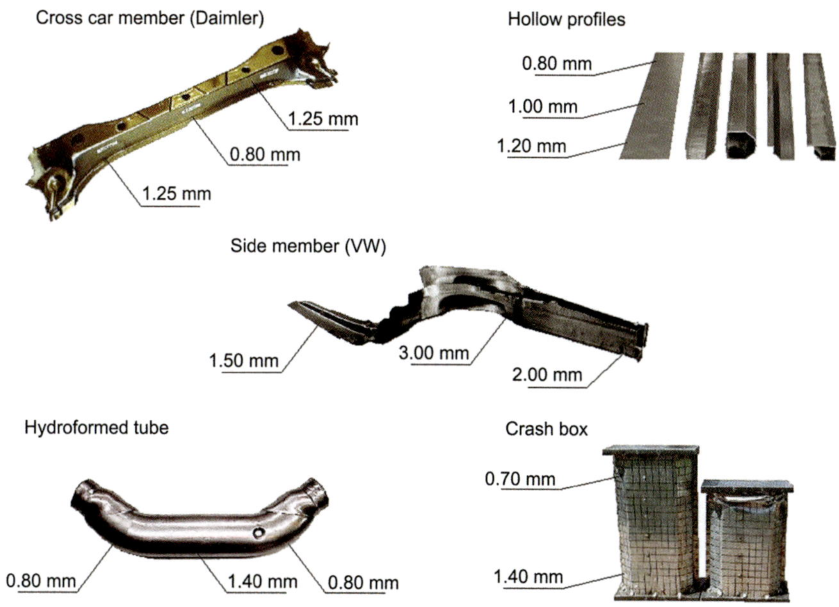

Fig. 15.27 Various applications of flexible rolling blanks. Source: Ref 15.10

REFERENCES

15.1 WorldAutoSteel, "FutureSteelVehicle (FSV)," Final Engineering Report, Steel Market Development Institute, May 2011, www.worldautosteel.org

15.2 M.Y. Demeri, J. Cao, and R. Venugopal, Process and Feedback Control for Manufacturing, *Metal Working: Sheet Forming,* Vol 14B, *ASM Handbook,* 2006, p 840–850

15.3 C.-W. Hsu, A.G. Ulsoy, and M.Y. Demeri, Development of Process Control in Sheet Metal Forming, *J. Mater. Process. Tech.,* Vol 127 (No. 3), 2002, p 361–368

15.4 J. Hengelhaupt and M. Vulcan, "Active Flexible Binder Control System for Robust Forming," Final Project (AMD 301) Report, Institute for Metal Forming Technology, Stuttgart, Aug 2006

15.5 R. Li and K.J. Weinman, Non-Symmetric Panel Forming of AA 6111-T4, Using Active Drawbeads in Sheet Metal Forming Technology, M.Y. Demeri, Ed., *TMS Proceedings,* Feb 28–Mar 4, 1999 (San Diego, CA), p 39–52

15.6 R. Li and K. Weinmann, Formability in Non-Symmetric Aluminum Panel Drawing using Active Drawbeads, *CIRP Ann.-Manuf. Techn.,* Vol 48 (No. 1), 1999, p 209–212

15.7 C. Du et al., Springback Control with Variable Binder Force—Experiments and FEA Simulation, *NUMIFORM 2004—Proc. of the 8th International Conference on Numerical Methods in Industrial Forming Processes,* AIP Conf. Proc. 712, 2004, p 970–976

15.8 M.Y. Demeri, "Flexible Binder Control System for Robust Forming," Final Project Report of AMD301, Contract # FC26-020R22910, USAMP/USCAR, Dec 10, 2006
15.9 K. Siegert and M. Vulcan, "Tool and Die Design for Deep Drawing AHSS," IFU Presentation, Great Designs in Steel, WorldAutoSteel, 2005, Livonia, Michigan
15.10 C. Greisert et al., Forming Behavior of Tailor Rolled Blanks, *Innovations in Processing and Manufacturing of Sheet Materials,* M.Y. Demeri, Ed., TMS, 2001, p 161–172

CHAPTER 16

Sustainability and Economics of Advanced High-Strength Steels

SUSTAINABILITY is a complex combination of an interdependent global system comprising of economic, social, political, and environmental entities whose goal is to ensure that human needs are met without compromising the needs of future generations or damaging the environment. It is influenced by actions and decisions taken by individuals, communities, societies, governments, and countries throughout the world. For materials, sustainability implies a link between depletion of limited resource, energy consumption, and protection of the environment. Materials are being consumed at incredible rates because of the increase in population and the growth of living standards. The supply of most materials comes from nonrenewable sources, which means that materials have to be managed in a sustainable way to ensure their availability for future generations. Improvement in processing, material substitution, light weighting, and recycling have contributed immensely to the cause of sustainability in the materials cycle.

The World Steel Association identifies eight indicators of sustainability (Ref 16.1):

- Greenhouse gas (GHG) emissions (tons CO_2/ton steel cast)
- Energy intensity (GJ/ton steel cast)
- Material efficiency (% of by-products reused)
- Environmental management systems (EMS) (% of employees)
- Lost time injury frequency rate (injuries/million hours worked)
- Employee training (training days/employee and year)
- Investment in new processes and products (% of revenue)
- Economic value distributed (% of revenue)

The first three indicators belong to the environmental area and are of interest in the production and use of advanced high-strength steels (AHSS). Greenhouse gas emissions include all emissions from any process to produce, use, and dispose of vehicles. Life cycle emissions for a vehicle are divided into four stages (Fig. 16.1): (1) material production, (2) vehicle manufacturing, (3) vehicle use, and (4) vehicle disposal.

There is a close connection, if not equivalency, between energy consumption and CO_2 emission. The energy consumed at each stage is indicative of the amount of emissions produced. For the material production stage, two approaches can be used to assess emissions. One is based on embodied energy and the other is based on the CO_2 footprint. Embodied energy of a material is the energy, in MJ, required to produce 1 kg (2 lb) of usable material (steel bars, plates, or sheet). The CO_2 footprint of a material is the amount of CO_2, in kg, emitted during the production of 1 kg (2 lb) of usable material.

16.1 Advanced High-Strength Steels and the Environment

The U.S. transportation sector consumes the largest percentage of petroleum and is the major contributor to GHG emissions in the United States. Approximately 98% of the CO_2 emissions come from petroleum fuels. The auto industry's vehicles, which include cars, light trucks, and heavy trucks, are responsible for the majority of GHG emissions (CO_2, CO, and NOx). Vehicle weight reduction to improve fuel efficiency has been adopted as a GHG emission reduction strategy. Many regulations in the United States and around the world, including the Kyoto Protocol of 1997, aim at reducing the amount of GHG emissions and have been adopted to reduce damage to the environment.

Steels, in general, and AHSS, in particular, are environmentally friendly and sustainable materials for these reasons:

- Steelmaking requires less energy than making other alternative materials.

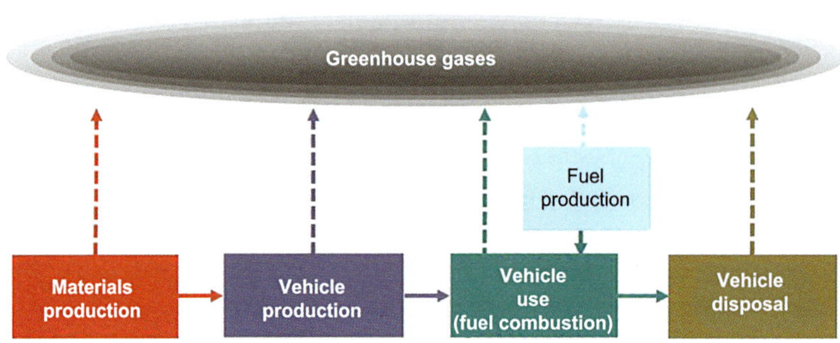

Fig. 16.1 Stages in life cycle emissions. Source: Ref 16.2

- Greenhouse gas emissions associated with making steel have been reduced.
- Using AHSS reduces the amount of steel produced per car, saving raw materials.
- Advanced high-strength steel can be produced on the same equipment currently used.
- Advanced high-strength steel reduces vehicle mass and leads to improved fuel economy and reduced emissions.
- End-of-life recycling reduces cost, energy, and raw material depletion.

Figure 16.2 shows a comparison of the carbon footprint for producing steel and three other alternate materials for lightweighting (aluminum, magnesium, and carbon fiber reinforced polymers). It is evident that producing steel generates CO_2 emissions that are 5 to 20 times lower than the other materials.

Reduction in CO_2 emissions are directly related to reductions in energy intensity for producing materials. The North American steel industry reduced the energy intensity by 27% and the CO_2 emissions by 33% per ton of steel produced between 1990 and 2010 (Fig. 16.3).

Emissions produced by the steel industry result mostly from the chemical reaction to reduce the iron ore to crude iron in the blast furnace. This is also the process that consumes the most energy. Since 1975, energy efficiency methods have led to reductions of approximately 50% of the energy required to produce a ton of crude steel.

There are three production routes for manufacturing crude steel: basic oxygen furnace (BOF), electric arc furnace (EAF), and open hearth furnace (OHF). To manufacture crude steel, BOF uses iron ore, coal, and

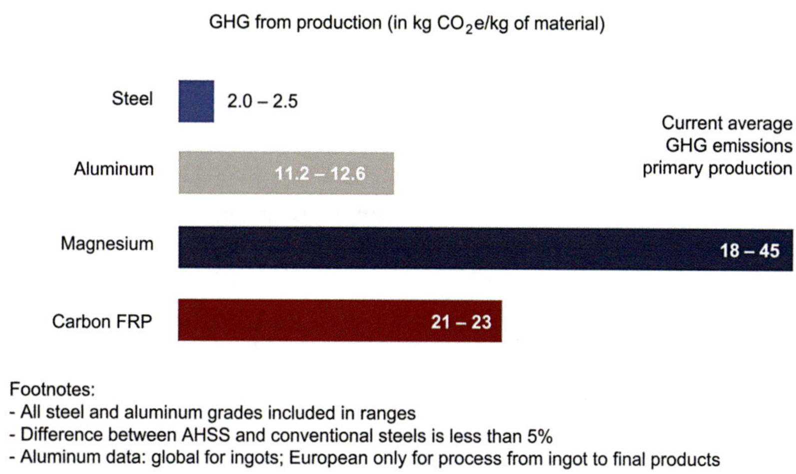

Fig. 16.2 Material production CO_2 emissions for steel and other alternate materials. GHG, greenhouse gas emissions; FRP, fiber-reinforced polymer. Source: Ref 16.3

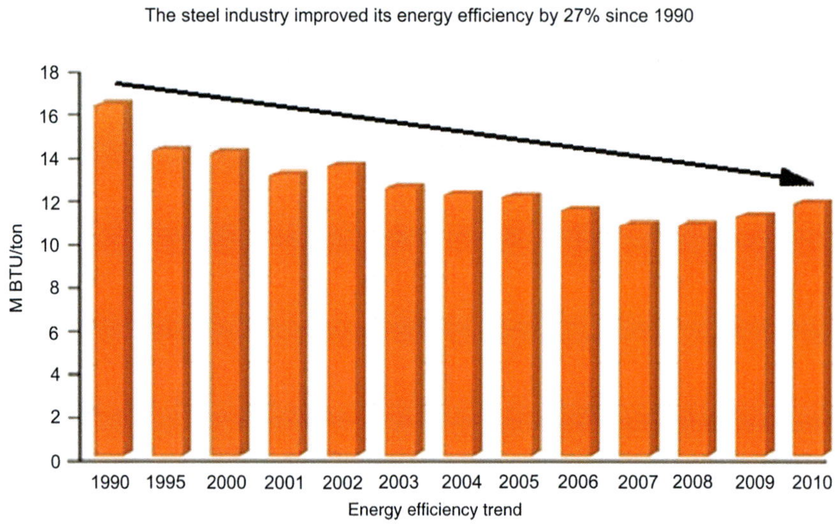

Fig. 16.3 Trend for energy consumption per ton of steel. Source: AUTO STEEL, Steel Market Development Institute

recycled steel; EAF uses only recycled steel. The OHF process is very energy intensive and is rarely used. The BOF process typically uses 1400 kg (3090 lb) of iron ore, 800 kg (1760 lb) of coal, 300 kg (660 lb) of limestone, and 120 kg (260 lb) of recycled steel to produce 1000 kg (2200 lb) of crude steel. The EAF process typically uses 880 kg (1940 lb) of recycled steel, 16 kg (35 lb) of coal and 64 kg (140 lb) of limestone to produce 1000 kg (2200 lb) of crude steel. About 70% of global steel production is based on BOF, 20% on EAF, and 1% on OHF. Based on the weighted average of the three production routes, emissions due to steel production amount to 1600 kg of CO_2/900 kg of crude steel, or 1.8 tons of CO_2/ton of crude steel (Ref 16.4).

A drastic reduction in low-emission steelmaking technology requires the development of breakthrough methods for crude steel production. Efforts to improve energy efficiency and reduce emissions in steelmaking are under development by an international team of collaborators. Nearly $1 billion has already been spent on research and development projects all around the world. Steelmaking technologies, which can potentially reduce CO_2 emissions by approximately 50%, have been identified. Work is underway to transfer these technologies to commercial implementation.

One of these projects is a three-year collaborative program between the U.S. Department of Energy (DOE), American Iron and Steel Institute (AISI), and the University of Utah to develop a hydrogen flash smelting process. The process produces iron by spraying iron ore directly into the furnace chamber using natural gas or hydrogen as a reducing agent to replace the energy-intensive coke oven and blast furnace process steps. This process has the potential to reduce energy consumption and CO_2 emissions by more than 50% (Ref 16.5). The energy needed and emissions produced, per ton of steel, are compared to those of aluminum (Table 16.1). Steelmaking uses less energy and has lower emissions than aluminum. The direct correlation between energy demand and GHG emissions is evident from Table 16.1.

The embodied energy approach can be used to assess CO_2 emissions during the material production stage of steel and other materials. Embodied energies are related to the thermodynamically required energies to reduce the metal oxide to metal. Iron is produced by the reduction of iron oxide, hematite (Fe_2O_3), by carbon. Aluminum is produced by the electrochemical reduction of bauxite (Al_2O_3), by liberating oxygen and depositing aluminum. The required energy to reduce the oxides to the metals is 5.5 MJ/kg for iron and 20.5 MJ/kg for aluminum. The embodied energies for the two metals are higher than those required to convert the ore to metals because they measure the total energy input per unit of usable output. The total energy includes the energy needed for the ore-to-metal conversion as well as the energy loss due to conversion process efficiency (38%) and the energy needed to support the plant infrastructure. Table 16.2 lists the energies required for conversion and the embodied energies for steel and aluminum.

The table shows that steel has lower energy requirements for extraction from ore and lower embodied energy than aluminum. Figure 16.4 shows a chart of embodied energy, MJ/kg, of basic metals. Steel is clearly the least energy demanding material compared to aluminum, magnesium, and titanium.

It is important to relate energy content per mass or volume to energy content per function. Advanced high-strength steels are characterized by their extremely high yield and high tensile strengths because they are mainly used in structural applications where high strength is specified to prevent failure. Figure 16.5 shows a strength-embodied energy per unit volume material chart. The chart can be used to select materials for a

Table 16.1 Energy and emissions for producing steel and aluminum

Material	Energy, GJ/ton	Global warming potential, ton CO_2 eq/ton	SO_2, eq kg/ton
Steel	21.6	2.01	5.12
Aluminum	155	11.06	50.43

Source: AUTOSTEEL, Steel Market Development Institute

Table 16.2 Energy required for ore conversion and embodied energy for steel and aluminum

Material	Material enthalpy of oxidation, MJ/kg	Embodied energy, MJ/kg
Carbon steel	5.5	17.1
Aluminum	20.5	194.0

Source: Ref 16.6

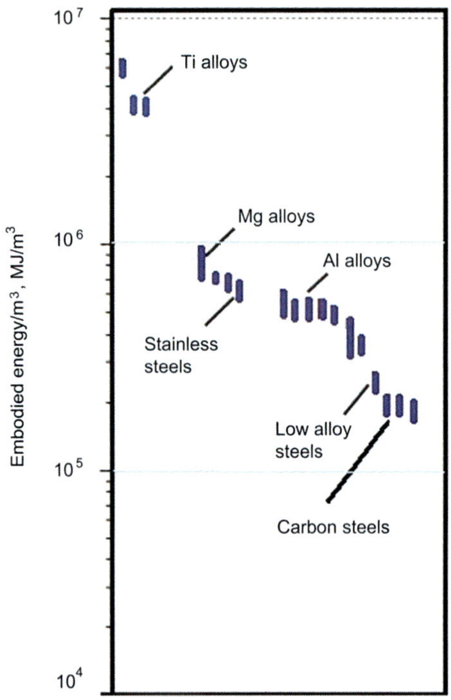

Fig. 16.4 Embodied energy chart of various metals. Source: Adapted from Ref 16.6

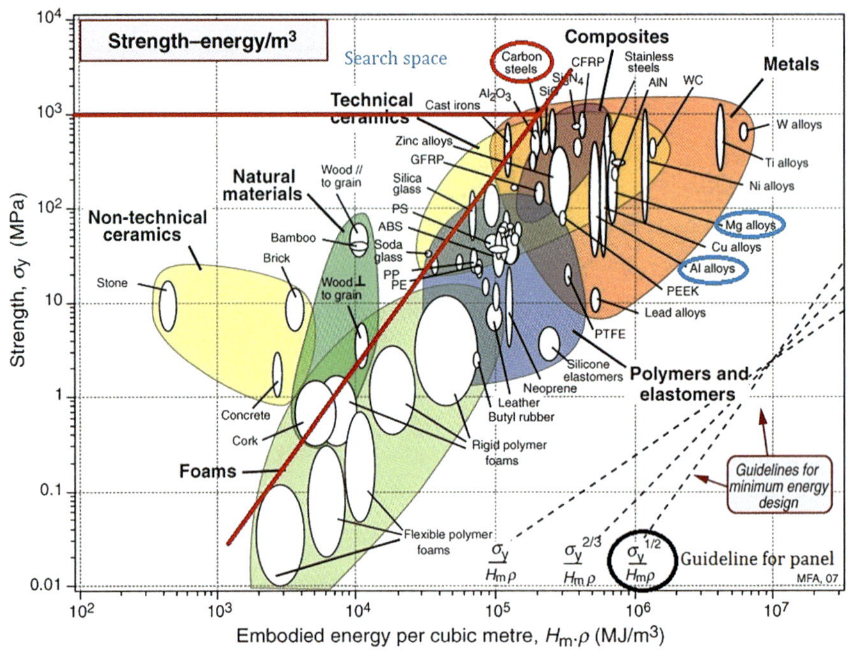

Fig. 16.5 Strength-embodied material property energy chart. Courtesy of Granta Design. Source: Ref 16.6

strong structure or strong panel with minimum embodied energy content. The following analysis derives an index for selecting the best material for a strong panel, loaded in bending, with minimum embodied energy content. The derivation follows Ashby's method (Ref 16.6).

Assume a flat panel has a specified length, *L*, and width, *w*, but a variable thickness, *t*. It is loaded in bending by a central load, *F*. The strength constraint requires that it must not yield under a bending moment, *M*, resulting from a design load. The objective is to minimize the embodied energy, *H*, of the material used to make the panel. The objective function is:

$$H = H_m m = H_m \, w \, t \, L\rho$$

where m is the mass of the panel, ρ is the density of the material, and H_m is the embodied energy/kg of the material. The maximum longitudinal stress in the panel occurs at the surface, at the greatest distance ($y_m = t/2$) from the neutral axis, and is caused by a moment, *M*:

$$\sigma_{max} = (M \, y_m)/I$$

where I is the second moment of inertia, $I = wt^3/12$.

Equating σ_{max} with the yield strength, σ_y, and assembling these results to eliminate t gives:

$$H = (6 \, M \, w)^{1/2} \, L[H_m \, \rho/\sigma]$$

The embodied energy is minimized by selecting materials with high values of the index:

$$(\sigma_y)^{1/2}/H_m\rho$$

The index can be used, in conjunction with Fig. 16.5, as a guideline for selecting the best material with the minimum energy design. The chart shows that the energy content at the same strength level of, for example, 1000 MPa (145 ksi), is lower for carbon steels than for aluminum or magnesium alloys.

Applying the selection guideline for the panel produces the search space indicated on the chart. The search space is defined by the optimized index guideline and the horizontal line defining the constraint that the yield strength of the panel $\sigma_y > 1000$ MPa. Results show that carbon steels, including AHSS, have the lowest embodied energy for the required yield strength.

Figure 16.6 shows a chart that compares the embodied energies per unit yield strength for various metals and alloys according to the equation:

Energy per unit strength = (Embodied energy × Density) / Yield strength

The chart shows clearly that steels, including AHSS, have the lowest embedded energy per unit yield strength.

Automotive companies have adopted strategies to minimize the impact of their vehicles and operations on climate change and the environment. Ford Motor Company, for example, has improved the fuel economy of their U.S. vehicles since 2006 by 16.8% and is set to reduce its facility CO_2 emissions by 30% per vehicle by 2025 compared to a 2010 baseline. The company's sustainability performance of its vehicles and facilities is listed in Table 16.3.

The importance of sustainable materials in the U.S. economy is reflected in an announcement, posted June 2012, by the Division of Materials Research of the National Science Foundation (NSF), listing grant opportunities for research proposals in the area of sustainability. The focus

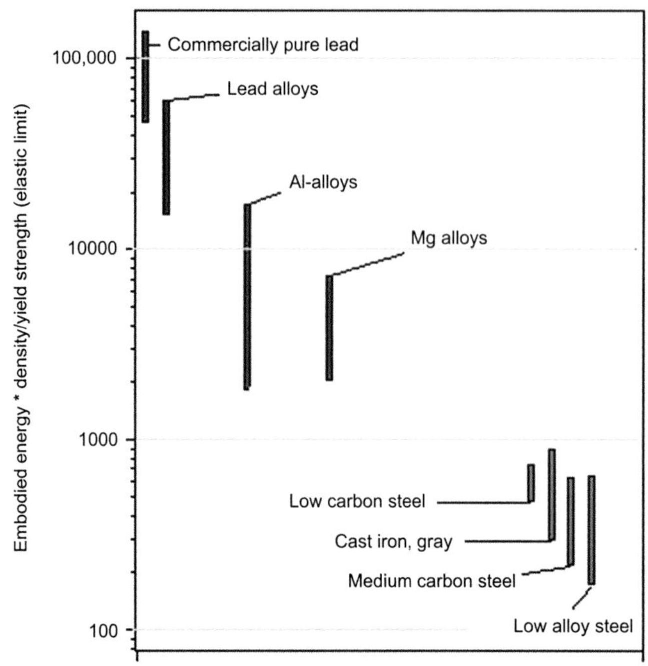

Fig. 16.6 Embodied energy per unit strength. Source: Adapted from Ref 16.6

Table 16.3 Sustainability performance of Ford Motor Company

	Year		
	2009	2010	2011
Fuel economy: U.S. corporate average combined car and truck fleet (miles per gallon)	27.1	26.9	27.8
CO_2 emissions: Worldwide facility CO_2 emissions per vehicle (metric tons)	1.07	1.01	0.91

Source: Ref 16.7

of the research is to reduce dependence on nonrenewable resources and improve the efficiency of industrial processes. Some of the proposed sustainability topics that have direct relevance to AHSS include:

- Enhancing recyclability
- Increasing conservation of natural resources
- Reducing carbon emissions
- Optimizing processing in terms of energy and materials usage
- Improving performance and increasing lifetime of materials

In conclusion, sustainability of AHSS includes the following aspects:

- Reducing consumption of steel
- Reducing pollution
- Recycling

16.2 Life Cycle Assessment

Life cycle assessment (LCA) is a method of accounting for the environmental performance of a product by assessing its total GHG emissions associated with its entire life cycle. For automobiles, this includes all emissions from any process to produce, use, and dispose of vehicles. Material production includes activities such as mining, smelting, refining, casting, rolling, extruding, shape casting, and galvanizing. Vehicle manufacturing includes activities such as forming, joining, and assembling. Vehicle use includes fuel and oil consumption during 200,000 km (124,000 miles) of travel. Table 16.4 lists the percentage of the GHG emissions for the four stages of the life cycle of a typical car.

The table shows that although the tailpipe emissions resulting from vehicle use account for about 85% of the total GHG emissions, vehicle production and vehicle manufacturing account for the remaining 15% of the total emissions. These emissions occur before a vehicle is driven. This is a relatively small percentage but it is important and should be included in the proposed U.S. fuel economy and emissions standards for 2017 to 2025. Emissions resulting from vehicle disposal are very small relative to the other life cycle stages.

The energy demands for a typical vehicle in each of the four life cycle stages are shown in Fig. 16.7. The energy fraction of the use phase is

Table 16.4 Life cycle greenhouse gas emissions for a typical car

Life cycle stage	Greenhouse gas (GHG) emissions, %
1. Material production	10.3
2. Vehicle manufacturing	4.3
3. Vehicle use	85.3
4. Vehicle disposal	0.1

Source: Compiled from Ref 16.8

85.3% and it dominates the life energy of the vehicle. The energy fraction for the material production stage is only 10.3% and its embodied energy is less dominating. If large changes in energy demand and CO_2 emissions are needed, the use phase should be first targeted. Although the energy fraction for the material production stage is only 10.3%, its impact on the total energy demand cannot be ignored.

A study commissioned by World Steel, the automotive group of the International Iron and Steel Institute (IISI), on the life cycle GHG emission assessment of automotive materials concluded that the life cycle GHG emissions attributable to AHSS and aluminum body-in-white (BIW) are essentially similar. Aluminum may offer higher weight savings potential (40 to 60%) than AHSS (10 to 20%), reducing fuel consumption and use-phase CO_2 emissions. This advantage, however, does not offset the increase in CO_2 emission during the material production phase, which is six times higher than steel. This is an energy penalty because aluminum has much higher embodied energy content than steel (Table 16.5).

Many studies have concluded that switching to aluminum from low-carbon steel does not reduce energy use and cost unless the use phase is substantially increased beyond the break-even point of approximately 150,000 km (93,000 miles). This assumes that the first generation of

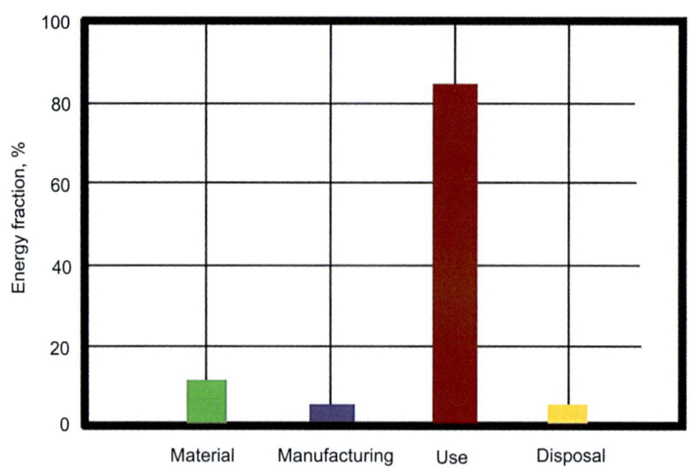

Fig. 16.7 Energy fraction consumed in life cycle stages of vehicle. Source: Adapted from Ref 16.6

Table 16.5 Embodied energy for steel and aluminum

Material	Embodied energy, MJ/kg	Embodied energy, GJ/m^3
Low-carbon steel	23.6	185
Aluminum alloy	194	523

Source: Ref 16.6

AHSS can be produced for the same price as traditional HSS. This assumption is valid because:

- First-generation AHSS are considered low-alloy steels and therefore keep expensive alloy content to a minimum.
- The intercritical annealing step in AHSS processing does not add additional costs to steel processing.

WorldAutoSteel has implemented the life cycle assessment (LCA) approach in the FutureSteelVehicle program where AHSS and optimized steel body structure designs were used to reduce vehicle weight by 35% and total life cycle emissions by nearly 70% at no additional cost.

In conclusion, AHSS benefit vehicles in the four stages of their life cycle:

1. Material production uses less energy and lower emissions.
2. Vehicle manufacturing produces lighter components. This reduces material usage, saves natural resources, and eliminates the energy and emissions associated with material production.
3. Vehicle use produces fuel savings because of the lighter vehicle weight.
4. Vehicle disposal by recycling reduces the consumption of raw materials and energy to produce the material. Steel can be recycled fully and indefinitely.

According to WorldAutoSteel, replacing conventional steel with AHSS in body structures of all cars produced worldwide would save 142 billion kg (156 million tons) of CO_2 equivalents. To put that in a LCA perspective, reducing vehicle weight by 117 kg (258 lb) would reduce its lifetime cycle emissions by 2000 kg (2.2 tons) of CO_2 equivalents. This means that for every 1 kg (2 lb) of AHSS used in a vehicle, 8 kg (18 lb) of CO_2 equivalents are saved over the total life cycle of the vehicle (Ref 16.9).

16.3 Recycling

Steel is the most recycled industrial material in the world with over 450 billion kg (500 million tons) of scrap steel recycled annually. In the U.S., 63 billion kg (70 million tons) of steel are recycled annually. Almost 50% of the current steel production is recycled, saving 33% of the energy needed to produce steel. Steel recycling is economic and environmental because it saves on the consumption of raw materials, minimizes energy consumption, and reduces GHG emissions. It is estimated that for every 900 kg (1 ton) of recycled steel, over 1400 kg (3085 lb) of iron ore, 740 kg (1630 lb) of coal, and 120 kg (265 lb) of limestone are saved. Figure 16.8 shows the life cycle of steel with the two sources of recycled steel:

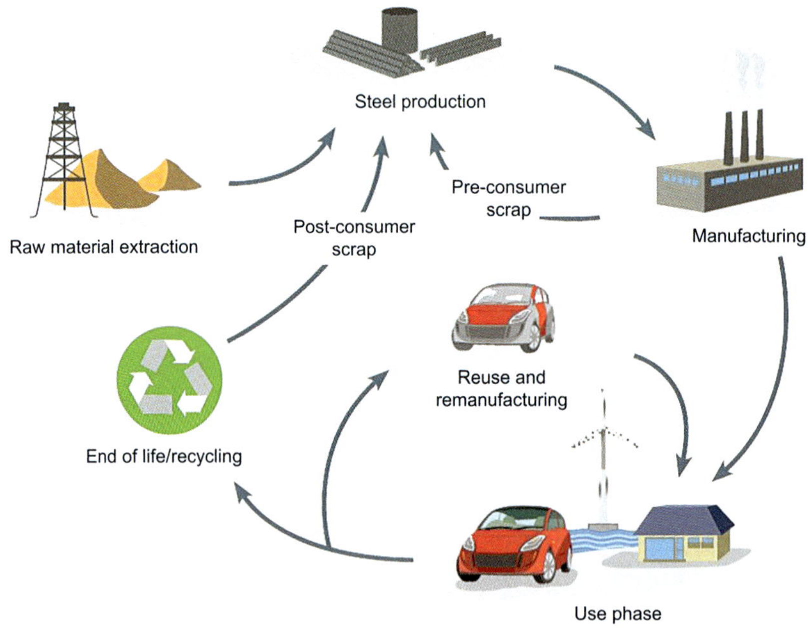

Fig. 16.8 Life cycle of steel. Source: Ref 16.10

excess material from steel production and downstream manufacturing (pre-consumer scrap), and steel at the end of a product's life (post-consumer scrap).

Steel has an endless life cycle and is easily recovered by magnetic separation. Recycling of vehicles at the end of their useful life may involve: (a) recycling the material back into the original product (slabs, bars, sheet, plate, etc.), and (b) downcycling into a lower-grade product because of contamination. Vehicles are the most recycled consumer product with more than 8 billion kg (14 million tons) of steel recycled every year from end-of-life vehicles. Approximately 75% of the weight of a typical vehicle is recycled; the remaining 25% is plastics, glass, and fabrics. The steel recycling rate is about 100% because it can be easily separated by magnetic techniques. Steel can be recycled indefinitely without degrading its properties and the infrastructure for recycling steel throughout the world is well established.

WorldAutoSteel and the steel producer ArcelorMittal are developing a process to recycle AHSS from the automotive market back into the basic oxygen furnace (BOF) steel production cycle. Because AHSS have high alloy content compared to traditional steels, recycling them causes unwanted alloys in steel production. Table 16.6 compares the chemistry of one type of AHSS with that of low-carbon steel.

Differences in the alloy content must be considered in recycling of AHSS so that the high alloy content of AHSS does not contaminate the

Table 16.6 Alloy content in one type of advanced high-strength steel and low-carbon steel

Material	Composition, wt%					
	Fe	C	Cr	Cb	Mn	Al
AHSS	bal	0.17	0.2	0.02	1.5	1.2
Low-carbon steel AK	bal	0.04	0.1	0.004	0.2	0.04

AK, aluminum killed. Source: Adapted from Ref 16.11

quality of the recycled steel. This means that AHSS must be separated from the normal scrap to prevent having an off-chemistry batch.

16.4 Economics of Advanced High-Strength Steels

Cost is a key incentive for any business and using AHSS helps automakers reduce costs because:

- Downgaging reduces the amount of steel used per product. If steel strength is increased from 500 to 1000 MPa (70 to 145 ksi), steel consumption would be reduced by half.
- Recycling saves raw materials.
- Recycling reduces consumption of energy and lowers emissions associated with steelmaking.
- Recycling recoups a portion of the cost of a vehicle.
- First-generation AHSS are considered low-alloy steels; hence, they could be produced for the same price as traditional HSS.
- Implementing thermal processing for first-generation AHSS is not difficult; no cost penalty is anticipated for this process.
- Steelmaking is an efficient process with 98% of the materials used to make crude steel converted to products and by-products (Ref 16.4).

For many applications, using a slightly more expensive higher-grade steel, such as AHSS, is cost neutral because less of it is used.

REFERENCES

16.1 World Steel Association, Sustainability Indicator, www.worldsteel.org, 2012

16.2 G. Coates, "Importance of Materials and Manufacturing Emissions for Future Vehicle Considerations," The Phoenix Group Presentation at Great Designs in Steel Seminar, May 16, 2012 (Livonia, MI), AUTOSTEEL

16.3 "Advanced High Strength Steels Applications Guidelines," Version 4.1, WorldAutoSteel, www.worldautosteel.org, June 2009

16.4 World Steel Association, "Sustainable Steel: At the Core of Green

Economy," publication at UN Conference R10+20, June 2012, www.worldsteel.org

16.5 "Energy Department Announces New Investments in Innovation Manufacturing Technologies," U.S. Department of Energy, June 12, 2012, www.energy.gov/articles

16.6 Granta Design, www.grantadesign.com

16.7 Blue Oval Connect Weekly Briefing, Ford Motor Co., July 13, 2012, www.blueovalconnect.com

16.8 D. Bhattacharya, *Developments in Advanced High Strength Steels,* Research and Development Center, Mittal Steel, East Chicago, IN, 2005

16.9 WorldAutoSteel Association, "Environmental Case Study: Automotive," www.worldautosteel.org

16.10 "Sustainable Steel at the Core of a Green Economy," World Steel Association Publication, 2012

16.11 ArcelorMittal Web site, www.arcelormittal.com

CHAPTER 17

Evolving Advanced High-Strength Steel Grades

THE FIRST GENERATION of advanced high-strength steels (AHSS) is ferritic-based and it includes dual-phase (DP), complex-phase (CP), transformation-induced plasticity (TRIP), and martensitic (MS) grades of steel. They were developed to extend the strength range of traditional HSS beyond the 800 MPa (115 ksi) limit. However, the increase in strength came at the expense of ductility, especially in MS steels, where the percent elongation drops to approximately 5%. The ductility problem was overcome with the development of the austenitic-based second generation of AHSS, which includes twinning-induced plasticity (TWIP), austenitic stainless steel (AUST SS), and lightweight induced-plasticity (L-IP). These steels are closely related to austenitic stainless steels and have both very high strength and excellent ductility. However, they contain high percentages of expensive alloying elements that result in higher costs and limited applications. Research is continuing to develop a third generation of AHSS that are less expensive but have good strength and high ductility. Table 17.1 summarizes the microstructural classes of conventional HSS and the three generations of AHSS.

17.1 Third-Generation Advanced High-Strength Steels

In 2007, researchers proposed to develop a third-generation AHSS with better strength and ductility than the first generation and at a lower cost than the second generation. Figure 17.1 shows the expected position, in

Table 17.1 Microstructural classes of conventional high-strength steels and three generations of advanced high-strength steels

Steel grade	Microstructure
HSS	Ferrite-based
Bake-hardenable (BH)	
High-strength, low-alloy (HSLA)	
First-generation AHSS	Ferrite-based
Dual-phase (DP)	
Complex-phase (CP)	
Bainitic	
Transformation-induced plasticity (TRIP)	
Martensitic (MS)	
Second-generation AHSS	Austenite-based
Austenitic stainless (AUST SS)	
Twinning-induced plasticity (TWIP)	
Lightweight-induced plasticity (L-IP)	
Others	
Third-generation AHSS (under development)	Multiphase: Composite (austenite and martensite), Ultra fine-grained ferrite, martensite, stabilized austenite

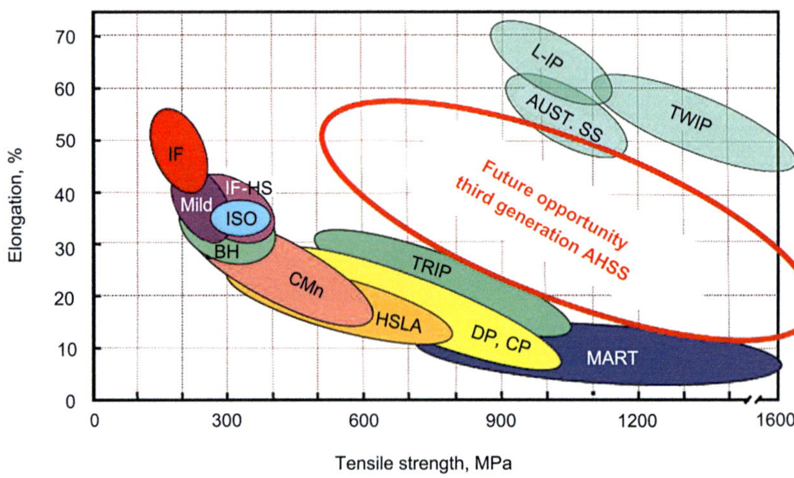

Fig. 17.1 Location of the third generation of advanced high-strength steel (AHSS) in the strength-ductility space. Source: Ref 17.1

the strength-ductility space, of the third-generation AHSS. It falls between the first- and second-generation bubbles and is expected to share the properties of both.

There are four key requirements to a viable third-generation AHSS:

- A high volume fraction of stable austenite
- Low carbon content so that weldability is not compromised
- Low alloy levels to reduce the cost of steel
- Processing consistent with current sheet steel production practice

Research is underway at university laboratories and industry research centers to develop new AHSS types and to expand the strength ranges of current steels. The extraordinary mechanical properties can be tailored and adjusted by alloying and processing. The introduction of steels with a microstructure consisting of at least two different components has led to a dramatic increase in the strength level without a decline in ductility. The key requirement for obtaining third-generation AHSS is the presence of significant amounts of stable retained austenite that transforms to hard martensite with progressive plastic deformation of the steel. This results in a high and sustained work-hardening rate of the material. Because austenite is not an equilibrium phase at room temperature, the challenge is to stabilize the high content of retained austenite in the final steel microstructure.

The critical part of the manufacturing of these steels is to control the processing conditions so that the microstructure and the strength-elongation relationship can be optimized. The two most important properties of any steel are strength to sustain loads and ductility to form parts. These properties can be controlled by quench and temper heat treatments for an optimal combination of strength and ductility. To address the cost problem, low-alloy steel compositions must be used to develop the third generation of AHSS.

The automotive industry has established an availability target of 2017 to 2025 for third-generation AHSS integration into vehicle structural designs to meet their lightweighting strategy of improving fuel economy and reducing CO_2 emissions. Figure 17.2 shows the location and established availability targets for integration of third-generation AHSS in structural designs and manufacturing of future vehicles.

Cooperative programs aimed at developing the metallurgy and microstructural designs needed for the third-generation AHSS have been established by the National Science Foundation (NSF), Alloy Stainless Prod-

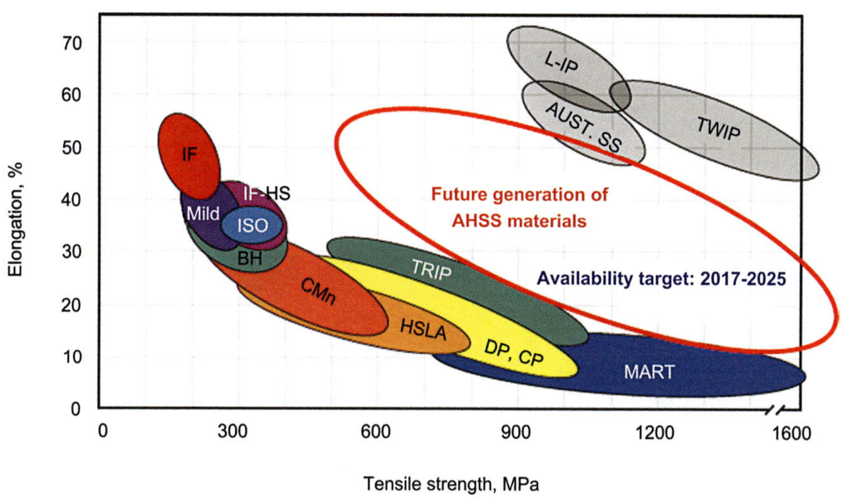

Fig. 17.2 Availability target for third-generation AHSS. Source: Ref 17.2

ucts (ASP), American Iron and Steel Institute (AISI), and nine universities. Different approaches and processing methods such as quenching and partitioning, double stabilization thermal cycle, and two-step austempering have been considered. Steel industry and academic research continue to develop new types and improve capabilities of AHSS. One example is the recent development of nanosteel AHSS. A review of the most promising processes are summarized subsequently.

17.2 Microstructure Design

The development of third-generation AHSS will require unique alloy/microstructure combinations to achieve the desired properties. The design and control of the microstructure determines the behavior and performance of AHSS. The basis of microstructural design of the third generation of AHSS is the assumption that the microstructure is a composite of two phases: one hard to give strength and one soft to give ductility.

The rule of mixtures can be used to manipulate and predict the strength and ductility of the composite microstructure. Assume a microstructure has two phases, (a) and (b). When a volume fraction of (a) is mixed with a volume fraction of (b) in the composite microstructure (c), the composite strength is given by the rule of mixtures as:

$$\sigma_c = f_a \sigma_a + f_b \sigma_b$$

The composite ductility is given by a similar equation:

$$\varepsilon_c = f_a \varepsilon_a + f_b \varepsilon_b$$

where σ is the strength, ε is the percent elongation, and f is the volume fraction of the constituents (a) and (b).

The equations can be solved if the volume fraction, strength, and ductility of each constituent are known.

Figure 17.3 shows a plot of the stress-strain curves for the composite and its components. The composite curve is obtained by applying the rule of mixtures to the components.

In the microstructure design for AHSS, the constituents are ferrite, austenite, and martensite. Their mechanical properties are listed in Table 17.2, which shows the high strength of martensite and the good ductility of austenite.

Based on the rule of mixtures concept, Matlock and Speer (Ref 17.3) predicted the strength-ductility relationships for steels with ferrite-martensite and austenite-martensite microstructures by varying the volume fraction of martensite. Figure 17.4 shows predicted strength-ductility relationships for two hypothetical steel microstructures, ferrite + martensite (DP steel) and stable austenite + martensite. The figure shows that

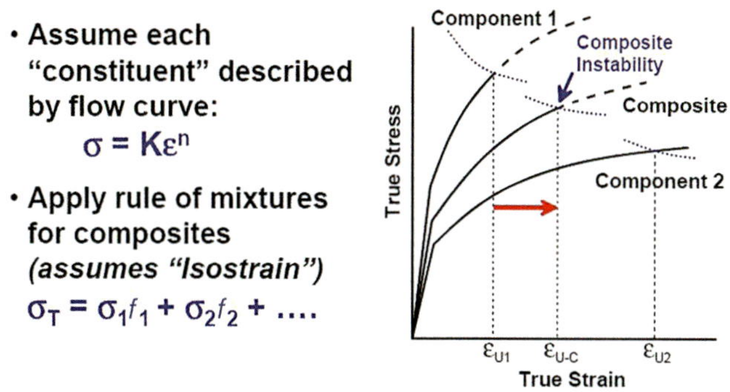

Fig. 17.3 Composite model for strength. Source: Ref 17.1

Table 17.2 Strength-ductility relationships for microstructure constituents

Constituent	Ultimate tensile strength, MPa (ksi)	Uniform true strain
Ferrite	300 (44)	0.30
Austenite	640 (93)	0.60
Martensite	2000 (290)	0.08

Source: Ref 17.3

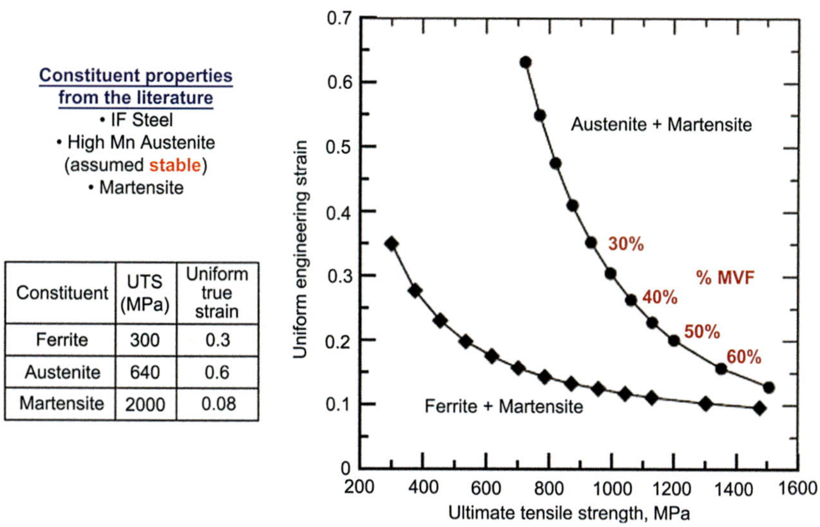

Fig. 17.4 Predicted strength/ductility relationships for two hypothetical steel microstructures. Source: Ref 17.1

strength increases with increase in the percent of martensite volume fraction (MVP). It also shows that the volume fraction of austenite has a stronger influence on ductility of the steel than the ferrite. Steel with a microstructure consisting of austenite and martensite has larger strain energy than steel consisting of ferrite and martensite; hence, the increase in uniform engineering strains of the austenite-martensite microstructure.

A superposition of the plots in Fig. 17.1 and 17.4 produces Fig. 17.5, which reveals that the prediction of the mechanical properties of steels with a ferrite-martensite microstructure fits well with the first-generation AHSS. It also reveals that the mechanical properties required for the third generation of AHSS could be realized by producing AHSS with austenite and martensite microstructure.

Computer modeling results have shown that the third-generation AHSS will include steel alloys with complex microstructures consisting of high-strength phases such as ultrafine-grained ferrite, martensite, or bainite. Significant amounts of high-ductility and work-hardening phases, such as retained austenite, will also be included in the third-generation AHSS.

17.3 Novel Processing Methods

As mentioned earlier, the key requirement for obtaining third-generation AHSS is the presence of significant amounts of retained austenite that transform to hard martensite, with progressive plastic deformation of the steel. This results in a high and sustained work-hardening rate of the material. However, austenite is not an equilibrium phase at room temperature and the challenge is to stabilize it in the final steel microstructure. This is done by controlling the processing conditions of quenching and tempering for obtaining high retained austenite content for optimal combination of strength and ductility.

The conventional AHSS processing routes are based on austempering and quench tempering processes. Novel AHSS processing routes include:

- Quenching and partitioning
- Double-stabilization thermal cycling

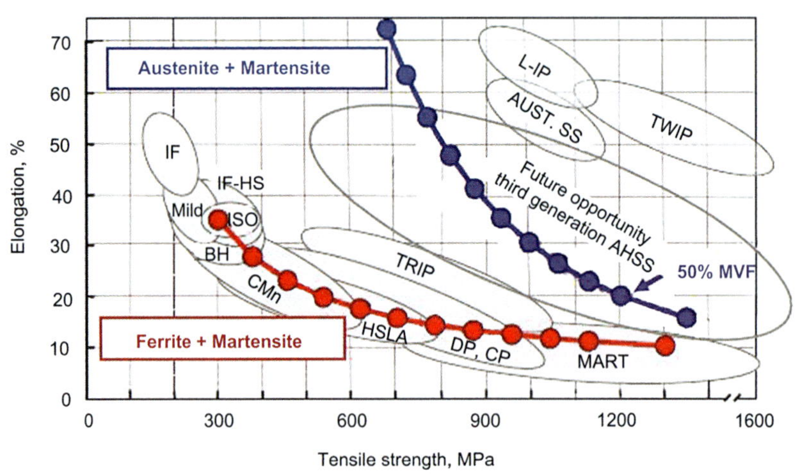

Fig. 17.5 Superposition of predicted microstructure design on strength-ductility plot. Source: Ref 17.1

Quenching and Partitioning (Q&P) Process. The Q&P process is a novel heat treatment process that uses thermal cycles to produce third-generation AHSS microstructure with high retained austenite. The process involves quenching austenite below the martensite start temperature, followed by a partitioning treatment to enrich the remaining austenite with carbon to make it stable at room temperature. The increased carbon content of the austenite is obtained through carbon depletion from the martensite and deposition into the austenite.

Obtaining high concentrations of retained austenite in the second generation of AHSS has been based on using substantial amounts of the expensive element manganese to stabilize the austenite and prevent it from transforming into ferrite at room temperature. Carbon is much less expensive than manganese and, in high concentrations, has the ability to stabilize austenite. Therefore, carbon has the potential to be an economical alternative for producing high fractions of retained austenite and creating the third generation of AHSS. However, high carbon content in steel produces iron carbide, which causes a reduction in the weldability of steel.

The Q&P process produces steels with high carbon concentration in the retained austenite but keeps the overall carbon content in the rest of the microstructure relatively low. This is achieved by using certain alloying elements and thermal treatment to suppress the formation of carbides and enrich austenite with carbon partitioned from martensite under paraequilibrium conditions. Paraequilibrium is a kinetically constrained equilibrium in which the diffusivity of substitutional elements is negligible compared to the fast diffusing interstitial element carbon. This means that under paraequilibrium conditions, substitutional alloys are not allowed to partition but carbon can redistribute between phases to reach equilibrium under the paraequilibrium constraint.

The Q&P process is schematically shown in Fig. 17.6; the figure also includes the development of microstructure in the heat treated steel. The four steps involved in the Q&P process are:

1. *Austenitizing:* A steel sample heated in the austenite region develops a fully austenitic microstructure. The carbon content of the initial alloy, C_i, is equal to the carbon content in the austenite, C_γ.
2. *Quenching:* The sample is quenched to a specific temperature, T_Q, to create certain volume fractions of austenite and martensite. The C_i is the same as C_γ and the carbon content in the martensite, C_m.
3. *Partitioning:* The sample is held at a partitioning temperature, T_P, to enrich austenite by allowing the carbon to diffuse from martensite regions to the surrounding austenite regions. This step increases the austenite stability and results in a higher austenite fraction. The C_γ is greater than C_i, and C_m is less than C_i.
4. *Cooling:* The sample is cooled from T_P to room temperature and additional martensite forms.

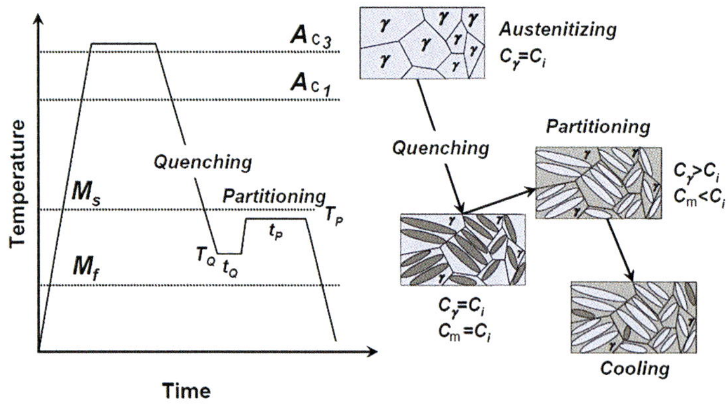

Fig. 17.6 Schematic illustration of the quenching and partitioning (Q&P) process. Source: Ref 17.4

The Q&P process has potential to produce third-generation AHSS provided that the alloy composition and the process times and temperatures for the partitioning treatment are optimized to produce the required fractions of retained austenite.

Double-Stabilization Thermal Cycle (DSTC) Process (Ref 17.5). Research at Case Western Reserve University (Cleveland, Ohio) was conducted on third-generation AHSS to determine the amount of retained austenite that is produced by the DSTC process. The objective is to produce steel that contains a high volume fraction of stable retained austenite and martensite with a sufficient carbon concentration to maintain a high strength level. As in the case of the Q&P process, carbide formation is avoided in the carbon partitioning process to allow as much carbon as possible to diffuse from martensite regions to surrounding austenite regions. Figure 17.7 shows a schematic of the DSTC five-step process:

1. *Austenitizing:* Steel is fully austenitized.
2. *Initial quenching:* Steel is cooled rapidly enough to prevent transformation to bainite en route to an initial hold temperature. The first stage of stabilization will occur at that point when the steel is briefly held just above its M_s temperature.
3. *Final quenching:* Steel is cooled to its final quench temperature to produce a specific volume fraction of the austenite to transform to martensite.
4. *Carbon partitioning:* Second stabilization occurs when the steel is aged at a higher temperature to allow partitioning of carbon from the martensite into the austenite under paraequilibrium, carbide-free conditions. This second stabilization will enable the austenite to have the required resistance to transformation to martensite.
5. *Air cooling:* Steel is cooled to room temperature.

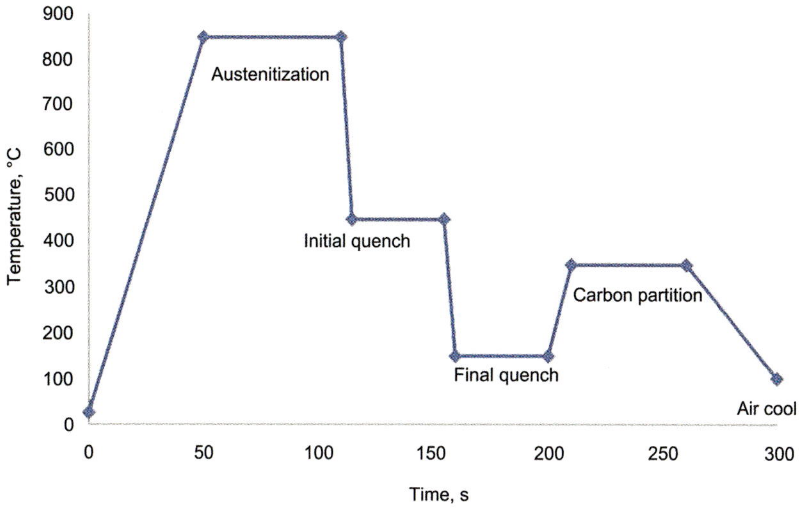

Fig. 17.7 Schematic of the double-stabilization thermal cycle (DSTC). Source: Ref 17.5

The DSTC process was used on steel with compositions of 0.29 wt% C, 3.99 wt% Mn, 2.12 wt% Si, and 1.50 wt% Al to produce microstructures containing more than 30 vol% austenite, 23 vol% ferrite, and 47 vol% martensite. Silicon and aluminum were included to suppress carbide formation. The processed steel attained a tensile strength of 1350 MPa (195 ksi). Electron backscatter diffraction (EBSD) phase maps showed that the retained austenite was uniformly distributed in the micron size range of 100 to 300 nm (1000 to 3000 Å).

Composite modeling is utilized to determine the fraction of stable retained austenite and martensite needed to meet the third-generation requirements for strength and ductility. More work is needed to optimize the steel composition and the heat treatment steps needed to maximize the amount of retained austenite in the steel (Ref 17.5, 17.6).

17.4 Nanosteels

Nanomaterials may be defined as materials containing 50% or more of particle sizes of at least one dimension in the range of 1 to 100 nm (10 to 1000 Å). At the nanomaterials level, properties are affected more by the laws of atomic physics than by the behavior of bulk materials. The small size feature of nanomaterials makes most of their atoms surface atoms with different coordination numbers, energy levels, electronic structure, and reactivity than interior atoms in the bulk material. This gives rise to completely different properties for nanomaterials.

Another aspect of crystalline nanomaterials is the contribution of their grain boundaries to strength. Grain boundaries provide strength to crystalline materials by impeding dislocation glide. For normal grain sizes, where D = 10–100 mm, the contribution of grain boundaries to yield strength is small because grain-boundary strengthening is found to be proportional to ($D^{-1/2}$), as described by the Hall-Petch equation. For nanomaterials, however, the grain size is D = 1–100 nm; hence, the contribution of grain boundaries to strength becomes significant. For example, assume that the strength of a 50 mm (500,000 Å) grain size DP steel has a yield strength of 350 MPa (50 ksi). Assume also that the grain size of the steel is refined to a smaller size of 50 nm (500 Å). According to the Hall-Petch equation, the yield strength of the nanosteel is 11,067 MPa (1,605 ksi). This number is very close to the theoretical strength of steel of 14,000 MPa (2,030 ksi). If the grains were made small enough, the yield strength would exceed the theoretical strength of the steel. This calculation assumes that grain boundaries and dislocation slip mechanisms exist and operate at the nanomaterial level.

Experimental and theoretical calculations predict that the number of pileup dislocations at grain boundaries decrease with decrease in grain size. No dislocation pileup is possible when the grain size reaches the critical value of 10 nm (100 Å). This is when the Hall-Petch equation breaks down and grain-boundary hardening ceases to exist (Ref 17.7).

Research at the University of Cambridge (United Kingdom) produced a strong and tough microstructure by using a low-temperature austempering process. In the heat treatment, the high-carbon austenite transforms to nanoscale thick bainitic microstructure with no carbides.

Most of the work on nanomaterials is considered research on advanced surface technology. New efforts are currently directed at developing bulk nanomaterials where grain/phase sizes are shrunk to the nanoscale level (1 to 100 nm).

The NanoSteel Company, Inc. announced the development of three classes of nanostructured, cold formable sheet AHSS with excellent combination of strength and ductility. Nanosteel properties result from using ultrafine, nanosized (<10 nm) particles, instead of islands of martensite, to strengthen the ferrite matrix in steel. The strength-elongation performances of the three classes of the nanosteels are: 950 MPa (138 ksi)/35%, 1200 MPa (174 ksi)/30%, and 1600 MPa (232 ksi)/15%, respectively. These steels are produced via conventional processes, avoid using expensive alloying elements, and are targeted for automotive structural applications.

Nanosteel location in the third-generation envelope of the ductility-strength diagram and its target date are shown in Fig. 17.8. The NanoSteel Company is promising a 2013 material target date and is stressing that the new material will have high strength, high ductility, and good performance. It will also be formed at ambient temperature and will use the existing infrastructure of steel.

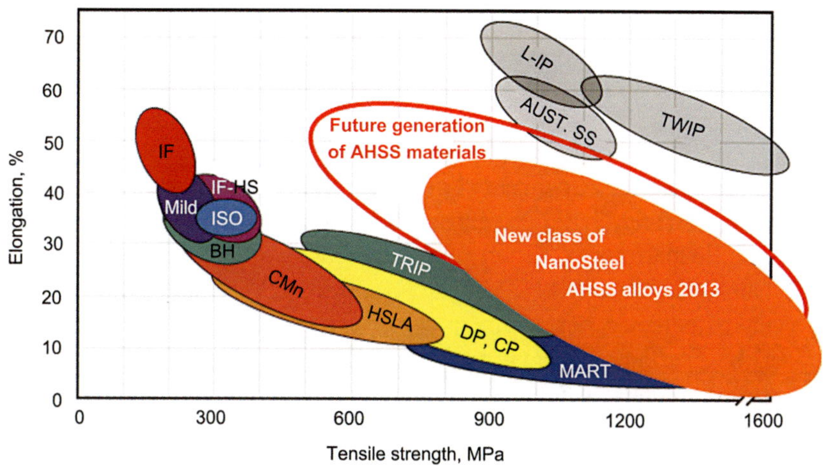

Fig. 17.8 Location of nanostructured AHSS and target date. Source: Ref 17.2

The NanoSteel Company is collaborating with General Motors (GM) to accelerate the final phases of the development of nanostructured AHSS for commercial applications. General Motor's support for nanosteel stems from its belief that the new steels offer a great potential for reducing vehicle weight and improving fuel economy without compromising performance and increasing cost.

The NanoSteel Company is not the only material developer that is involved in producing nanosteels; ArcelorMittal has indicated that they are developing nanosteel that would reduce vehicle weight to match the weight of aluminum cars but at a lower cost. Neither NanoSteel nor ArcelorMital are disclosing any of the technical aspects of nanosteels.

17.5 Integrated Computational Materials Engineering Approach

Integrated Computational Materials Engineering (ICME) is a computationally-driven engineering system approach that can be used to accelerate materials development by integrating processing, manufacturing, and design to produce optimized products.

Department of Energy has recently (first quarter 2013) awarded $6 million to a consortium of Steel Market Development Institute (SMDI), Auto Steel Partnership (ASP), and United States Automotive Materials Partnership (USAMP) for a four-year research and development project to utilize the ICME approach to accelerate the development of third-generation AHSS. Participants in the project include car companies, suppliers, universities and a national lab. Designing third-generation AHSS involves starting from atomic and nano scale models of these steels and ending with performance evaluation of components in actual cars.

REFERENCES

17.1 R.H. Wagoner, Advanced High Strength Steel Workshop, Oct 22–23, 2006 (Arlington, VA), sponsored by the National Science Foundation, Department of Energy, and the Auto/Steel Partnership

17.2 Nano-Structured AHSS, The NanoSteel Company, Inc., www.nanosteelco.com/markets/automotive/new-class-of-steel

17.3 D.K. Matlock and J.G. Speer, Third Generation of AHSS: Microstructure Design Concepts, Arunansu Haldar, Ed., *Microstructure and Texture in Steels and Other Materials,* Springer, London, 2009, p 185–205

17.4 E. De Moor, "Assessment of Quenching and Partitioning as a Fundamentally New Way of Producing Advanced High Strength Martensitic Steel Grades with Improved Ductility," Ph.D. dissertation, Gent University, 2009

17.5 H. Qu, AHSS Through Paraequilibrium Carbon Partitioning and Austenite Stabilization, master's thesis, Case Western Reserve University, Jan 2011

17.6 G.M. Michal, AHSS through Paraequilibrium Carbon Partitioning and Austenite Stabilization, *Proc. of 2009 NSF Engineering Research and Innovation Conference,* June 22–25, 2009 (Honolulu, HI), National Science Foundation, 2009

17.7 M. Ashby, P. Ferreira, and D. Schodek, *Nanomaterials, Nanotechnologies and Design,* Elsevier, 2009

Index

A

D

G

H

I

J

K

L

M

N

O

P

Q

R

S

U

V

W

Y